MONOGRAPHS ON STATISTICS AND APPLIED PROBABILITY

General Editors

V. Isham, N. Keiding, T. Louis, N. Reid, R. Tibshirani, and H. Tong

1 Stochastic Population Models in Ecology and Epidemiology *M.S. Barlett* (1960)
2 Queues *D.R. Cox and W.L. Smith* (1961)
3 Monte Carlo Methods *J.M. Hammersley and D.C. Handscomb* (1964)
4 The Statistical Analysis of Series of Events *D.R. Cox and P.A.W. Lewis* (1966)
5 Population Genetics *W.J. Ewens* (1969)
6 Probability, Statistics and Time *M.S. Barlett* (1975)
7 Statistical Inference *S.D. Silvey* (1975)
8 The Analysis of Contingency Tables *B.S. Everitt* (1977)
9 Multivariate Analysis in Behavioural Research *A.E. Maxwell* (1977)
10 Stochastic Abundance Models *S. Engen* (1978)
11 Some Basic Theory for Statistical Inference *E.J.G. Pitman* (1979)
12 Point Processes *D.R. Cox and V. Isham* (1980)
13 Identification of Outliers *D.M. Hawkins* (1980)
14 Optimal Design *S.D. Silvey* (1980)
15 Finite Mixture Distributions *B.S. Everitt and D.J. Hand* (1981)
16 Classification *A.D. Gordon* (1981)
17 Distribution-Free Statistical Methods, 2nd edition *J.S. Maritz* (1995)
18 Residuals and Influence in Regression *R.D. Cook and S. Weisberg* (1982)
19 Applications of Queueing Theory, 2nd edition *G.F. Newell* (1982)
20 Risk Theory, 3rd edition *R.E. Beard, T. Pentikäinen and E. Pesonen* (1984)
21 Analysis of Survival Data *D.R. Cox and D. Oakes* (1984)
22 An Introduction to Latent Variable Models *B.S. Everitt* (1984)
23 Bandit Problems *D.A. Berry and B. Fristedt* (1985)
24 Stochastic Modelling and Control *M.H.A. Davis and R. Vinter* (1985)
25 The Statistical Analysis of Composition Data *J. Aitchison* (1986)
26 Density Estimation for Statistics and Data Analysis *B.W. Silverman* (1986)
27 Regression Analysis with Applications *G.B. Wetherill* (1986)
28 Sequential Methods in Statistics, 3rd edition
G.B. Wetherill and K.D. Glazebrook (1986)
29 Tensor Methods in Statistics *P. McCullagh* (1987)
30 Transformation and Weighting in Regression
R.J. Carroll and D. Ruppert (1988)
31 Asymptotic Techniques for Use in Statistics
O.E. Bandorff-Nielsen and D.R. Cox (1989)
32 Analysis of Binary Data, 2nd edition *D.R. Cox and E.J. Snell* (1989)
33 Analysis of Infectious Disease Data *N.G. Becker* (1989)
34 Design and Analysis of Cross-Over Trials *B. Jones and M.G. Kenward* (1989)
35 Empirical Bayes Methods, 2nd edition *J.S. Maritz and T. Lwin* (1989)

36 Symmetric Multivariate and Related Distributions
K.T. Fang, S. Kotz and K.W. Ng (1990)
37 Generalized Linear Models, 2nd edition *P. McCullagh and J.A. Nelder* (1989)
38 Cyclic and Computer Generated Designs, 2nd edition
J.A. John and E.R. Williams (1995)
39 Analog Estimation Methods in Econometrics *C.F. Manski* (1988)
40 Subset Selection in Regression *A.J. Miller* (1990)
41 Analysis of Repeated Measures *M.J. Crowder and D.J. Hand* (1990)
42 Statistical Reasoning with Imprecise Probabilities *P. Walley* (1991)
43 Generalized Additive Models *T.J. Hastie and R.J. Tibshirani* (1990)
44 Inspection Errors for Attributes in Quality Control
N.L. Johnson, S. Kotz and X. Wu (1991)
45 The Analysis of Contingency Tables, 2nd edition *B.S. Everitt* (1992)
46 The Analysis of Quantal Response Data *B.J.T. Morgan* (1992)
47 Longitudinal Data with Serial Correlation—A State-Space Approach
R.H. Jones (1993)
48 Differential Geometry and Statistics *M.K. Murray and J.W. Rice* (1993)
49 Markov Models and Optimization *M.H.A. Davis* (1993)
50 Networks and Chaos—Statistical and Probabilistic Aspects
O.E. Barndorff-Nielsen, J.L. Jensen and W.S. Kendall (1993)
51 Number-Theoretic Methods in Statistics *K.-T. Fang and Y. Wang* (1994)
52 Inference and Asymptotics *O.E. Barndorff-Nielsen and D.R. Cox* (1994)
53 Practical Risk Theory for Actuaries
C.D. Daykin, T. Pentikäinen and M. Pesonen (1994)
54 Biplots *J.C. Gower and D.J. Hand* (1996)
55 Predictive Inference—An Introduction *S. Geisser* (1993)
56 Model-Free Curve Estimation *M.E. Tarter and M.D. Lock* (1993)
57 An Introduction to the Bootstrap *B. Efron and R.J. Tibshirani* (1993)
58 Nonparametric Regression and Generalized Linear Models
P.J. Green and B.W. Silverman (1994)
59 Multidimensional Scaling *T.F. Cox and M.A.A. Cox* (1994)
60 Kernel Smoothing *M.P. Wand and M.C. Jones* (1995)
61 Statistics for Long Memory Processes *J. Beran* (1995)
62 Nonlinear Models for Repeated Measurement Data
M. Davidian and D.M. Giltinan (1995)
63 Measurement Error in Nonlinear Models
R.J. Carroll, D. Rupert and L.A. Stefanski (1995)
64 Analyzing and Modeling Rank Data *J.J. Marden* (1995)
65 Time Series Models—In Econometrics, Finance and Other Fields
D.R. Cox, D.V. Hinkley and O.E. Barndorff-Nielsen (1996)
66 Local Polynomial Modeling and its Applications *J. Fan and I. Gijbels* (1996)
67 Multivariate Dependencies—Models, Analysis and Interpretation
D.R. Cox and N. Wermuth (1996)
68 Statistical Inference—Based on the Likelihood *A. Azzalini* (1996)
69 Bayes and Empirical Bayes Methods for Data Analysis
B.P. Carlin and T.A Louis (1996)

70 Hidden Markov and Other Models for Discrete-Valued Time Series
I.L. Macdonald and W. Zucchini (1997)
71 Statistical Evidence—A Likelihood Paradigm *R. Royall* (1997)
72 Analysis of Incomplete Multivariate Data *J.L. Schafer* (1997)
73 Multivariate Models and Dependence Concepts *H. Joe* (1997)
74 Theory of Sample Surveys *M.E. Thompson* (1997)
75 Retrial Queues *G. Falin and J.G.C. Templeton* (1997)
76 Theory of Dispersion Models *B. Jørgensen* (1997)
77 Mixed Poisson Processes *J. Grandell* (1997)
78 Variance Components Estimation—Mixed Models, Methodologies and Applications
P.S.R.S. Rao (1997)
79 Bayesian Methods for Finite Population Sampling
G. Meeden and M. Ghosh (1997)
80 Stochastic Geometry—Likelihood and computation
O.E. Barndorff-Nielsen, W.S. Kendall and M.N.M. van Lieshout (1998)
81 Computer-Assisted Analysis of Mixtures and Applications—
Meta-analysis, Disease Mapping and Others *D. Böhning* (1999)
82 Classification, 2nd edition *A.D. Gordon* (1999)
83 Semimartingales and their Statistical Inference *B.L.S. Prakasa Rao* (1999)
84 Statistical Aspects of BSE and vCJD—Models for Epidemics
C.A. Donnelly and N.M. Ferguson (1999)
85 Set-Indexed Martingales *G. Ivanoff and E. Merzbach* (2000)
86 The Theory of the Design of Experiments *D.R. Cox and N. Reid* (2000)
87 Complex Stochastic Systems
O.E. Barndorff-Nielsen, D.R. Cox and C. Klüppelberg (2001)
88 Multidimensional Scaling, 2nd edition *T.F. Cox and M.A.A. Cox* (2001)
89 Algebraic Statistics—Computational Commutative Algebra in Statistics
G. Pistone, E. Riccomagno and H.P. Wynn (2001)
90 Analysis of Time Series Structure—SSA and Related Techniques
N. Golyandina, V. Nekrutkin and A.A. Zhigljavsky (2001)
91 Subjective Probability Models for Lifetimes
Fabio Spizzichino (2001)
92 Empirical Likelihood *Art B. Owen* (2001)
93 Statistics in the 21st Century
Adrian E. Raftery, Martin A. Tanner, and Martin T. Wells (2001)
94 Accelerated Life Models: Modeling and Statistical Analysis
Vilijandas Bagdonavičius and Mikhail Nikulin (2001)
95 Subset Selection in Regression, Second Edition *Alan Miller* (2002)
96 Topics in Modelling of Clustered Data
Marc Aerts, Helena Geys, Geert Molenberghs, and Louise M. Ryan (2002)
97 Components of Variance *D.R. Cox and P.J. Solomon* (2002)
98 Design and Analysis of Cross-Over Trials, 2nd Edition
Byron Jones and Michael G. Kenward (2003)
99 Extreme Values in Finance, Telecommunications, and the Environment
Bärbel Finkenstädt and Holger Rootzén (2003)
100 Statistical Inference and Simulation for Spatial Point Processes
Jesper Møller and Rasmus Plenge Waagepetersen (2004)

Statistical Inference and Simulation for Spatial Point Processes

Jesper Møller
Rasmus Plenge Waagepetersen

CHAPMAN & HALL/CRC

A CRC Press Company
Boca Raton   London   New York   Washington, D.C.

Library of Congress Cataloging-in-Publication Data

Møller, Jesper.
 Statistical inference and simulation for spatial point processes / Jesper Møller and Rasmus Plenge Waagepetersen.
    p. cm.
 Includes bibliographical references and index.
 ISBN 1-58488-265-4 (alk. paper)
    1. Point processes. 2. Spatial analysis (Statistics) I. Waagepetersen, Rasmus Plenge. II. Title.

QA274.42.M65 2003
519.2′3—dc22   2003058463

This book contains information obtained from authentic and highly regarded sources. Reprinted material is quoted with permission, and sources are indicated. A wide variety of references are listed. Reasonable efforts have been made to publish reliable data and information, but the author and the publisher cannot assume responsibility for the validity of all materials or for the consequences of their use.

Neither this book nor any part may be reproduced or transmitted in any form or by any means, electronic or mechanical, including photocopying, microfilming, and recording, or by any information storage or retrieval system, without prior permission in writing from the publisher.

The consent of CRC Press LLC does not extend to copying for general distribution, for promotion, for creating new works, or for resale. Specific permission must be obtained in writing from CRC Press LLC for such copying.

Direct all inquiries to CRC Press LLC, 2000 N.W. Corporate Blvd., Boca Raton, Florida 33431.

Trademark Notice: Product or corporate names may be trademarks or registered trademarks, and are used only for identification and explanation, without intent to infringe.

Visit the CRC Press Web site at www.crcpress.com

© 2004 by Chapman & Hall/CRC

No claim to original U.S. Government works
International Standard Book Number 1-58488-265-4
Library of Congress Card Number 2003058463
Printed in the United States of America 1 2 3 4 5 6 7 8 9 0
Printed on acid-free paper

Contents

Preface		xiii
Acknowledgments		xv
1	**Examples of spatial point patterns**	1
2	**Introduction to point processes**	7
	2.1 Point processes on $\mathbb{R}^d$	7
	2.2 Marked point processes and multivariate point processes	8
	2.3 Unified framework	8
	2.3.1 Characterisation using void events	9
	2.3.2 Characterisation using the generating functional	9
	2.3.3 The standard proof	10
	2.4 Space-time processes	10
3	**Poisson point processes**	13
	3.1 Basic properties	13
	3.1.1 Definitions	13
	3.1.2 Existence and independent scattering property	15
	3.1.3 Constructions of stationary Poisson processes	17
	3.2 Further results	20
	3.2.1 Slivnyak-Mecke's theorem	20
	3.2.2 Superpositioning and thinning	22
	3.2.3 Simulation of Poisson processes	24
	3.2.4 Densities for Poisson processes	24
	3.3 Marked Poisson processes	25
	3.3.1 Random independent displacements of the points in a Poisson process	27
	3.3.2 Multivariate Poisson processes and random labelling	27
4	**Summary statistics**	29
	4.1 First and second order properties	29
	4.1.1 Basic definitions and results	30

		4.1.2 The second order reduced moment measure	32
	4.2	Summary statistics	33
		4.2.1 Second order summary statistics	33
		4.2.2 Directional K-functions	34
		4.2.3 Summary statistics based on interpoint distances	35
	4.3	Nonparametric estimation	36
		4.3.1 Nonparametric estimation of intensity functions	36
		4.3.2 Nonparametric estimation of K and L	37
		4.3.3 Edge correction	39
		4.3.4 Envelopes for summary statistics	40
		4.3.5 Nonparametric estimation of g	44
		4.3.6 Nonparametric estimation of F, G, and J-functions	46
	4.4	Summary statistics for multivariate point processes	47
		4.4.1 Definitions and properties	48
		4.4.2 The stationary case	50
		4.4.3 Nonparametric estimation	51
	4.5	Summary statistics for marked point processes	53
5	**Cox processes**		**57**
	5.1	Definition and simple examples	57
	5.2	Basic properties	60
	5.3	Neyman-Scott processes as Cox processes	61
	5.4	Shot noise Cox processes	62
		5.4.1 Shot noise Cox processes as cluster processes	63
		5.4.2 Relation to marked point processes	64
		5.4.3 Examples	64
		5.4.4 Summary statistics	66
	5.5	Approximate simulation of SNCPs	68
	5.6	Log Gaussian Cox processes	72
		5.6.1 Conditions on the covariance function	73
		5.6.2 Summary statistics	75
	5.7	Simulation of Gaussian fields and LGCPs	76
	5.8	Multivariate Cox processes	78
		5.8.1 Summary statistics	78
		5.8.2 Multivariate log Gaussian Cox processes	79
		5.8.3 Multivariate shot noise Cox processes	80
6	**Markov point processes**		**81**
	6.1	Finite point processes with a density	81
		6.1.1 Papangelou conditional intensity and stability conditions	83
	6.2	Pairwise interaction point processes	84
		6.2.1 Definitions and properties	84

CONTENTS ix

		6.2.2	Examples of pairwise interaction point processes	85
	6.3	Markov point processes		88
		6.3.1	Definition and characterisation	88
		6.3.2	Examples	91
		6.3.3	A spatial Markov property	93
	6.4	Extensions of Markov point processes to $\mathbb{R}^d$		94
		6.4.1	Infinite Gibbs point processes	94
		6.4.2	Summary statistics	96
	6.5	Inhomogeneous Markov point processes		97
		6.5.1	First order inhomogeneity	98
		6.5.2	Thinning of homogeneous Markov point processes	98
		6.5.3	Transformation of homogeneous Markov point processes	98
	6.6	Marked and multivariate Markov point processes		99
		6.6.1	Finite marked and multivariate point processes with a density	99
		6.6.2	Definition and characterisation of marked and multivariate Markov point processes	100
		6.6.3	Examples of marked and multivariate Markov point processes	101
		6.6.4	Summary statistics for multivariate Markov point processes	104
7	**Metropolis-Hastings algorithms**			**107**
	7.1	Description of algorithms		107
		7.1.1	Metropolis-Hastings algorithms for the conditional case of point processes with a density	108
		7.1.2	Metropolis-Hastings algorithms for the unconditional case	112
		7.1.3	Simulation of marked and multivariate point processes with a density	115
	7.2	Background material for Markov chains		118
		7.2.1	Irreducibility and Harris recurrence	119
		7.2.2	Aperiodicity and ergodicity	121
		7.2.3	Geometric and uniform ergodicity	122
	7.3	Convergence properties of algorithms		125
		7.3.1	The conditional case	125
		7.3.2	The unconditional case	128
		7.3.3	The case of marked and multivariate point processes	132
8	**Simulation-based inference**			**135**
	8.1	Monte Carlo methods and output analysis		135

	8.1.1	Ergodic averages	136
	8.1.2	Assessment of convergence	136
	8.1.3	Estimation of correlations and asymptotic variances	137
	8.1.4	Subsampling	139
8.2	Estimation of ratios of normalising constants		140
	8.2.1	Setting and assumptions	140
	8.2.2	Exponential family models	141
	8.2.3	Importance sampling	142
	8.2.4	Bridge sampling and related methods	144
	8.2.5	Path sampling	145
8.3	Approximate likelihood inference using MCMC		146
	8.3.1	Some basic ingredients in likelihood inference	146
	8.3.2	Estimation and maximisation of log likelihood functions	147
8.4	Monte Carlo error		149
8.5	Distribution of estimates and hypothesis tests		150
8.6	Approximate missing data likelihoods		151
	8.6.1	Importance, bridge, and path sampling for missing data likelihoods	152
	8.6.2	Derivatives and approximate maximum likelihood estimates	153
	8.6.3	Monte Carlo EM algorithm	154

9 Inference for Markov point processes — 157

9.1	Maximum likelihood inference		158
	9.1.1	Likelihood functions for Markov point processes	158
	9.1.2	Conditioning on the number of points	161
	9.1.3	Asymptotic properties of maximum likelihood estimates	161
	9.1.4	Monte Carlo maximum likelihood	162
	9.1.5	Examples	163
9.2	Pseudo likelihood		171
	9.2.1	Pseudo likelihood functions	171
	9.2.2	Practical implementation of pseudo likelihood estimation	174
	9.2.3	Consistency and asymptotic normality of pseudo likelihood estimates	176
	9.2.4	Relation to Takacs-Fiksel estimation	177
	9.2.5	Time-space processes	178
9.3	Bayesian inference		179

10 Inference for Cox processes — 181

10.1	Minimum contrast estimation	182
10.2	Conditional simulation and prediction	184
	10.2.1 Conditional simulation for Neyman-Scott processes	185
	10.2.2 Conditional simulation for SNCPs	186
	10.2.3 Conditional simulation for LGCPs	190
10.3	Maximum likelihood inference	192
	10.3.1 Likelihood inference for a Thomas process	192
	10.3.2 Likelihood inference for a Poisson-gamma process	197
	10.3.3 Likelihood inference for LGCPs	199
10.4	Bayesian inference	200
	10.4.1 Bayesian inference for cluster processes	204

11 Birth-death processes and perfect simulation 205
11.1	Spatial birth-death processes	205
	11.1.1 General definition and description of spatial birth-death processes	206
	11.1.2 General algorithms	207
	11.1.3 Simulation of spatial point processes with a density	209
	11.1.4 A useful coupling construction in the locally stable and constant death rate case	211
	11.1.5 Ergodic averages for spatial birth-death processes	214
11.2	Perfect simulation	216
	11.2.1 General CFTP algorithms	217
	11.2.2 Propp-Wilson's CFTP algorithm	220
	11.2.3 Propp-Wilson's monotone CFTP algorithm	221
	11.2.4 Perfect simulation of continuum Ising models	223
	11.2.5 Read-once algorithm	225
	11.2.6 Dominated CFTP	227
	11.2.7 Clans of ancestors	232
	11.2.8 Empirical findings	233
	11.2.9 Other perfect simulation algorithms	236

Appendices 237

A History, bibliography, and software 239
A.1	Brief history	239
A.2	Brief bibliography	240
A.3	Software	240

B Measure theoretical details 241
B.1	Preliminaries	241
B.2	Formal definition of point processes	241
B.3	Some useful conditions and results	243

C	**Moment measures and Palm distributions**	**247**
	C.1 Moment measures	247
	C.1.1 Moment measures in a general setting	247
	C.1.2 The second order reduced moment measure	248
	C.2 Campbell measures and Palm distributions	248
	C.2.1 Campbell measures and Palm distributions in a general setting	248
	C.2.2 Palm distributions in the stationary case	251
	C.2.3 Interpretation of $\mathcal{K}$ and G as Palm expectations	252
D	**Perfect simulation of SNCPs**	**253**
E	**Simulation of Gaussian fields**	**257**
F	**Nearest-neighbour Markov point processes**	**261**
	F.1 Definition and characterisation	261
	F.2 Examples	263
	F.3 Connected component Markov point processes	265
G	**Results for spatial birth-death processes**	**269**
	G.1 Jump processes	269
	G.2 Coupling constructions	270
	G.3 Detailed balance	272
	G.4 Ergodicity properties	274
References		**279**
Subject index		**293**
Notation index		**299**

Preface

Spatial point processes are used to model point patterns where the points typically are positions or centres of objects in a two- or three-dimensional region. The points may be decorated with marks (such as sizes or types of the objects) whereby marked point processes are obtained. The areas of applications are manifold and include astronomy, ecology, forestry, geography, image analysis, and spatial epidemiology. For more than 30 years spatial point processes have been a major area of research in spatial statistics, see e.g. Ripley (1977, 1981), Diggle (1983), Cressie (1993), Stoyan & Stoyan (1994), Stoyan, Kendall & Mecke (1995), and van Lieshout (2000). We expect that research in spatial point processes will continue to be of importance as new technology makes huge amounts of spatial point process data available and new applications emerge.

Some of the earliest applications of computational methods in statistics are related to spatial point processes, cf. the historical account in Section A.1. In the last decade computational methods, and particularly Markov Chain Monte Carlo (MCMC) methods, have undergone major developments. However, general textbooks on MCMC and statistical applications contain very little material associated with spatial point processes. The recent survey papers by Geyer (1999) and Møller & Waagepetersen (2003) and the book by van Lieshout (2000) contain material on MCMC methods for spatial point processes, but a more comprehensive book which collects and unifies recent theoretical advances and shows examples of applications in simulation-based inference for spatial point process seems missing.

This book aims at filling this gap; it is concerned with simulation-based inference for spatial point processes, with an emphasis on MCMC methods. It should be accessible for a fairly general readership interested in spatial models, including senior undergraduate students and Ph.D. students in statistics, experienced statisticians, and applied probabilists. Moreover, researchers working with complex stochastic systems in e.g. environmental statistics, ecology, and materials science may benefit from reading the parts of the book concerned with the statistical aspects.

A substantial part of the book consists of statistical methodology, simulation studies, and examples of statistical analysis of the different datasets introduced in Chapter 1. The book provides further a detailed treatment of the mathematical theory for spatial point processes and simulation-based methods. We have sought to make the book as self-contained as possible, and give proofs of almost all mathematical results in the book. Thereby the reader can be acquainted with the basic mathematical techniques for point processes. The general mathematical theory for point processes is heavily based on measure theory. In order to make the book accessible to a wide audience, we mainly restrict attention to point processes on $\mathbb{R}^d$, whereby the mathematical treatment is simplified. However, most results generalise in an obvious way to more general state spaces. Measure theoretical details are confined to Appendices B–C. Readers with no or very limited knowledge on measure theory (or who are less interested in the measure theoretical details) can read the book without consulting these appendices, since only a very few proofs (marked with †) rely on such knowledge.

The book is organised as follows. The first part (Chapters 1–4) deals with background material for point processes, including Poisson point processes and nonparametric methods based on summary statistics. The second part (Chapters 5–6) concerns Cox and Markov point processes which we consider the most useful model classes in data analysis. The third part (Chapters 7–9) treats both some general background material on MCMC methods, particularly Monte Carlo methods related to statistical inference as well as theoretical issues such as stability properties of Markov chains, and simulation-based inference for Cox and Markov point process models. The fourth part (Chapter 11 – Appendix E) deals with some more specialised topics and technical issues, including recent advances of perfect simulation for spatial point processes. Finally, Appendices A–G provide a brief history and bibliography, some comments on software, and various technical topics.

Acknowledgments

We have been supported by the European Union's TMR network "Statistical and Computational Methods for the Analysis of Spatial Data. ERB-FMRX-CT96-0095," by the Centre for Mathematical Physics and Stochastics (MaPhySto), funded by a grant from the Danish National Research Foundation, and by the Danish Natural Science Research Council. We are grateful to Adrian J. Baddeley, Kasper K. Berthelsen, Anders Brix, Peter J. Diggle, Marie-Colette van Lieshout, Shigeru Mase, Linda Stougaard Nielsen, Antti Penttinen, Jakob Gulddahl Rasmussen, Dietrich Stoyan, and Robert L. Wolpert for their useful comments and assistance with computer code and data. Our exposition of the subjects of spatial point processes and simulation-based inference has further been much inspired by Julian Besag, Charles J. Geyer, Eva Vedel Jensen, Wilfrid S. Kendall, and Gareth Roberts. Our absorption in the work on the book has demanded much patience and tolerance from our families to whom we owe a special thank you. In particular, we dedicate this book to Søren, Peter, and Katrine.

Jesper Møller
Rasmus Plenge Waagepetersen
Aalborg, Denmark.

CHAPTER 1

Examples of spatial point patterns

This chapter introduces some examples of spatial point patterns which are used throughout this book for illustrative purposes. They have been chosen both because of simplicity and because they are flexible enough to show Markov chain Monte Carlo (MCMC) applications of many point process models. Other examples of applications can be found in the references given in this book, particularly many of those mentioned in Section A.2.

EXAMPLE 1.1 (*Hickory trees*) Figure 1.1 shows the positions of 85 hickory trees in an 120 by 120m square subset of the Bormann research plot in Duke Forest in Durham, North Carolina. This data set which served as an example in Wolpert & Ickstadt (1998) is a subset of a much larger data set including 7859 trees of 38 species, which was initially collected in 1951–52 by Bormann and updated in 1974, 1982, 1989, and 1993 by N. Christensen, R. Peet and members of the Duke University and University of North Carolina botany departments in a continuing study of forest maturation (Bormann 1953, Christensen 1977). The trees in Figure 1.1 seem to aggregate, possibly due to environmental conditions which we later model by an unobserved intensity process (Chapters 5 and 10).

EXAMPLE 1.2 (*Amacrine cells*) The points in Figure 1.2 show amacrine cell centres observed in a rectangular area within the retinal ganglion cell layer of a rabbit. There are 152 "on" cells and 142 "off" cells, corresponding to cells which process "light-on" and "light-off" information. The observation window is scaled to $[0, 1.6065] \times [0, 1]$. Compared to the point pattern in Figure 1.1, the two point patterns in Figure 1.2 seem more regular, possibly due to a repulsive behaviour between the points. Scientific hypotheses of interest are whether the two types of points can be regarded as realisations of independent point processes, and whether the division of cells into "on" respective "off" cells is completely random given the positions of the cells. For details, see Wässle, Boycott & Illing (1981) and Diggle (1986).

The amacrine cell data is an example of a multitype (or bivariate) point pattern, and it is analysed in Diggle (1986) and van Lieshout & Baddeley (1999) using nonparametric summary statistics (Chapter 4).

The univariate pattern of the "on" cell centres is analysed in several papers, including Diggle & Gratton (1984) who use an ad hoc method for estimation, Ogata & Tanemura (1984) and Särkkä (1993) who use likelihood and pseudo likelihood methods (Chapter 9), and Heikkinen & Penttinen (1999) and Berthelsen & Møller (2003) who use a nonparametric Bayesian MCMC setting (Chapter 9).

EXAMPLE 1.3 (*Mucous membrane cells*) The point pattern in Figure 1.3 shows cell centres in a cross section of the mucous membrane of the stomach of a rat. The rectangular observation window is scaled to $[0, 1] \times [0, 0.7]$. The cells are divided into two classes: ECL cells and other

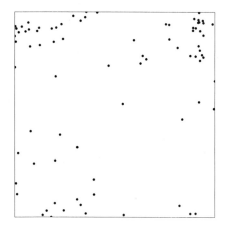

Figure 1.1 *Positions of hickory trees.*

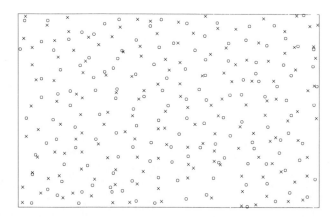

Figure 1.2 *Amacrine "on" (crosses) and "off" (circles) cell centres.*

EXAMPLES OF SPATIAL POINT PATTERNS

cells for which, respectively, 86 and 807 cell centres are observed. An additional feature compared with Example 1.2 is the increasing trend in the intensity from top to bottom so that more points occur in the bottom part of the window. A hypothesis of interest is whether the spatially varying intensities of ECL cells and other cells are proportional. A similar data set is examined in Nielsen (2000), using a particular type of inhomogeneous Markov point process obtained by transforming a homogeneous Markov point process (Chapters 6 and 9).

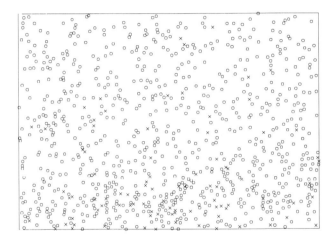

Figure 1.3 *Mucous membrane ECL (crosses) and other (circles) cell centres.*

EXAMPLE 1.4 (*Norwegian spruces*) Figure 1.4 shows 134 discs, where the centres are the positions of Norwegian spruces observed in a rectangular window of size 56×38m, and the radii are the stem diameters multiplied by 5. As discussed in Penttinen, Stoyan & Henttonen (1992) and Goulard, Särkkä & Grabarnik (1996) the "influence zone" of a tree is about five times the stem diameter. The discs can be modelled by a so-called marked point process (Section 2.2). The data has been analysed in Fiksel (1984a), Penttinen et al. (1992), Stoyan et al. (1995), Goulard et al. (1996), and Møller & Waagepetersen (2003) using different Markov marked point process models (Chapters 6 and 9).

EXAMPLE 1.5 (*Weed*) In this example the points are positions of 976 *Trifolium* spp. (clover) and 406 *Veronica* spp. (speedwell) weed plants observed within 45 metal frames on a Danish barley field. The 45 frames are of size 30×20cm and organised in 9 groups, each containing 5 frames, where the vertical and horizontal distances between two neighbouring

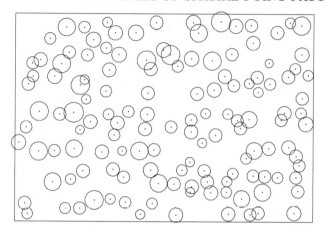

Figure 1.4 *Norwegian spruces. The radii of the discs equal 5 times the stem diameters.*

groups are 1 and 1.5m, respectively, see Figure 1.5. The size of the experimental area is 7.5 × 5m, where the longest side is parallel with the ploughing direction. The weed plant positions are shown in Figure 1.6 where we have rotated the design 90° and omitted some space between the frames. Note the trend in the point pattern of *Trifolium* spp. weed plants: in general more plants occur in the upper frames, i.e. the frames to the left in Figure 1.6.

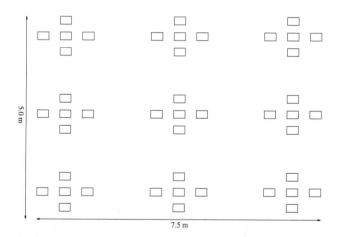

Figure 1.5 *Sampling design for weed data.*

EXAMPLES OF SPATIAL POINT PATTERNS

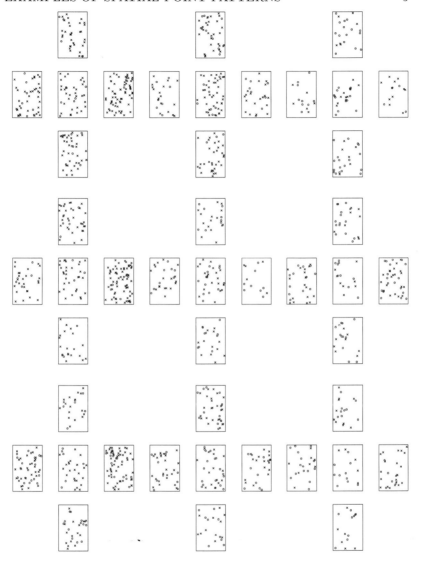

Figure 1.6 *Positions of weed plants when the design is rotated 90° (some space between the frames is omitted). Crosses correspond to Trifolium spp. and circles to Veronica spp.*

This bivariate point pattern is a subset of a much larger dataset where weed plants observed at eight different dates are considered (Figure 1.6 shows the observation at date 30 May 1996). The entire dataset is analysed by different Cox process models (Chapters 5 and 10) in

Brix & Møller (2001) and Brix & Chadoeuf (2000). See also Møller & Waagepetersen (2003).

CHAPTER 2

Introduction to point processes

This chapter gives an informal introduction to point process and to marked point processes with points in $\mathbb{R}^d$ (the d-dimensional Euclidian space). It also states the definitions and notation used throughout the book. The rigorous mathematical framework for point processes on a general state space is given in Appendix B, which can be skipped by readers who are unfamiliar with or less interested in measure theoretical details. Section A.1 describes the history of spatial point processes, and Section A.2 contains several references for further reading.

2.1 Point processes on $\mathbb{R}^d$

A *spatial point process* X is a random countable subset of a space S. Throughout this book, unless otherwise stated, we will always assume that $S \subseteq \mathbb{R}^d$. Often S will be a d-dimensional box or the whole of $\mathbb{R}^d$, but it could also be e.g. the $(d-1)$-dimensional unit sphere. In practice we observe only the points contained in a bounded observation window $W \subseteq S$. In Examples 1.2–1.4, $d = 2$ and W is either a rectangle or a union of rectangles.

We restrict attention to point processes X whose realisations are locally finite subsets of S. Formally, for any subset $x \subseteq S$, let $n(x)$ denote the cardinality of x, setting $n(x) = \infty$ if x is not finite. Then x is said to be *locally finite*, if $n(x_B) < \infty$ whenever $B \subseteq S$ is bounded, where

$$x_B = x \cap B$$

is the restriction of a point configuration x to B (similarly X_B is the restriction of X to B). Thus X takes values in the space defined by

$$N_{\mathrm{lf}} = \{x \subseteq S : n(x_B) < \infty \text{ for all bounded } B \subseteq S\}.$$

Elements of N_{lf} are called *locally finite point configurations*, and they will be denoted by $x, y, \ldots$, while $\xi, \eta, \ldots$ will denote points in S. The empty point configuration is denoted $\emptyset$. We shall abuse notation and write $x \cup \xi$ for $x \cup \{\xi\}$, $x \setminus \eta$ for $x \setminus \{\eta\}$, etc., when $x \in N_{\mathrm{lf}}$ and $\xi, \eta \in S$.

2.2 Marked point processes and multivariate point processes

Let Y be a point process on $T \subseteq \mathbb{R}^d$ (the reason for shift in notation will become apparent in Section 2.3). Given some space M, if a random "mark" $m_\xi \in M$ is attached to each point $\xi \in Y$, then

$$X = \{(\xi, m_\xi) : \xi \in Y\}$$

is called a *marked point process* with points in T and *mark space* M. In this book we mainly consider the cases where M is either a finite set or a subset of $\mathbb{R}^p$, $p \geq 1$. For more general cases of marked point processes, see Stoyan & Stoyan (1994), Schlather (2001), and the references therein.

For the disc process in Example 1.4, $M = (0, \infty)$, where (ξ, m_ξ) is identified with the disc with centre ξ and radius m_ξ. Similarly, we may obtain marked point processes with other kinds of geometric objects (line segments, ellipses, etc.) which can be identified with points in $\mathbb{R}^p$. Such processes are called *germ-grain models* where ξ (the germ) specifies the location of the object m_ξ (the grain).

Another simple example is a *multitype point process*, where $M = \{1, \ldots, k\}$ and the marks specify k different types of points (e.g. different types of cells or weed plants as in Examples 1.2, 1.3, and 1.5 where $k = 2$). This is equivalent to a k-dimensional *multivariate point process*, that is a tuple $(X_1, \ldots, X_k)$ of point processes $X_1, \ldots, X_k$ corresponding to the k different types of points.

2.3 Unified framework for point processes and marked point processes

In this book we distinguish between point processes on $S \subseteq \mathbb{R}^d$ and marked point processes with points in $T \subseteq \mathbb{R}^d$. However, both types of processes as well as point processes on non-Euclidian spaces can be handled in a unified framework where S is a general metric space, see Appendix B (for a marked point process as in Section 2.2, $S = T \times M$). We denote the metric $d(\cdot, \cdot)$, and for $\xi \in S$ and $r \geq 0$, we let

$$b(\xi, r) = \{\eta \in S : d(\xi, \eta) \leq r\}$$

denote the closed ball in S with centre ξ and radius r. Examples of metrics are given in Appendix B. When $S \subseteq \mathbb{R}^d$, we let always, unless otherwise stated, $d(\xi, \eta) = \|\xi - \eta\|$ be the usual Euclidian distance.

For the mathematical treatment of point processes we equip S and N_{lf} with appropriate σ-algebras denoted $\mathcal{B}$ and $\mathcal{N}_{\text{lf}}$, respectively (for details, see Section B.2). Readers unfamiliar with measure theory may just think of $\mathcal{B}$ and $\mathcal{N}_{\text{lf}}$ as very large families of subsets of S and N_{lf}. We shall often just write $B \subseteq S$ and $F \subseteq N_{\text{lf}}$ when we mean $B \in \mathcal{B}$ and $F \in \mathcal{N}_{\text{lf}}$.

Appendix B contains some useful characterisation and uniqueness re-

UNIFIED FRAMEWORK

sults for spatial point processes in a general setting. An informal account is given below when $S \subseteq \mathbb{R}^d$. We also briefly explain how a technique called the standard proof becomes useful.

2.3.1 Characterisation using void events

Let
$$\mathcal{B}_0 = \{B \in \mathcal{B} : B \text{ is bounded}\}.$$
For a point process X on S consider the *count function*
$$N(B) = n(X_B)$$
which is the random number of points falling in $B \subseteq S$. Sets of the form $F_B = \{x \in N_{\text{lf}} : n(x_B) = 0\}$ with $B \in \mathcal{B}_0$ are called *void events*. Note that $X \in F_B$ if and only if $N(B) = 0$. Under mild conditions (which will be satisfied throughout the book, cf. Theorem B.1 in Section B.3) the *distribution* of X (i.e. $P(X \in F)$, $F \in \mathcal{N}_{\text{lf}}$) is determined by its *void probabilities* defined by
$$v(B) = P(N(B) = 0), \quad B \in \mathcal{B}_0.$$

Similarly, the distribution of a multivariate point process $(X_1, \ldots, X_k)$ with $X_1, \ldots, X_k$ defined on $T \subseteq \mathbb{R}^d$ and corresponding count functions $N_1, \ldots, N_k$, is determined by the void probabilities
$$v(B_1, \ldots, B_k) = P(N_1(B_1) = 0, \ldots, N_k(B_k) = 0)$$
for bounded $B_1, \ldots, B_k \subseteq T$.

Consider
$$\nu(E) = \mathbb{E} \sum_{\xi \in X} \mathbf{1}[(\xi, X) \in E], \quad E \subseteq S \times N_{\text{lf}},$$
where $\mathbf{1}[\cdot]$ denotes the indicator function. Then ν is a *measure*, meaning that $\nu(\cdot) \geq 0$, $\nu(\emptyset) = 0$, and $\nu(\cup_{i=1}^{\infty} E_i) = \sum_{i=1}^{\infty} \nu(E_i)$ for disjoint $E_1, E_2, \ldots \subseteq S \times N_{\text{lf}}$. In fact ν is uniquely determined by its values on sets of the form $E = B \times F$ where $B \in \mathcal{B}_0$ and F is a void event. Thus ν is an example of a measure defined in terms of a certain expectation, where the measure is determined by its values on a small class of sets. We shall often consider such measures and characterisations, which are treated in more detail in Appendix B.

2.3.2 Characterisation using the generating functional

The *generating functional* for a point process X on S is defined by
$$G_X(u) = \mathbb{E} \prod_{\xi \in X} u(\xi) \qquad (2.1)$$

for functions $u : S \to [0,1]$ with $\{\xi \in S : u(\xi) < 1\}$ bounded. For $B \in \mathcal{B}_0$ and $u(\xi) = t^{\mathbf{1}[\xi \in B]}$ with $0 \le t \le 1$, $G_X(u) = \mathbb{E} t^{N(B)}$ is the generating function for $N(B)$. Hence the distribution of X is uniquely determined by G_X.

The generating functional can be a very useful tool, see e.g. Daley & Vere-Jones (1988), but we shall only use it a few times in this book. By putting additional assumptions on X, $G_X(u)$ can be defined for more general functions u, see Daley & Vere-Jones (1988) and Proposition 3.3 in Section 3.1.2.

2.3.3 The standard proof

Suppose we are given two expressions in terms of integrals or expectations of an arbitrary nonnegative function h defined on some space H and that we want to establish the equality of these expressions. For instance, we shall later consider the expressions $\mathbb{E} \sum_{\xi \in X} h(\xi)$ and $\int h(\xi)\rho(\xi)\mathrm{d}\xi$ where $H = S$ and $\rho : S \to [0,\infty)$ is a certain function called the intensity function. Assume that the two expressions become measures whenever h is an indicator function; e.g. both $\nu_1(A) = \mathbb{E} \sum_{\xi \in X} \mathbf{1}[\xi \in A]$ and $\nu_2(A) = \int_A \rho(\xi)\mathrm{d}\xi$ are measures for $A \subseteq S$. Then it suffices to verify the equality for every indicator function (for more details see the remark to Theorem 16.11 in Billingsley 1995). We shall refer to this technique as the *standard proof*.

2.4 Space-time processes

A spatial point pattern observed within some bounded window $W \subseteq \mathbb{R}^d$ is usually a realisation of a continuous-time process observed at a fixed time. Sometimes point patterns are observed at discrete times for a continuous-time process. One example is the weed data (Example 1.5) where the full space-time dataset is analysed in Brix & Møller (2001). Other examples include the space-time modelling of localised cases of gastro-intestinal infections (Brix & Diggle 2001), forest trees (Rathbun & Cressie 1994b), geological data (Fiksel 1984b), sand dunes (Møller & Sørensen 1994), and earthquake occurrences (Ogata 1998, Zhung, Ogata & Vere-Jones 2002). See also Schoenberg, Brillinger & Guttorp (2002) and the references therein.

For many applications only one realisation of a spatial point process is available and information concerning the underlying space-time process is missing. This is also the case for the statistical analyses considered in this book, and the theory we present will accordingly be restricted to this situation. Statistical inference for space-time processes is often

SPACE-TIME PROCESSES

simpler than for spatial point processes, see Daley & Vere-Jones (1988, 2003) and Section 9.2.5.

Space-time processes and Monte Carlo methods provide indispensable tools for simulation-based inference of spatial point process models. For example, even for the Poisson process, which is a simple and fundamental point process model studied in detail in Chapter 3, most statistics have intractable or unknown distributions. The Poisson point process (and some related point process models) can easily be simulated, but most point process models are so complicated that direct simulation is infeasible. Instead simulations can be obtained from discrete or continuous time Markov chains with a specified stationary distribution. Chapters 7–11 treat these points in detail.

CHAPTER 3

Poisson point processes

Poisson point processes play a fundamental role. They serve as a tractable model class for "no interaction" or "complete spatial randomness" in spatial point patterns. They also serve as reference processes when summary statistics are studied (Chapter 4) and when more advanced point process models are constructed (Chapters 5–6).

In this chapter we study some of the most important properties of Poisson processes. Section 3.1 contains basic definitions and results, particularly concerning existence and construction of Poisson processes. Section 3.2 contains some further important results, including the useful Slivnyak-Mecke theorem and results for superpositioning, thinning, simulation, and densities for Poisson processes. Section 3.3 deals with marked Poisson processes.

Further material on Poisson processes can be found in Daley & Vere-Jones (1988), Kingman (1993), and Stoyan et al. (1995).

3.1 Basic properties

3.1.1 Definitions

We start by considering Poisson point processes defined on a space $S \subseteq \mathbb{R}^d$ and specified by a so-called *intensity function* $\rho : S \to [0, \infty)$ which is *locally integrable*, i.e. $\int_B \rho(\xi) \mathrm{d}\xi < \infty$ for all bounded $B \subseteq S$. This is by far the most important case for applications.

REMARK 3.1 In the definition below of a Poisson process we use only the *intensity measure* μ given by

$$\mu(B) = \int_B \rho(\xi) \mathrm{d}\xi, \quad B \subseteq S. \tag{3.1}$$

This measure is *locally finite*, i.e. $\mu(B) < \infty$ for bounded $B \subseteq S$, and *diffuse*, i.e. $\mu(\{\xi\}) = 0$ for $\xi \in S$. For readers familiar with measure theory it should be obvious how many of the definitions and results in this chapter extend to a general metric space S as in Appendix B and a locally finite diffuse measure μ on S.

Before defining a Poisson process with intensity function ρ, we define a related process:

DEFINITION 3.1 Let f be a density function on a set $B \subseteq S$, and let $n \in \mathbb{N}$ (where $\mathbb{N} = \{1, 2, 3, \ldots\}$). A point process X consisting of n i.i.d. points with density f is called a *binomial point process* of n points in B with density f. We write $X \sim$ binomial(B, n, f) (where $\sim$ means "distributed as").

Note that B in Definition 3.1 has volume $|B| > 0$, since $\int_B f(\xi) \mathrm{d}\xi = 1$. In the simplest case, $|B| < \infty$ and each of the n i.i.d. points follows Uniform(B), the uniform distribution on B, i.e. $f(\xi) = 1/|B|$ is the uniform density on B.

DEFINITION 3.2 A point process X on S is a *Poisson point process* with intensity function ρ if the following properties are satisfied (where μ is given by (3.1)):

(i) for any $B \subseteq S$ with $\mu(B) < \infty$, $N(B) \sim$ po$(\mu(B))$, the Poisson distribution with mean $\mu(B)$ (if $\mu(B) = 0$ then $N(B) = 0$);

(ii) for any $n \in \mathbb{N}$ and $B \subseteq S$ with $0 < \mu(B) < \infty$, conditional on $N(B) = n$, $X_B \sim$ binomial(B, n, f) with $f(\xi) = \rho(\xi)/\mu(B)$.

We then write $X \sim$ Poisson(S, ρ).

For any bounded $B \subseteq S$, μ determines the expected number of points in B,
$$\mathbb{E}N(B) = \mu(B).$$
Heuristically, $\rho(\xi)\mathrm{d}\xi$ is the probability for the occurrence of a point in an infinitesimally small ball with centre ξ and volume $\mathrm{d}\xi$.

DEFINITION 3.3 If ρ is constant, the process Poisson(S, ρ) is called a *homogeneous Poisson process* on S with *rate* or *intensity* ρ; else it is said to be an *inhomogeneous Poisson process* on S. Moreover, Poisson$(S, 1)$ is called the *standard Poisson point process* or *unit rate Poisson process* on S.

Examples of simulated homogeneous and inhomogeneous Poisson point processes are shown in Figure 3.1. For a homogeneous Poisson process on $\mathbb{R}^d$, $\rho^{-1/d}$ is a scale parameter, since $X \sim$ Poisson$(\mathbb{R}^d, 1)$ implies that $\{\rho^{-1/d}\xi : \xi \in X\} \sim$ Poisson$(\mathbb{R}^d, \rho)$.

A homogeneous Poisson point process is both stationary and isotropic in the following sense.

DEFINITION 3.4 A point process X on $\mathbb{R}^d$ is *stationary* if its distribution

BASIC PROPERTIES

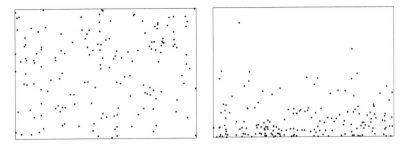

Figure 3.1 *Simulation of homogeneous (left) and inhomogeneous (right) Poisson processes on $S = [0,1] \times [0, 0.7]$. In both cases the expected number of points is 150. For the inhomogeneous Poisson process, $\rho(\xi_1, \xi_2) \propto \exp(-10.6\xi_2)$, $(\xi_1, \xi_2) \in S$.*

is invariant under translations, that is, the distribution of $X+s = \{\xi+s : \xi \in X\}$ is the same as that of X for any $s \in \mathbb{R}^d$. It is *isotropic* if its distribution is invariant under rotations about the origin in $\mathbb{R}^d$, i.e. the distribution of $\mathcal{O}X = \{\mathcal{O}\xi : \xi \in X\}$ is the same as that of X for any rotation $\mathcal{O}$ around the origin.

3.1.2 Existence and independent scattering property

We shall often use the following expansion for the Poisson process:

PROPOSITION 3.1 **(i)** $X \sim \text{Poisson}(S, \rho)$ if and only if for all $B \subseteq S$ with $\mu(B) = \int_S \rho(\xi) \mathrm{d}\xi < \infty$ and all $F \subseteq N_{\mathrm{lf}}$,

$$P(X_B \in F)$$
$$= \sum_{n=0}^{\infty} \frac{\exp(-\mu(B))}{n!} \int_B \cdots \int_B \mathbf{1}[\{x_1, \ldots, x_n\} \in F] \prod_{i=1}^{n} \rho(x_i) \mathrm{d}x_1 \cdots \mathrm{d}x_n \quad (3.2)$$

where the integral for $n = 0$ is read as $\mathbf{1}[\emptyset \in F]$.

(ii) If $X \sim \text{Poisson}(S, \rho)$, then for functions $h : N_{\mathrm{lf}} \to [0, \infty)$ and $B \subseteq S$ with $\mu(B) < \infty$,

$$\mathbb{E}[h(X_B)]$$
$$= \sum_{n=0}^{\infty} \frac{\exp(-\mu(B))}{n!} \int_B \cdots \int_B h(\{x_1, \ldots, x_n\}) \prod_{i=1}^{n} \rho(x_i) \mathrm{d}x_1 \cdots \mathrm{d}x_n. \quad (3.3)$$

Proof. (i) follows immediately from (i) and (ii) in Definition 3.2. (i) and the standard proof imply (ii). □

If S is bounded (or just $\int_S \rho(\xi)\mathrm{d}\xi < \infty$), the existence of Poisson(S,ρ) follows easily from Proposition 3.1 (with $B=S$): let $n(X) \sim \mathrm{po}(\mu(S))$ and $X|n(X)=n \sim$ binomial $(S,n,\rho/\mu(S))$ for $n \in \mathbb{N}$. For an unbounded S, existence of Poisson(S,ρ) is verified in the following theorem.

THEOREM 3.1 $X \sim$ Poisson(S,ρ) exists and is uniquely determined by its void probabilities

$$v(B) = \exp(-\mu(B)), \quad \text{bounded } B \subseteq S. \tag{3.4}$$

Proof. Let $\xi \in S$ be an arbitrary point and set $B_i = \{\eta \in S : i-1 \le \|\eta-\xi\| < i\}$ for $i \in \mathbb{N}$. Clearly, S is a disjoint union of the bounded B_i. Let $X = \cup_1^\infty X_i$ where $X_i \sim$ Poisson(B_i,ρ_i), $i=1,2,\ldots$, are independent, and where ρ_i is the restriction of ρ to B_i. Then for bounded $B \subseteq S$,

$$P(X \cap B = \emptyset) = \prod_{i=1}^\infty P(X_i \cap B = \emptyset) = \prod_{i=1}^\infty \exp(-\mu(B \cap B_i))$$
$$= \exp\left(-\sum_{i=1}^\infty \mu(B \cap B_i)\right) = \exp(-\mu(B)).$$

This is the void probability for a Poisson process with intensity measure μ, so the existence and uniqueness follow from Section 2.3 (or Theorem B.1 in Section B.3). □

REMARK 3.2 In fact, "bounded" in (3.4) can be replaced by "compact" (Matheron 1975) or by "finite union of rectangles" (Rényi's theorem, see e.g. Kingman 1993). Theorem 3.1 extends immediately to independent Poisson processes: if X_1 and X_2 are independent Poisson processes on S with intensity measures μ_1 and μ_2, the distribution of (X_1,X_2) is uniquely determined by the void probabilities

$$P(X_1 \cap A = \emptyset, X_2 \cap B = \emptyset) = \exp(-\mu_1(A) - \mu_2(B)) \tag{3.5}$$

for bounded $A,B \subseteq S$.

By Theorem 3.1, Definition 3.2 states more than needed for characterising a Poisson process. However, as seen in the proof of Theorem 3.1 and many places in the sequel, (i) and (ii) in Definition 3.2 provide an easy construction of a Poisson process. Sometimes in the literature, (ii) is replaced by the condition that $N(B_1),\ldots,N(B_n)$ are independent for disjoint sets $B_1,\ldots,B_n \subseteq S$ and $n \ge 2$. This property is called *independent scattering*. It can be extended as follows.

PROPOSITION 3.2 If X is a Poisson process on S, then $X_{B_1}, X_{B_2},\ldots$ are independent for disjoint sets $B_1,B_2,\ldots \subseteq S$.

Proof. We need only to verify that $X_{B_1}, \ldots, X_{B_n}$ are independent for disjoint sets $B_1, \ldots, B_n \subseteq S$ and $n \geq 2$. This follows straightforwardly from Proposition 3.1 when $n = 2$ and $B_1, B_2 \subseteq S$ are bounded and disjoint (with $B = B_1 \cup B_2$), and hence by induction also when $B_1, \ldots, B_n \subseteq S$ are bounded and disjoint. Hence $X_{B_1}, X_{B_2}, \ldots$ are independent for bounded disjoint $B_i \subseteq S$, $i = 1, 2, \ldots$. Since subsets of S are countable unions of bounded subsets of S, the assertion is now seen to hold for any disjoint $B_1, \ldots, B_n \subseteq S$ and $n \in \mathbb{N}$. □

Proposition 3.2 explains the terminology of *no interaction* and *complete spatial randomness* in the Poisson process.

We conclude by deriving the generating functional for a Poisson process, noticing that in this case the definition (2.1) of $G_X(u)$ extends to any function $u : S \to [0, 1]$, setting $\exp(-\infty) = 0$.

PROPOSITION 3.3 If $X \sim \text{Poisson}(S, \rho)$ then

$$G_X(u) = \exp\left(-\int_S (1 - u(\xi))\rho(\xi)\mathrm{d}\xi\right) \quad (3.6)$$

for functions $u : S \to [0, 1]$.

Proof. We verify this only in the special case when $u(\xi) = 1 - \sum_{i=1}^{\infty}(1 - a_i)\mathbf{1}[\xi \in C_i]$, where $0 \leq a_i \leq 1$ and the $C_i \subseteq S$ are bounded and pairwise disjoint. (The general case follows then by standard arguments and the monotone convergence theorem from measure theory.) By Proposition 3.2, $N(C_i) \sim \text{po}(\mu(C_i))$, $i = 1, 2, \ldots$, are independent. Hence,

$$G_X(u) = \mathbb{E}\prod_{i=1}^{\infty} a_i^{N(C_i)} = \prod_{i=1}^{\infty} \mathbb{E} a_i^{N(C_i)} = \prod_{i=1}^{\infty} \exp(-(1 - a_i)\mu(C_i))$$

$$= \exp\left(-\sum_{i=1}^{\infty}(1 - a_i)\mu(C_i)\right) = \exp\left(-\int (1 - u(\xi))\rho(\xi)\mathrm{d}\xi\right).$$

□

3.1.3 Constructions of stationary Poisson processes

A construction of a Poisson process with a given intensity function is given in the proof of Theorem 3.1. In the stationary case Quine & Watson (1984) give an easier construction using a shift to polar coordinates. This construction is given in the following proposition, letting

$$\omega_d = \pi^{d/2}/\Gamma(1 + d/2), \quad \sigma_d = 2\pi^{d/2}/\Gamma(d/2),$$

denote the volume and surface area of the d-dimensional unit ball.

PROPOSITION 3.4 Let $S_1, U_1, S_2, U_2, \ldots$ be mutually independent where each U_i is uniformly distributed on $\{u \in \mathbb{R}^d : \|u\| = 1\}$ (the $(d-1)$-dimensional unit sphere), and each $S_i \sim \text{Exp}(\rho \omega_d)$ is exponentially distributed with mean $1/(\rho \omega_d)$ for a $\rho > 0$. Let $R_0 = 0$ and $R_i^d = R_{i-1}^d + S_i$, $i = 1, 2, \ldots$. Then $X = \{R_1 U_1, R_2 U_2, \ldots\} \sim \text{Poisson}(\mathbb{R}^d, \rho)$.

Proof. By Theorem 3.1 we have to verify that $v(B) = \exp(-\rho|B|)$ for any bounded $B \subset \mathbb{R}^d$. This follows if for any $r > 0$, (i) $N(b(0,r)) \sim \text{po}(\rho \omega_d r^d)$ and (ii) conditional on $N(b(0,r)) = n$, we have that $\xi_1, \ldots, \xi_n$ are distributed as n uniform points on $b(0,r)$ ordered according to Euclidian distance. Because if $B \subseteq b(0,r)$ then by (i) and (ii),

$$P(N(B) = 0)$$
$$= \sum_{n=0}^{\infty} P(N(B) = 0 | N(b(0,r)) = n) P(N(b(0,r)) = n)$$
$$= \sum_{n=0}^{\infty} (|b(0,r) \setminus B|/|b(0,r)|)^n \exp(-\rho|b(0,r)|)(\rho|b(0,r)|)^n/n!$$
$$= \exp(\rho|b(0,r) \setminus B| - \rho|b(0,r)|) = \exp(-\rho|B|).$$

We verify now (i) and (ii).[†] For a fixed $n \in \mathbb{N}$, the joint density of $(S_1, U_1, \ldots, S_n, U_n)$ is

$$g_1(s_1, u_1, \ldots, s_n, u_n) = (\rho \omega_d)^n \exp(-\rho \omega_d r_n^d) \sigma_d^{-n}$$

where $r_n = (\sum_{j=1}^n s_j)^{1/d}$ and the density is with respect to the n-fold product measure of Lebesgue measure on $(0, \infty)$ times surface measure on the d-dimensional unit ball. The conditional probability of $N(b(0,r)) = n$ given $(S_1, U_1, \ldots, S_n, U_n) = (s_1, u_1, \ldots, s_n, u_n)$ is

$$\mathbf{1}[r_n \leq r] \exp(-\rho \omega_d (r^d - r_n^d)),$$

since $N(b(0,r)) = n$ is equivalent to $R_n \leq r$ and $S_{n+1} > r^d - R_n^d$. The joint density of $(S_1, U_1, \ldots, S_n, U_n, \mathbf{1}[N(b(0,r)) = n])$ therefore satisfies

$$g_2(s_1, u_1, \ldots, s_n, u_n, 1) = \mathbf{1}[r_n \leq r] (\rho \omega_d)^n \sigma_d^{-n} \exp(-\rho \omega_d r^d).$$

Integrating $g_2(s_1, u_1, \ldots, s_n, u_n, 1)$ over $(s_1, u_1, \ldots, s_n, u_n)$ we obtain

$$P(N(b(0,r)) = n) = \exp(\rho \omega_d r^d)(\rho \omega_d r^d)^n / n!,$$

and so (i) is verified. Thus the conditional density of $(R_1, U_1, \ldots, R_n, U_n)$

[†] Readers who are unfamiliar with measure theory may prefer to skip this part of the proof.

BASIC PROPERTIES

given $N(b(0,r)) = n$ is

$$g_3(r_1, u_1, \ldots, r_n, u_n) =$$

$$n! \sigma_d^{-n} r^{-nd} \mathbf{1}[0 \leq r_1 \leq r_2 \leq \cdots \leq r_n \leq r] \prod_{i=1}^{n} dr_i^{d-1}$$

where $\sigma_d r^d / d = \omega_d r^d$ is the volume of $b(0,r)$. Transforming from polar coordinates (R_i, U_i) to $X_i = R_i U_i$ (see e.g. Proposition 2.8 in Jensen 1998) we further obtain that the conditional density of $(\xi_1, \ldots, \xi_n)$ given $N(b(0,r)) = n$ is

$$g_4(\xi_1, \ldots, \xi_n) = n!(\omega_d r^d)^{-n} \mathbf{1}[\|\xi_1\| \leq \|\xi_2\| \leq \cdots \leq \|\xi_n\| \leq r].$$

Thereby (ii) is verified. □

When $d = 1$ it seems more natural to apply the following proposition where points on the real line are ordered as usual.

PROPOSITION 3.5 Let $\ldots, T_{-2}, T_{-1}, T_1, T_2, \ldots$ be mutually independent and $\text{Exp}(\rho)$-distributed for a $\rho > 0$. Let $\xi_0 = 0$, and for $i = 1, 2, \ldots$, let $\xi_i = \xi_{i-1} + T_i$ and $\xi_{-i} = \xi_{-i+1} - T_{-i}$. Then $X = \{\ldots, \xi_{-2}, \xi_{-1}, \xi_1, \xi_2, \ldots\}$ $\sim \text{Poisson}(\mathbb{R}, \rho)$.

Proof. This is verified along similar lines as in the proof of Proposition 3.4. □

REMARK 3.3 The point $\xi_0 = 0$ serves as a starting point when generating $\xi_1, \xi_2, \ldots$ and $\xi_{-1}, \xi_{-2}, \ldots$. By stationarity, an arbitrary point $\xi_0 \in \mathbb{R}^d$ can be chosen. Note that the exponential distribution has *no memory* in the sense that for $a > 0$ and $t > 0$,

$$T \sim \text{Exp}(a) \Rightarrow T - t | T > t \sim \text{Exp}(a).$$

So the construction of $\text{Poisson}(I, \rho)$ on an interval $I = (a, b)$, where either a or b or both are finite, is similar to that of Proposition 3.5, taking either $\xi_0 = a$ or $\xi_0 = b$.

The interval $[\xi_{-1}, \xi_1]$ which contains ξ_0 is longer than a typical interval $[\xi_{i-1}, \xi_i]$. Specifically, the length of a typical interval is $\text{Exp}(\rho)$-distributed, while $\xi_1 - \xi_{-1}$ is distributed as the sum of two independent $\text{Exp}(\rho)$-distributed random variables. This seemingly "paradox" is due to sampling bias: Consider a stationary point process with points $\ldots < \eta_{-1} < \eta_0 < \eta_1 < \ldots$ on the real line and which is independent of a random real variable Z, and suppose that we sample the interval $[\eta_{k-1}, \eta_k]$ which contains Z. Let L_0 denote the length of this "point sampled" interval and let L be the length of a typical interval (i.e. the distribution of L is given by the so-called Palm distribution, see e.g.

Stoyan et al. 1995 and Appendix C). Since larger intervals are more likely to get sampled, L_0 will on the average be greater than L. In fact, it is not hard to verify that for functions $k : (0, \infty) \to (0, \infty)$,

$$\mathbb{E}k(L)/\mathbb{E}L = E(k(L_0)/L_0),$$

so

$$\mathbb{E}L_0 = \mathbb{E}(L^2)/\mathbb{E}L \geq (\mathbb{E}L)^2/\mathbb{E}L = \mathbb{E}L.$$

Even a stronger result holds, namely $P(L > l) \leq P(L_0 > l)$ for all $t \geq 0$ (Mecke 1999).

REMARK 3.4 Proposition 3.5 extends to the inhomogeneous case by *transforming a standard Poisson process*. For simplicity suppose that $\rho(\xi) > 0$ for all $\xi \in \mathbb{R}$ and that

$$H(t) = \int_t^\infty \rho(s) \mathrm{d}s$$

is finite for all $t \in \mathbb{R}$. Then $H(t)$ is strictly decreasing. Let $X = \{H^{-1}(\eta) : \eta \in Y\}$ where $Y \sim \text{Poisson}(\mathbb{R}, 1)$. Then $X \sim \text{Poisson}(\mathbb{R}, \rho)$, since

$$P(X \cap A = \emptyset) = P(Y \cap H(A) = \emptyset) = \exp(-|H(A)|) = \exp\left(-\int_A \rho(\xi) \mathrm{d}\xi\right)$$

for bounded $A \subset \mathbb{R}$. This construction is easily modified to the case where $\rho(\xi) \geq 0$.

For a fixed $\xi_0 \in \mathbb{R}^d$, enumerate the points in Y as $\cdots > \eta_2 > \eta_1 > \eta_{-1} > \eta_{-2} > \cdots$ so that $\eta_1 \geq \eta_0 > \eta_{-1}$ where $\eta_0 = H(\xi_0)$, and let $\xi_i = H^{-1}(\eta_{-i})$. Then the differences $T_i = \xi_i - \xi_{i-1}$ and $T_{-i} = \xi_{-i+1} - \xi_{-i}$, $i = 1, 2, \ldots$, are mutually independent but not exponentially distributed (unless of course $\rho(\cdot)$ is constant). The sequence $\ldots, \xi_{-2}, \xi_{-1}, \xi_0, \xi_1, \xi_2, \ldots$ is a Markov chain which can be viewed as the arrival times of a *birth process* with birth rate ρ.

3.2 Further results

Poisson processes are very tractable for mathematical analysis. This section collects several useful results.

3.2.1 Slivnyak-Mecke's theorem

The simplest way of characterising a Poisson point process is by its void probabilities (3.4). A less well known but very useful characterisation of a Poisson process is provided by the following theorem called the *Slivnyak-Mecke theorem* (Mecke 1967).

FURTHER RESULTS

THEOREM 3.2 If $X \sim \text{Poisson}(S, \rho)$, then for functions $h : S \times N_{\text{lf}} \to [0, \infty)$,

$$\mathbb{E} \sum_{\xi \in X} h(\xi, X \setminus \xi) = \int_S \mathbb{E} h(\xi, X) \rho(\xi) \mathrm{d}\xi \qquad (3.7)$$

(where the left hand side is finite if and only if the right hand side is finite).

Proof. Consider first the case where

$$\sum_{\xi \in X} h(\xi, X \setminus \xi)$$

only depends on X through X_B for some $B \subseteq S$ with $\int_B \rho(\xi)\mathrm{d}\xi < \infty$. Applying Proposition 3.1 we obtain

$$\mathbb{E} \sum_{\xi \in X} h(\xi, X \setminus \xi) =$$

$$\sum_{n=1}^{\infty} \frac{\exp(-\mu(B))}{n!} \int_B \cdots \int_B \sum_{i=1}^n h(x_i, \{x_1, \ldots, x_n\} \setminus x_i) \prod_{i=1}^n \rho(x_i) \mathrm{d}x_1 \cdots \mathrm{d}x_n$$

$$= \sum_{n=1}^{\infty} \frac{\exp(-\mu(B))}{(n-1)!} \int_B \cdots \int_B h(x_n, \{x_1, \ldots, x_{n-1}\}) \prod_{i=1}^n \rho(x_i) \mathrm{d}x_1 \cdots \mathrm{d}x_n$$

$$= \int_S \mathbb{E} h(\xi, X) \rho(\xi) \mathrm{d}\xi.$$

Consider next the general case.[†] If we can prove (3.7) when $h(\xi, x) = \mathbf{1}[(\xi, x) \in F]$, $F \subseteq S \times N_{\text{lf}}$, is an indicator function, then by the standard proof, (3.7) is satisfied for any nonnegative function h. The left and right sides in (3.7) both become measures on $S \times N_{\text{lf}}$ when we consider indicator functions $h(\xi, x) = \mathbf{1}[(\xi, x) \in F]$, $F \subseteq S \times N_{\text{lf}}$, and they are both equal to $\mathbb{E} N(B) = \int_B \rho(\xi) \mathrm{d}\xi$ if $h(\xi, x) = \mathbf{1}[(\xi, x) \in B \times N_{\text{lf}}]$ for bounded $B \subseteq S$. By Lemma B.4 in Section B.3, we thus just need to verify (3.7) when h is of the form

$$h(\xi, x) = \mathbf{1}[\xi \in C, n(x_A) = 0], \quad \text{bounded } A, C \subseteq S. \qquad (3.8)$$

But this follows from the first part of the proof, since $B = A \cup C$ is bounded and ρ is locally integrable. □

By induction we easily obtain the following *extended Slivnyak-Mecke theorem.*

[†] Readers who are unfamiliar with measure theory may prefer to skip this part of the proof.

THEOREM 3.3 If $X \sim \mathrm{Poisson}(S, \rho)$, then for any $n \in \mathbb{N}$ and any function $h : S^n \times N_{\mathrm{lf}} \to [0, \infty)$,

$$\mathbb{E} \sum_{\xi_1,\ldots,\xi_n \in X}^{\neq} h(\xi_1, \ldots, \xi_n, X \setminus \{\xi_1, \ldots, \xi_n\})$$
$$= \int_S \cdots \int_S \mathbb{E} h(\xi_1, \ldots, \xi_n, X) \prod_{i=1}^n \rho(\xi_i) \mathrm{d}\xi_1 \cdots \mathrm{d}\xi_n \quad (3.9)$$

where the $\neq$ over the summation sign means that the n points $\xi_1, \ldots, \xi_n$ are pairwise distinct.

Proof. For $n = 1$, (3.7) and (3.9) obviously agree. For $n \geq 2$, consider

$$\tilde{h}(\xi, x) = \sum_{\xi_2,\ldots,\xi_n \in x}^{\neq} h(\xi, \xi_2, \ldots, \xi_n, x \setminus \{\xi_2, \ldots, \xi_n\}).$$

By (3.7) and the induction hypothesis, the left hand side in (3.9) is equal to

$$\mathbb{E} \sum_{\xi \in X} \tilde{h}(\xi, X \setminus \xi) = \int_S \mathbb{E}\tilde{h}(\xi, X) \rho(\xi) \mathrm{d}\xi$$
$$= \int_S \int_S \cdots \int_S \mathbb{E} h(\xi, \xi_2, \ldots, \xi_n, X) \rho(\xi_2) \cdots \rho(\xi_n) \mathrm{d}\xi_2 \cdots \mathrm{d}\xi_n \rho(\xi) \mathrm{d}\xi$$

which agrees with the right side in (3.9). □

3.2.2 Superpositioning and thinning

We now define two basic operations for point processes.

DEFINITION 3.5 A disjoint union $\cup_{i=1}^{\infty} X_i$ of point processes $X_1, X_2, \ldots,$ is called a *superposition*.

DEFINITION 3.6 Let $p : S \to [0, 1]$ be a function and X a point process on S. The point process $X_{\mathrm{thin}} \subseteq X$ obtained by including $\xi \in X$ in X_{thin} with probability $p(\xi)$, where points are included/excluded independently of each other, is said to be an *independent thinning* of X with *retention probabilities* $p(\xi)$, $\xi \in S$. Formally, we can set

$$X_{\mathrm{thin}} = \{\xi \in X : R(\xi) \leq p(\xi)\}$$

where $R(\xi) \sim \mathrm{Uniform}[0, 1]$, $\xi \in S$, are mutually independent and independent of X.

As shown in the following two propositions, the class of Poisson processes is closed under both superpositioning and independent thinning.

FURTHER RESULTS

PROPOSITION 3.6 *If* $X_i \sim \text{Poisson}(S, \rho_i)$, $i = 1, 2, \ldots$, *are mutually independent and* $\rho = \sum \rho_i$ *is locally integrable, then with probability one,* $X = \cup_{i=1}^{\infty} X_i$ *is a disjoint union, and* $X \sim \text{Poisson}(S, \rho)$.

Proof. The first part can easily be verified using Proposition 3.1 and considering two independent Poisson processes within a bounded ball B which expands to S; it is also verified in the Disjointness Lemma in Kingman (1993). The second part follows from Theorem 3.1: for bounded $B \subseteq S$,

$$P(X_B = \emptyset) = \prod_{i=1}^{\infty} P(X_i \cap B = \emptyset) = \prod_{i=1}^{\infty} \exp(-\mu_i(B)) = \exp(-\mu(B))$$

where $\mu_i(B) = \int_B \rho_i(\xi) d\xi$. □

PROPOSITION 3.7 *Suppose that* $X \sim \text{Poisson}(S, \rho)$ *is subject to independent thinning with retention probabilities* $p(\xi)$, $\xi \in S$, *and let*

$$\rho_{\text{thin}}(\xi) = p(\xi)\rho(\xi), \quad \xi \in S.$$

Then X_{thin} *and* $X \setminus X_{\text{thin}}$ *are independent Poisson processes with intensity functions* ρ_{thin} *and* $\rho - \rho_{\text{thin}}$, *respectively.*

Proof. Let μ_{thin} be given by $\mu_{\text{thin}}(B) = \int \rho_{\text{thin}}(\xi) d\xi$. By (3.5), we only need to verify that

$$P(X_{\text{thin}} \cap A = \emptyset, (X \setminus X_{\text{thin}}) \cap B = \emptyset) = \exp(-\mu_{\text{thin}}(A) - \mu(B) + \mu_{\text{thin}}(B)) \quad (3.10)$$

for bounded $A, B \subseteq S$. By Proposition 3.1 and Definition 3.6, for any bounded $C \subseteq S$,

$$P(X_{\text{thin}} \cap C = \emptyset) = \sum_{n=0}^{\infty} \frac{1}{n!} e^{-\mu(C)} \left(\int_C (1 - p(\xi))\rho(\xi) d\xi \right)^n$$
$$= \exp(-\mu_{\text{thin}}(C)).$$

By symmetry,

$$P((X \setminus X_{\text{thin}}) \cap C = \emptyset) = \exp(-(\mu - \mu_{\text{thin}})(C)).$$

The result now follows by Proposition 3.2: for bounded $A, B \subseteq S$,

$P(X_{\text{thin}} \cap A = \emptyset, (X \setminus X_{\text{thin}}) \cap B = \emptyset)$
$= P(X \cap A \cap B = \emptyset) P(X_{\text{thin}} \cap (A \setminus B) = \emptyset) P((X \setminus X_{\text{thin}}) \cap (B \setminus A) = \emptyset)$
$= \exp(-\mu(A \cap B) - \mu_{\text{thin}}(A \setminus B) - (\mu - \mu_{\text{thin}})(B \setminus A))$

whereby (3.10) is obtained. □

The following corollary shows that inhomogeneous Poisson processes often can be derived by thinning a homogeneous Poisson process.

COROLLARY 3.1 Suppose that $X \sim \text{Poisson}(\mathbb{R}^d, \rho)$ where the intensity function ρ is bounded by a finite constant c. Then X is distributed as an independent thinning of $\text{Poisson}(\mathbb{R}^d, c)$ with retention probabilities $p(\xi) = \rho(\xi)/c$.

Proof. Follows immediately from Proposition 3.7. □

3.2.3 Simulation of Poisson processes

Simulation of a Poisson point process $X \sim \text{Poisson}(\mathbb{R}^d, \rho)$ within a bounded set $B \subset \mathbb{R}^d$ is usually easy.

Consider first the homogeneous case where $\rho(\xi) = \rho_0 > 0$ is constant for all $\xi \in B$. For $d = 1$, if B is an interval, we may simulate the restriction to B of the Poisson process considered in Proposition 3.5. For $d \geq 2$, we consider three cases:

1. If $B = b(0, r)$ is a ball, simulation of X_B is straightforward by Proposition 3.4: simulate $S_1, \ldots, S_m$ and $U_1, \ldots, U_{m-1}$, where m is given by that $R_{m-1} \leq r < R_m$, and return $X_B = \{R_1 U_1, \ldots, R_{m-1} U_{m-1}\}$. This *radial simulation procedure* is due to Quine & Watson (1984).

2. If $B = [0, a_1] \times \cdots \times [0, a_d]$ is a box, then by (i)–(ii) in Definition 3.2, we may first generate the number of points $N(B) \sim \text{po}(\rho_0 a_1 \cdots a_d)$ (see e.g. Ripley, 1987), and second the locations of the $N(B)$ independent and uniformly distributed points in B. An alternative is to apply Proposition 3.5, where we exploit that by (i)–(ii) in Definition 3.2,

    - the point process consisting of the first coordinates $\xi^{(1)}$ of the points $\xi = (\xi^{(1)}, \xi^{(2)}, \ldots, \xi^{(d)}) \in X_B$ is $\text{Poisson}([0, a_1], \rho_0 a_2 \cdots a_d)$,
    - the remaining components $(\xi^{(2)}, \ldots, \xi^{(d)})$ of such points are independent and uniformly distributed on $[0, a_2] \times \cdots \times [0, a_d]$.

3. If B is bounded but neither a ball nor a box, we simply simulate X on a ball or box B_0 containing B, and disregard the points falling in $B_0 \setminus B$.

Suppose next that X_B is inhomogeneous with $\rho(\xi), \xi \in B$, bounded by a constant $\rho_0 > 0$. Then we may first generate a homogeneous Poisson process Y on B with intensity ρ_0. Next we obtain X_B as an independent thinning of Y_B with retention probabilities $p(\xi) = \rho(\xi)/\rho_0$, $\xi \in B$, cf. Proposition 3.7. The right plot in Figure 3.1 is generated in this way. For $d = 1$, an alternative is the transformation method considered in Remark 3.4 (this method does not require ρ to be bounded on B).

3.2.4 Densities for Poisson processes

If X_1 and X_2 are two point processes defined on the same space S, then X_1 is *absolutely continuous* with respect to X_2 (or more precisely the

MARKED POISSON PROCESSES

distribution of X_1 is absolutely continuous with respect to the distribution of X_2) if and only if $P(X_2 \in F) = 0$ implies that $P(X_1 \in F) = 0$ for $F \subseteq N_{\text{lf}}$. Equivalently, by the Radon-Nikodym theorem (see e.g. p. 422 in Billingsley 1995), there exists a function $f : N_{\text{lf}} \to [0, \infty]$ so that

$$P(X_1 \in F) = \mathbb{E}\big[\mathbf{1}[X_2 \in F] f(X_2)\big], \quad F \subseteq N_{\text{lf}}.$$

We call f a *density* for X_1 with respect to X_2.

The following proposition shows that Poisson processes are not always absolutely continuous with respect to each other; but they are always absolutely continuous with respect to the standard Poisson process if we let S be bounded.

PROPOSITION 3.8

(i) For any numbers $\rho_1 > 0$ and $\rho_2 > 0$, Poisson($\mathbb{R}^d, \rho_1$) is absolutely continuous with respect to Poisson($\mathbb{R}^d, \rho_2$) if and only if $\rho_1 = \rho_2$.

(ii) For $i = 1, 2$, suppose that $\rho_i : S \to [0, \infty)$ so that $\mu_i(S) = \int_S \rho_i(\xi) d\xi$ is finite, and that $\rho_2(\xi) > 0$ whenever $\rho_1(\xi) > 0$. Then Poisson(S, ρ_1) is absolutely continuous with respect to Poisson(S, ρ_2), with *density*

$$f(x) = \exp(\mu_2(S) - \mu_1(S)) \prod_{\xi \in x} \rho_1(\xi)/\rho_2(\xi) \quad (3.11)$$

for finite point configurations $x \subset S$ (taking $0/0 = 0$).

Proof. (i) The "if" part is trivial. Assume that $\rho_1 \neq \rho_2$. Consider disjoint subsets $A_i \subseteq \mathbb{R}^d$ with $|A_i| = 1$, $i = 1, 2, \ldots$, and let $F = \{x \in N_{\text{lf}} : \lim_{m \to \infty} \sum_{i=1}^m n(x \cap A_i)/m = \rho_1\}$. Under both Poisson($\mathbb{R}^d, \rho_1$) and Poisson($\mathbb{R}^d, \rho_2$), the $N(A_i)$ are i.i.d. but with means ρ_1 and ρ_2, respectively. Hence, by the strong law of large numbers, $P(X_1 \in F) = 1$ but $P(X_2 \in F) = 0$.

(ii) If Y is Poisson(S, ρ_2), we have to verify that $P(X \in F) = \mathbb{E}\big[\mathbf{1}[Y \in F] f(Y)\big]$ for $F \subseteq N_{\text{lf}}$. This is straightforwardly derived from Proposition 3.1. □

We consider densities of point processes in more detail in Chapter 6.

3.3 Marked Poisson processes

Consider now a marked point process $X = \{(\xi, m_\xi) : \xi \in Y\}$ with points in T and mark space M, cf. Section 2.2 or the general setting in Section B.2.

DEFINITION 3.7 Suppose that Y is Poisson(T, ϕ), where ϕ is a locally integrable intensity function, and conditional on Y, the marks $\{m_\xi :$

$\xi \in Y$} are mutually independent. Then X is a *marked Poisson process*. If the marks are identically distributed with a common distribution Q, then Q is called the *mark distribution*.

The marks can e.g. be integers, real numbers, geometric objects, or even themselves point processes. If $M = \{1, \ldots, k\}$, then X is called a *multitype Poisson process*. If $M = \{K \subseteq S : K \text{ compact}\}$, then the so-called *Boolean model* is obtained (more details can be found in Molchanov 1997).

A simulation of a Boolean model is shown in Figure 3.2 (left). For comparison with the spruces data in Example 1.4 we use $T = [0, 56] \times [0, 38]$ and sample the marks from the empirical distribution of the disc radii in Figure 1.4 (see also Figure 4.13 in Section 4.5). The large number of overlapping discs in the simulation indicates that a Boolean model is not appropriate for the spruces data. In Examples 9.1 and 9.2 in Sections 9.1.4 we consider more appropriate Markov point process models.

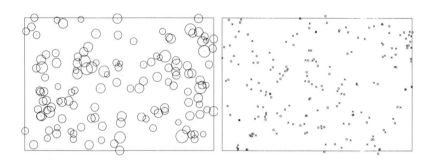

Figure 3.2 *Left: simulation of a homogeneous Boolean model conditional on 134 points and with disc radii sampled from the empirical distribution of the disc radii in Figure 1.4. Right: simulation of a bivariate Poisson process where circles are given by random displacements of the crosses using a bivariate standard normal distribution. In both plots the simulation window is $[0, 56] \times [0, 38]$.*

In the rest of this section we consider the case $M \subseteq \mathbb{R}^p$, $p \geq 1$.

PROPOSITION 3.9 Let X be a marked Poisson process with $M \subseteq \mathbb{R}^p$ and where, conditional on Y, each mark m_ξ has a discrete or continuous density p_ξ which does not depend on $Y \setminus \xi$. Let $\rho(\xi, m) = \phi(\xi) p_\xi(m)$. Then

(i) $X \sim \text{Poisson}(T \times M, \rho)$;

(ii) if the density on M defined by $\kappa(m) = \int_T \rho(\xi, m) \mathrm{d}\xi$ is locally in-

MARKED POISSON PROCESSES

tegrable, then $\{m_\xi : \xi \in Y\} \sim \text{Poisson}(M, \kappa)$ (where we refer to Remark 3.1 for the discrete case of M).

Proof. (i) follows directly from Proposition 3.1.

(ii) Let B be a bounded Borel set in M. Then

$$P(\{m_\xi : \xi \in Y\} \cap B = \emptyset) = 1 - P(X \cap (T \times B) \neq \emptyset) = \exp(-\int_B \kappa(m) \mathrm{d}m)$$

where the last equality follows from (i). The result now follows from Theorem 3.1. □

Note that not all $\text{Poisson}(T \times M, \rho)$-processes are marked Poisson processes, since Definition 3.7 requires local integrability of $\phi(\xi)$ given by $\int_M \rho(\xi, m) \mathrm{d}m$ in the continuous case or by $\sum_{\xi \in M} \rho(\xi, m)$ in the discrete case. Likewise, it is easy to construct marked Poisson processes which are not Poisson processes, since the conditional distribution of m_ξ for a marked Poisson process may depend on $Y \setminus \xi$.

3.3.1 Random independent displacements of the points in a Poisson process

Suppose that $Y^* = \{\xi + m_\xi : \xi \in Y\}$ is obtained by *random independent displacements* of the points in $Y \sim \text{Poisson}(\mathbb{R}^d, \rho)$ where conditional on Y, the m_ξ are independent and each distributed according to a density p_ξ on $\mathbb{R}^d$ which does not depend on $Y \setminus \xi$. Letting X^* be the marked Poisson point process with marked points $(\xi, \xi + m_\xi)$, it follows from (ii) in Proposition 3.9 that Y^* is a Poisson process with intensity function

$$\rho^*(\eta) = \int \rho(\xi) p_\xi(\eta - \xi) \mathrm{d}\xi$$

provided ρ^* is locally integrable. If ρ is constant and p_ξ does not depend on ξ, then Y^* becomes a stationary Poisson process with intensity $\rho^* = \rho$.

Figure 3.2 (right) shows a simulation of a bivariate Poisson process, where one type of points (the circles) are random displacements of the other type of points (the crosses), which follow a $\text{Poisson}([0, 56] \times [0, 38], 100)$-process. The displacements are given by independent bivariate standard normal distributions.

3.3.2 Multivariate Poisson processes and random labelling

By a *multivariate Poisson process* (or a *bivariate Poisson process* if $k = 2$) is usually meant that each X_i is a stationary Poisson process on $\mathbb{R}^d$ with intensity $0 < \rho_i < \infty$ for $i = 1, \ldots, k$, see e.g. Diggle (1983). An example of a bivariate Poisson process with positive correlation between the components is (Y, Y^*) where Y is a stationary Poisson process and Y^* is obtained by random independent displacements of the points in Y. A construction of a bivariate Poisson process with negative correlation between the components is given in Brown, Silverman & Milne (1981).

Consider now a multitype process X with $M = \{1, \ldots, k\}$. This can be identified with a multivariate point process $(X_1, \ldots, X_k)$, cf. Section 2.2. We have equivalence between the following two properties: (i) $P(m_\xi = i | Y = y) = p_\xi(i)$ depends only on ξ for realisations y of Y and $\xi \in y$; (ii) $(X_1, \ldots, X_k)$ is a multivariate Poisson process with independent components $X_i \sim \text{Poisson}(T, \rho_i)$ where $\rho_i(\xi) = \phi(\xi)p_\xi(i)$, $i = 1, \ldots, k$. That (i) implies (ii) follows by induction using Proposition 3.7, since if e.g. $k = 2$, we can think of $X_1 = \{\xi \in Y : m_\xi = 1\}$ as an independent thinning of Y with retention probability $p_\xi(1)$, $\xi \in Y$. That (ii) implies (i) is a consequence of the well-known result that we obtain a multinomial distribution when conditioning on the sum of k independent Poisson variables.

A common hypothesis for marked point processes $\{(\xi, m_\xi) : \xi \in Y\}$ is that of *random labelling* which means that conditional on Y, the marks m_ξ are mutually independent and the distribution of m_ξ does not depend on Y. For a multitype Poisson process, for example, random labelling means (i) above with $p_\xi(i) = p(i)$ not depending on the location ξ.

CHAPTER 4

Summary statistics

This chapter surveys the most commonly used summary statistics for point processes and multitype point processes in the homogeneous as well as the inhomogeneous case. Exploratory analysis for spatial point patterns and the validation of fitted models are often based on nonparametric estimates of various summary statistics. For example, the initial step of a point process analysis often consists of looking for discrepancies with a Poisson model using nonparametric estimates of summary statistics.

Section 4.1 deals with characteristics describing the first and second order properties of a point process, particularly intensity and pair correlation functions and second order reduced moment measures. Section 4.2 focuses on summary statistics for second order properties (such as the so-called K and L-functions, and including the anisotropic case) and summary statistics based on interpoint distances (such as the so-called F, G, and J-functions). Section 4.3 concerns nonparametric estimation of summary statistics and contains some examples of applications; many other examples are given in the other chapters of this book. Sections 4.4–4.5 concern summary statistics for multivariate and marked point processes. The more technical part related to this chapter is deferred to Appendix C.

Further material on summary statistics can be found in Stoyan & Stoyan (1994, 2000), Baddeley & Gill (1997), Ohser & Mücklich (2000), and the references therein. The application of many of the summary statistics in this chapter requires some sort of stationarity assumption. We therefore consider point processes (or marked point processes) with points in $\mathbb{R}^d$. However, in some cases we can replace $\mathbb{R}^d$ with a proper subset of $\mathbb{R}^d$ or an arbitrary space S, cf. Appendix C.

4.1 First and second order properties of a point process

Throughout this section X denotes a point process on $S = \mathbb{R}^d$.

4.1.1 Basic definitions and results

The first and second order properties of the random count variables $N(B)$ for $B \subseteq S$ are described by the so-called intensity measure and second order factorial moment measure defined as follows.

DEFINITION 4.1 The *intensity measure* μ on $\mathbb{R}^d$ is given by
$$\mu(B) = \mathbb{E}N(B), \quad B \subseteq \mathbb{R}^d,$$
and the *second order factorial moment measure* $\alpha^{(2)}$ on $\mathbb{R}^d \times \mathbb{R}^d$ by
$$\alpha^{(2)}(C) = \mathbb{E} \sum_{\xi,\eta \in X}^{\neq} \mathbf{1}[(\xi, \eta) \in C], \quad C \subseteq \mathbb{R}^d \times \mathbb{R}^d.$$

In Appendix C we consider in more detail moment measures of any order on a general space S. Note that
$$\mathbb{E}\big(N(B_1)N(B_2)\big) = \alpha^{(2)}(B_1 \times B_2) + \mu(B_1 \cap B_2), \quad B_1, B_2 \subseteq \mathbb{R}^d, \quad (4.1)$$
so $\alpha^{(2)}$ and μ determine the second order moments of the random variables $N(B)$, $B \subseteq \mathbb{R}^d$.

DEFINITION 4.2 If the intensity measure μ can be written as
$$\mu(B) = \int_B \rho(\xi) \mathrm{d}\xi, \quad B \subseteq \mathbb{R}^d,$$
where ρ is a nonnegative function, then ρ is called the *intensity function*. If ρ is constant, then X is said to be *homogeneous* or *first order stationary* with *intensity* ρ; otherwise X is said to be *inhomogeneous*.

Definition 4.2 is in agreement with Definition 3.3. Heuristically, $\rho(\xi)\mathrm{d}\xi$ is the probability for the occurrence of a point in an infinitesimally small ball with centre ξ and volume $\mathrm{d}\xi$. For a homogeneous point process, ρ is the mean number of points per unit volume.

DEFINITION 4.3 If the second order factorial moment measure $\alpha^{(2)}$ can be written as
$$\alpha^{(2)}(C) = \int\int \mathbf{1}[(\xi,\eta) \in C] \rho^{(2)}(\xi,\eta) \mathrm{d}\xi \mathrm{d}\eta, \quad C \subseteq \mathbb{R}^d \times \mathbb{R}^d,$$
where $\rho^{(2)}$ is a nonnegative function, then $\rho^{(2)}$ is called the *second order product density*.

Intuitively, $\rho^{(2)}(\xi,\eta) \mathrm{d}\xi \mathrm{d}\eta$ is the probability for observing a pair of points from X occuring jointly in each of two infinitesimally small balls with centres ξ, η and volumes $\mathrm{d}\xi, \mathrm{d}\eta$. In order to study whether a point

FIRST AND SECOND ORDER PROPERTIES 31

process deviates from the Poisson process, it is useful to normalise the second order product density $\rho^{(2)}(\xi,\eta)$ by dividing with $\rho(\xi)\rho(\eta)$.

DEFINITION 4.4 *If both ρ and $\rho^{(2)}$ exist, the* pair correlation function *is defined by*

$$g(\xi,\eta) = \frac{\rho^{(2)}(\xi,\eta)}{\rho(\xi)\rho(\eta)}$$

where we take $a/0 = 0$ for $a \geq 0$.

The g-function is widely used in astronomy and astrophysics (Peebles 1974, Besag 1977b, Kerscher 2000). For a Poisson process we immediately obtain from the extended Slivnyak-Mecke Theorem 3.3 (Section 3.2) that we can take $\rho^{(2)}(\xi,\eta) = \rho(\xi)\rho(\eta)$ so that $g(\xi,\eta) = 1$. If for example $g(\xi,\eta) > 1$, this indicates that pair of points are more likely to occur jointly at the locations ξ,η than for a Poisson process with the same intensity function as X.

If X is stationary, g becomes translation invariant, i.e. $g(\xi,\eta) = g(\xi - \eta)$.[†] Note that there exist cases where g is translation invariant but ρ is inhomogeneous, see e.g. Section 5.6.

PROPOSITION 4.1 *Suppose that X has intensity function ρ and second order product density $\rho^{(2)}$. Then for functions $h_1 : \mathbb{R}^d \to [0,\infty)$ and $h_2 : \mathbb{R}^d \times \mathbb{R}^d \to [0,\infty)$,*

$$\mathbb{E} \sum_{\xi \in X} h_1(\xi) = \int h_1(\xi)\rho(\xi)\mathrm{d}\xi \qquad (4.2)$$

and

$$\mathbb{E} \sum_{\xi,\eta \in S}^{\neq} h_2(\xi,\eta) = \int\int h_2(\xi,\eta)\rho^{(2)}(\xi,\eta)\mathrm{d}\xi\mathrm{d}\eta. \qquad (4.3)$$

Proof. By the standard proof these equalities follow directly from Definitions 4.1–4.3. □

A more general version of these results is given by (C.1) in Section C.1.1.

PROPOSITION 4.2 *Suppose that X has intensity function ρ and second order product density $\rho^{(2)}$, and that X_{thin} is an independent thinning of X with retention probabilities $p(\xi)$, $\xi \in \mathbb{R}^d$. Then the intensity function and second order product density of X_{thin} are given by*

[†] Strictly speaking this holds for all $\xi,\eta \in \mathbb{R}^d \setminus A$ where $A \subset \mathbb{R}^d$ is of zero volume. Note that we abuse notation and denote by g also the function which describes how the pair correlation depends on differences.

$\rho_{\text{thin}}(\xi) = p(\xi)\rho(\xi)$ and $\rho_{\text{thin}}^{(2)}(\xi,\eta) = p(\xi)p(\eta)\rho^{(2)}(\xi,\eta)$, and the pair correlation function is *invariant under independent thinning*: $g = g_{\text{thin}}$.

Proof. Using a notation as in Definition 3.6, recall that $R(\xi) \sim$ Uniform $([0,1])$, $\xi \in \mathbb{R}^d$, are mutually independent and independent of X. So for $B \subseteq \mathbb{R}^d$,

$$\mathbb{E}\, n(X_{\text{thin}} \cap B) = \mathbb{E}\left[\mathbb{E}\left[\sum_{\xi \in X} \mathbf{1}[\xi \in B, R(\xi) \leq p(\xi)]\,\Big|\, X\right]\right]$$

$$= \mathbb{E}\sum_{\xi \in X} p(\xi)\mathbf{1}[\xi \in B] = \int_B p(\xi)\rho(\xi)\,\mathrm{d}\xi$$

where the last equality follows from (4.2). Hence $\rho_{\text{thin}}(\xi) = p(\xi)\rho(\xi)$. It follows in the same way that $\rho_{\text{thin}}^{(2)}(\xi,\eta) = p(\xi)p(\eta)\rho^{(2)}(\xi,\eta)$, and so

$$g_{\text{thin}}(\xi,\eta) = \rho_{\text{thin}}^{(2)}(\xi,\eta)/(\rho_{\text{thin}}(\xi)\rho_{\text{thin}}(\eta)) = \rho^{(2)}(\xi,\eta)/(\rho(\xi)\rho(\eta))$$
$$= g(\xi,\eta).$$

$\square$

4.1.2 The second order reduced moment measure

The second order factorial moment measure can sometimes be expressed in terms of the intensity function and the following measure $\mathcal{K}$ on $\mathbb{R}^d$.

DEFINITION 4.5 Suppose that X has intensity function ρ and that the measure

$$\mathcal{K}(B) = \frac{1}{|A|}\mathbb{E}\sum_{\xi,\eta \in X}^{\neq} \frac{\mathbf{1}[\xi \in A, \eta - \xi \in B]}{\rho(\xi)\rho(\eta)}, \quad B \subseteq \mathbb{R}^d, \quad (4.4)$$

does not depend on the choice of $A \subseteq \mathbb{R}^d$ with $0 < |A| < \infty$, where we take $a/0 = 0$ for $a \geq 0$. Then X is said to be *second order intensity reweighted stationary* and $\mathcal{K}$ is called the *second order reduced moment measure*.

Stationarity of X implies second order intensity reweighted stationarity.[†] If the pair correlation function exists and is invariant under trans-

---

[†] This follows by some elementary measure theoretical properties: In the stationary case where ρ is constant,

$$\nu(A) = |A|\rho^2 \mathcal{K}(B) = \mathbb{E}\sum_{\xi,\eta \in X}^{\neq} \mathbf{1}[\xi \in A, \eta - \xi \in B]$$

is seen to be a translation invariant measure for $A \subseteq \mathbb{R}^d$ when B is fixed, and so ν is proportional to Lebesgue measure on $\mathbb{R}^d$.

SUMMARY STATISTICS

lations, then we have second order intensity reweighted stationarity and

$$\mathcal{K}(B) = \int_B g(\xi) \mathrm{d}\xi, \quad B \subseteq \mathbb{R}^d. \tag{4.5}$$

This follows by applying (4.3) to the right hand side in (4.4). Note that $\mathcal{K}$ is in general not translation invariant (even if X is stationary).

When the intensity ρ is constant we have that

$$\alpha^{(2)}(B_1 \times B_2) = \rho^2 \int_{B_1} \mathcal{K}(B_2 - \xi) \mathrm{d}\xi.$$

In the stationary case $\rho\mathcal{K}(B)$ can be interpreted as the conditional expectation of the number of further points in B given that X has a point at the origin. This interpretation is given a precise meaning in Section C.2.3.

The measure $\mathcal{K}$ is used in Sections 4.2.1–4.2.2 for constructing various summary statistics. A useful property, which holds for $\mathcal{K}$ and hence the related summary statistics, is that $\mathcal{K}$ is invariant under *independent thinning*:

PROPOSITION 4.3 *If X_{thin} is an independent thinning of a second order reweighted stationary point process X, then X_{thin} is second order reweighted stationary, and X_{thin} and X have the same $\mathcal{K}$-measure.*

Proof. This follows along similar lines as in the proof of Proposition 4.2. □

4.2 Summary statistics

In this section we consider various summary statistics for a point process X on $\mathbb{R}^d$: the so-called K, L, g-functions in Section 4.2.1, directional K-functions in Section 4.2.2, and F, G, J-functions in Section 4.2.3.

4.2.1 Second order summary statistics

In applications we consider estimates of $\mathcal{K}(B)$ for a class of test sets B such as balls. Thereby second order summary statistics are obtained as described below.

DEFINITION 4.6 *The K and L-functions for a second order reweighted stationary point process are defined by*

$$K(r) = \mathcal{K}(b(0, r)), \quad L(r) = (K(r)/\omega_d)^{1/d}, \quad r > 0.$$

This definition, which extends the definition of *Ripley's K-function* (Ripley 1976, 1977) for the stationary case to the case of second order intensity reweighted stationarity, is due to Baddeley, Møller & Waagepetersen (2000). In the stationary case, $\rho K(r)$ is the expected number of further points within distance r from the origin given that X has a point at the origin.

The K and L-functions are in one-to-one correspondence, and in applications the L-function is often used instead of the K-function. One reason is that L is the identity for a Poisson process. In general, at least for small values of r, $L(r) - r > 0$ indicates *aggregation* or *clustering* at distances less than r, and $L(r) - r < 0$ *regularity* at distances less than r. This may be due to certain latent processes (Chapter 5) or *attraction* or *repulsion* between the points (Chapter 6). Moreover, for a homogeneous Poisson process, the transformation $K \to L$ is variance stabilising when K is estimated by nonparametric methods (Besag 1977b).

If $\mathcal{K}$ is invariant under rotations, then $\mathcal{K}$ is determined by K. This is the case if X is isotropic, cf. (4.4), or if $g(\xi, \eta) = g(\|\xi - \eta\|)$ is isotropic,[†] cf. (4.5). If g is isotropic, then by (4.5),

$$K(r) = \sigma_d \int_0^r t^{d-1} g(t) \mathrm{d}t. \qquad (4.6)$$

This shows the close relationship between K and g. Many statisticians seem more keen on using the L or K-function than g (see e.g. Ripley 1981 and Diggle 1983), possibly because it is simpler to estimate K than g, cf. Section 4.3.2. However, since K is a cumulative function, it is usually easier to interpret a plot of g than a plot of K. We shall therefore also consider estimates of g when it is isotropic. In general, at least for small values of r, $g(r) > 1$ indicates *aggregation* or *clustering* at distances r, and $g(r) < 1$ *regularity* at such distances.

It should be noticed that very different point process models can share the same K-function, see Baddeley & Silverman (1984) and Baddeley, Møller & Waagepetersen (2000). In Section 5.4 we show that different shot noise Cox processes can have the same pair correlation function.

4.2.2 Directional K-functions

Test sets B other than balls may be considered for $\mathcal{K}(B)$. For example, in the planar case $d = 2$, in order to investigate for a possible *anisotropy*, *directional K-functions* can be constructed as follows (see also Stoyan & Stoyan 1994 and Brix & Møller 2001).

[†] Here and in the following we abuse notation and denote by g also the function which describes how the pair correlation depends on interpoint distances.

SUMMARY STATISTICS

For $-\pi/2 \leq \varphi < \pi/2$, $\varphi < \psi \leq \varphi + \pi$, and $r > 0$, define

$$K(\varphi, \psi, r) = \mathcal{K}(B(\varphi, \psi, r)) \qquad (4.7)$$

where

$$B(\varphi, \psi, r) = \{t(\cos v, \sin v) : 0 \leq t \leq r, \phi \leq v \leq \psi \text{ or } \phi + \pi \leq v \leq \psi + \pi\}$$

is the union of the two sectors of $b(0, r)$ with the angles of the first sector between φ and ψ, and the angles of the second sector between $\varphi + \pi$ and $\psi + \pi$, see Figure 4.1. If X is isotropic, then $K(\varphi, \psi, r) = K(r)(\psi - \varphi)/\pi$ for all (φ, ψ).

Figure 4.1 $B(-\pi/6, \pi/6, r)$ (light grey).

4.2.3 Summary statistics based on interpoint distances

We now consider three summary statistics based on interpoint distances.

DEFINITION 4.7 Assume that X is stationary. The *empty space function* F is the distribution function of the distance from the origin (or another fixed point in $\mathbb{R}^d$) to the nearest point in X, i.e.

$$F(r) = P(X \cap b(0, r) \neq \emptyset), \quad r > 0. \qquad (4.8)$$

The *nearest-neighbour function* G is

$$G(r) = \frac{1}{\rho |A|} \mathbb{E} \sum_{\xi \in X \cap A} \mathbf{1}[(X \setminus \xi) \cap b(\xi, r) \neq \emptyset], \quad r > 0, \qquad (4.9)$$

for an arbitrary set $A \subset \mathbb{R}^d$ with $0 < |A| < \infty$. Finally, the *J-function* is defined by

$$J(r) = (1 - G(r))/(1 - F(r)) \quad \text{for } F(r) < 1. \qquad (4.10)$$

It is not obvious how to extend the definitions of F, G, and J to the nonstationary case. The empty space function is also called the *spherical contact distribution function*. By stationarity, (4.9) does not depend on the choice of A. As indicated by its name, G can be interpreted as the distribution function of the distance from a typical point in X to the nearest-neighbour in X, see Section C.2.3. The J-function was suggested by van Lieshout & Baddeley (1996).

For a stationary Poisson process on $\mathbb{R}^d$ with intensity $\rho < \infty$,

$$F(r) = G(r) = 1 - \exp(-\rho \omega_d r^d) \quad \text{and} \quad J(r) = 1 \quad \text{for } r > 0,$$

where the equality for G follows from the Slivnyak-Mecke Theorem 3.2 (Section 3.2). For other kind of models, closed form expressions of F, G, J are rarely known. In general, at least for small values of $r > 0$, $F(r) < G(r)$ (or $J(r) < 1$) indicates *aggregation* or *clustering*, and $F(r) > G(r)$ (or $J(r) > 1$) *regularity*. Bedford & van den Berg (1997) show that $J = 1$ does not imply that X is a stationary Poisson process.

4.3 Nonparametric estimation

We now turn to nonparametric estimation of summary statistics. In this section we let X be a spatial point process on $\mathbb{R}^d$ with intensity function ρ; if X is stationary, ρ is assumed to be a constant with $0 < \rho < \infty$. Whenever needed we assume that the pair correlation function g or the measure $\mathcal{K}$ exists. We confine ourself to the case where a single point pattern $X_W = x$ is observed in a bounded window $W \subset \mathbb{R}^d$ with $|W| > 0$, and discuss nonparametric estimation of $\rho, \mathcal{K}, K, L, g, F, G$, and J. Extensions to replicated point patterns and to marked point processes are sometimes obvious; see Diggle, Lange & Beneš (1991), Baddeley, Moyeed, Howard & Boyde (1993), Diggle, Mateu & Clough (2000), Schlather (2001), and the references therein. *Higher order summary statistics* can be introduced as well, but the corresponding nonparametric estimators may be less stable if the number of points observed is not sufficiently large; see Peebles & Groth (1975), Stoyan & Stoyan (1994), Møller, Syversveen & Waagepetersen (1998), and Schladitz & Baddeley (2000).

4.3.1 Nonparametric estimation of intensity functions

In the homogeneous case, a natural unbiased estimate of the intensity is

$$\hat{\rho} = n(x)/|W|. \tag{4.11}$$

This is in fact the maximum likelihood estimate if X is a homogeneous Poisson process.

In the inhomogeneous case, a nonparametric *kernel* estimate of the intensity function is

$$\hat{\rho}_b(\xi) = \sum_{\eta \in x} k_b(\xi - \eta)/c_{W,b}(\eta), \quad \xi \in W \tag{4.12}$$

NONPARAMETRIC ESTIMATION

(Diggle 1985). Here k_b is a kernel with band width $b > 0$, i.e. $k_b(\xi) = k(\xi/b)/b^d$ where k is a given density function, and

$$c_{W,b}(\eta) = \int_W k_b(\xi - \eta)\mathrm{d}\xi$$

is an *edge correction factor*. The estimate (4.12) is usually sensitive to the choice of b, while the choice of k is less important. In our examples where $d = 2$ we use a product kernel given by $k(\xi) = e(\xi_1)e(\xi_2)$ for $\xi = (\xi_1, \xi_2) \in \mathbb{R}^2$, where

$$e(u) = (3/4)(1 - |u|)\mathbf{1}[|u| \leq 1], \quad u \in \mathbb{R}, \qquad (4.13)$$

is the *Epanečnikov kernel*.

LEMMA 4.1 $\int_W \hat{\rho}_b(\xi)\mathrm{d}\xi$ is an unbiased estimate of $\mu(W)$.

Proof. We have that

$$\mathbb{E}\int_W \sum_{\eta \in X_W} k_b(\xi - \eta)/c_{W,b}(\eta)\mathrm{d}\xi = \int_W \mathbb{E}\sum_{\eta \in X_W} k_b(\xi - \eta)/c_{W,b}(\eta)\mathrm{d}\xi$$

$$= \int_W \int_W (k_b(\xi - \eta)/c_{W,b}(\eta))\rho(\eta)\mathrm{d}\eta\mathrm{d}\xi = \int_W \rho(\eta)\mathrm{d}\eta = \mu(W)$$

using (4.2) in the second equality. □

EXAMPLE 4.1 (*Intensity function for mucous cell data*) Figure 4.2 shows nonparametric estimates of the intensity function for the point pattern in Example 1.3 when the information of point type is ignored. The band widths used are $b = 0.08, 0.12, 0.16,$ and 0.20. The choice $b = 0.16$ seems to provide a suitable impression of the large scale variation in the point pattern.

4.3.2 Nonparametric estimation of K and L

For nonparametric estimation of $\mathcal{K}$, it is useful to establish the following lemma where we write $W_\xi = \{\eta + \xi : \eta \in W\}$ for the translate of W by $\xi \in \mathbb{R}^d$.

LEMMA 4.2 Suppose that $|W \cap W_\xi| > 0$ for all $\xi \in B$, and that X is second order intensity reweighted stationary. Then

$$\sum_{\xi,\eta \in x}^{\neq} \frac{\mathbf{1}[\eta - \xi \in B]}{\rho(\xi)\rho(\eta)|W \cap W_{\eta-\xi}|} \qquad (4.14)$$

is an unbiased estimate of $\mathcal{K}(B)$.

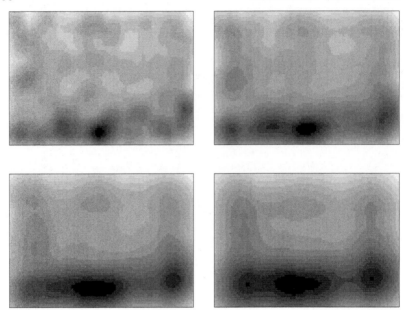

Figure 4.2 *Nonparametric estimates of the intensity function for the mucous membrane cell data obtained with band widths* $b = 0.08$ *(upper left),* 0.12 *(upper right),* 0.16 *(lower left), and* 0.20 *(lower right).*

Proof. For simplicity assume that X has a translation invariant pair correlation function g so that we can use (4.3) with $\rho^{(2)}(\xi,\eta) = \rho(\eta)\rho(\xi)g(\eta-\xi)$. In the general case the proof is similar, using (C.2) instead of (4.3). Letting $h(\xi,\eta) = \mathbf{1}[\eta - \xi \in B]/(\rho(\xi)\rho(\eta)|W \cap W_{\eta-\xi}|)$ in (4.3) we get

$$\mathbb{E} \sum_{\xi,\eta \in X_W}^{\neq} \frac{\mathbf{1}[\eta - \xi \in B]}{\rho(\xi)\rho(\eta)|W \cap W_{\eta-\xi}|}$$

$$= \int\int \frac{\mathbf{1}[\xi \in W, \eta \in W, \eta - \xi \in B]}{|W \cap W_{\eta-\xi}|} g(\eta - \xi) \mathrm{d}\eta \mathrm{d}\xi$$

$$= \int\int \frac{\mathbf{1}[\xi \in W \cap W_{-\tilde{\eta}}, \tilde{\eta} \in B]}{|W \cap W_{\tilde{\eta}}|} g(\tilde{\eta}) \mathrm{d}\xi \mathrm{d}\tilde{\eta}$$

$$= \mathcal{K}(B)$$

using (4.5) and that $|W \cap W_{\tilde{\eta}}| = |W \cap W_{-\tilde{\eta}}|$ in the last equality. $\square$

The condition on B in Lemma 4.2 means that if e.g. $B = b(0,r)$ and W is rectangular, it is required that r be smaller than the smallest side in W. The condition can be weakened: for instance, if the pair correlation

NONPARAMETRIC ESTIMATION

exists, it suffices to assume that

$$|\{\xi \in B : |W \cap W_\xi| = 0\}| = 0. \qquad (4.15)$$

We illustrate later in Example 4.3 what this means for the special design of the weed plants.

Lemma 4.2 provides an unbiased estimate of $\mathcal{K}(B)$ provided ρ is known. In practice ρ is not known, so $\rho(\xi)\rho(\eta)$ in (4.14) must be replaced by an estimate $\widehat{\rho(\xi)\rho(\eta)}$. The combined estimate

$$\hat{\mathcal{K}}(B) = \sum_{\xi,\eta \in x}^{\neq} \frac{\mathbf{1}[\eta - \xi \in B]}{\widehat{\rho(\xi)\rho(\eta)}|W \cap W_{\eta-\xi}|} \qquad (4.16)$$

is then biased. The estimate of $L(r)$ obtained from transforming that of $K(r)$ is in general biased as well.

In fact unbiasedness is usually unobtainable for many estimators in spatial statistics, but instead they are often *ratio-unbiased*, i.e. of the form $\hat{\theta} = Y/Z$ where $\theta = \mathbb{E}Y/\mathbb{E}Z$. For example, in the homogeneous case, if $\widehat{\rho(\xi)\rho(\eta)} = \widehat{\rho^2}$ is unbiased, then (4.16) is ratio-unbiased. Stoyan & Stoyan (2000) discuss various possibilities for the homogeneous case: one possibility is to transform the estimate in (4.11) to obtain $n(x)^2/|W|^2$ as an estimate of ρ^2; an alternative is

$$\widehat{\rho^2} = n(x)(n(x) - 1)/|W|^2 \qquad (4.17)$$

which is unbiased for a Poisson process. For the inhomogeneous case, Baddeley et al. (2000) propose to use $\widehat{\rho(\xi)\rho(\eta)} = \bar{\rho}_b(\xi)\bar{\rho}_b(\eta)$ where

$$\bar{\rho}_b(\xi) = \sum_{\eta \in x \setminus \xi} \kappa_b(\xi - \eta)/c_{W,b}(\eta), \quad \xi \in W, \qquad (4.18)$$

is a slight modification of (4.12). Baddeley et al. (2000) show that for an inhomogeneous Poisson processes, $\bar{\rho}_b(\xi)$ is less biased than $\hat{\rho}_b(\xi)$ when $\xi \in x$ is a data point. They argue that this is also the case for a point process with pair correlation function $g \geq 1$, while the picture is less clear if $g \leq 1$.

4.3.3 Edge correction

The weight $1/|W \cap W_{\eta-\xi}|$ in (4.14) and (4.16) is an *edge correction factor* which we use because of its simplicity and general applicability; see also the discussion in Stoyan & Stoyan (1994). However, numerous other edge correction factors have been suggested in the literature, see e.g. Stoyan & Stoyan (1994, 2000) and Ohser & Mücklich (2000).

A simpler alternative is based on *minus sampling*: For $r > 0$, let

$$W_{\ominus r} = \{\xi \in W : b(\xi, r) \subseteq W\}$$

denote the set of points in W with a distance to the boundary of W which is greater than r. Then

$$\sum_{\xi\in x,\eta\in x\cap W_{\ominus r}}^{\neq} \frac{\mathbf{1}[\eta-\xi\in B]}{\rho(\xi)\rho(\eta)}$$

is an unbiased estimate of $\mathcal{K}(B)$ (which of course becomes biased when we replace $\rho(\xi)\rho(\eta)$ by an estimate). This is called the *simple border correction estimate* or *reduced-sample estimate* of $\mathcal{K}(B)$. There is a loss of information in this estimate, since some pairs of points are excluded. In contrast the sum in (4.16) includes all observed pairs of points. However, given sufficient data, one may prefer the reduced-sample estimate if in (4.16) very large weights $1/|W\cap W_{\eta-\xi}|$ are given to point pairs $\{\xi,\eta\}\subseteq x$ with $\eta-\xi\in B$.

4.3.4 Envelopes for summary statistics

A plot of a nonparametric estimate of a summary statistic like $L(r)$ may be supplied by a *confidence interval* for each value of r. In the following we consider the L-function, but the techniques apply on any of the summary statistics considered in this chapter.

Consider a simple hypothesis H. Confidence intervals and other distributional characteristics associated with the nonparametric estimate $\hat{L}(r)$ can be obtained by a *bootstrap* using simulation under H. For a given distance $r>0$, let $T_0(r)=T(X,r)$ denote $\hat{L}(r)$ obtained from the point process X observed within the window W. Let $T_1(r)=T(X_1,r),\ldots,T_n(r)=T(X_n,r)$ be obtained from i.i.d. simulations $X_1,\ldots,X_n$ under H. From the empirical distribution of $T_1(r),\ldots,T_n(r)$ we can estimate any quantile for the distribution of $T_0(r)$ under H, and we can do this with any desired precision if n is large enough. Notice that although $T_1(r),\ldots,T_n(r)$ are i.i.d., the random vectors $(T_1(r),\ldots,T_n(r))$ considered for different values of $r>0$ are dependent. So some caution should be taken when we compare the results for different values of r.

If the computation of $T_i(r)$, $i=1,\ldots,n$, is time consuming (this is e.g. the case for the weed data in Example 1.5), the following envelopes may be used where n is small (e.g. $n=39$ as exemplified below). Let

$$T_{\min}(r)=\min\{T_1(r),\ldots,T_n(r)\} \text{ and } T_{\max}(r)=\max\{T_1(r),\ldots,T_n(r)\}. \tag{4.19}$$

Under H,

$$P(T_0(r)<T_{\min}(r))=P(T_0(r)>T_{\max}(r))\le 1/(n+1) \tag{4.20}$$

with equality if $T_0(r),T_1(r),\ldots,T_n(r)$ are almost surely different. The bounds $T_{\min}(r)$ and $T_{\max}(r)$ are called the $100/(n+1)\%$-lower and the

NONPARAMETRIC ESTIMATION

$100n/(n+1)\%$-upper *envelope* at the distance $r > 0$. For example, if we let $n = 39$, we obtain a 2.5% lower and a 97.5% upper envelope. Again some caution should be taken when we interpret a plot of $(T_0(r), T_{\min}(r), T_{\max}(r))$ for different values of r.

The same techniques are often used when instead of a simple hypothesis the fit of an estimated model is checked; several examples are considered in the sequel. Then $T_1(r), \ldots, T_n(r)$ are simulated under the estimated model, so they may be dependent on X and only approximately distributed as $T_0(r)$. In some cases it is possible to condition on a sufficient statistic so that a simple hypothesis is obtained. For example, for a homogeneous Poisson point process model, the conditional distribution of the process on W given the observed number of points is simply a binomial process with a uniform density for the points.

EXAMPLE 4.2 (*L-function for Norwegian spruces*) The left plot in Figure 4.3 shows the estimated L-function for the Norwegian spruces in Example 1.4. The intensity is assumed to be constant and the squared intensity is estimated using (4.17). The figure also shows the average and envelopes obtained from L-functions estimated from 39 simulations under a homogeneous Poisson process with expected number of points equal to the observed number of points. The values of $\hat{L}(r) - r$ for $r < 5$ indicate repulsion.

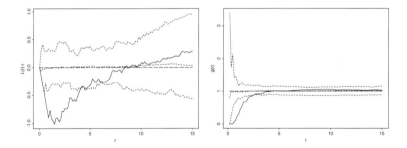

Figure 4.3 *Left: Estimated $(L(r) - r)$-function for spruces (solid line), average and envelopes calculated from 39 simulations of an homogeneous Poisson process (dashed lines), and theoretical value of $L(r) - r$ for a Poisson process (long-dashed line). Right: as left but for the g-function (see Example 4.6).*

EXAMPLE 4.3 (*L-function for weed data*) Assuming a constant intensity for the *Veronica* spp. weed data in Figure 1.6, an estimate and envelopes for the L-function are shown in the left plot in Figure 4.4. The squared intensity is estimated by (4.17) where W is the union of the 45 observation frames. The plot clearly indicates aggregation. The condition on

B in Lemma 4.2 implies that r should be smaller than 20cm but the weaker condition (4.15) only requires that $r < (30^2 + 60^2)^{1/2} \approx 67$cm. The envelopes for L are wide for $r > 25$cm due to few observed interpoint distances between 25cm and 50cm, see also the envelopes for the g-function in the right plot in Figure 4.4. The g-function is further considered in Example 4.7.

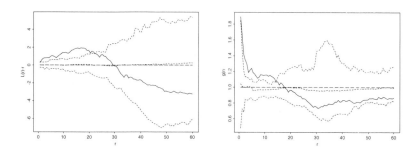

Figure 4.4 *Left: estimate of L for Veronica spp. weed data (solid line), average and envelopes calculated from 39 simulations of an homogeneous Poisson process (dashed lines), and theoretical value of L for a Poisson process (long-dashed line). Right: as left but for the g-function estimated using k_5^* (see Example 4.7).*

EXAMPLE 4.4 (*L-function for mucous cell data*) The estimated $(L(r) - r)$-function for the point pattern in Example 1.3 (where we ignore the classification of the points) is shown in Figure 4.5 (left). The intensity estimate $\bar{\rho}_b$ is used with $b = 0.16$, see Example 4.1. Envelopes are calculated from simulations of an inhomogeneous Poisson process with intensity function $\bar{\rho}_b$. The bias of the nonparametric estimate of L is apparent from the average of the simulated nonparametric estimates. The behaviour of $\hat{L}(r) - r$ for $r < 0.05$ indicates some repulsion between the cells.

The right plot in Figure 4.5 shows the estimated $(L(r) - r)$-function and envelopes under the assumption of stationarity. The apparent clustering is due to the inhomogeneity of the point pattern.

The right plot in Figure 4.6 shows another estimate of the $(L(r) - r)$-function. The nonparametric estimate of the intensity function is now replaced by a parametric estimate, where the log parametric intensity function is a fourth order polynomial in the second spatial coordinate, see the left plot in Figure 4.6. The five coefficients in the polynomial are estimated using maximum likelihood, assuming that the point pattern is a realisation of an inhomogeneous Poisson process (see Section 9.2

NONPARAMETRIC ESTIMATION

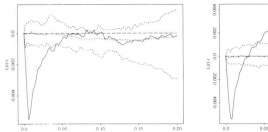
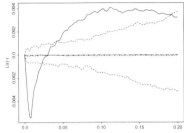

Figure 4.5 *Left: Estimated $(L(r) - r)$-function for mucous cell data (solid line), average and envelopes calculated from 39 simulations of an inhomogeneous Poisson process (dashed lines), and theoretical value of $L(r) - r$ for a Poisson process (long-dashed line). Right: as left but now under the assumption of homogeneity.*

where the pseudo likelihood function coincides with the likelihood function when a Poisson process is considered). Comparing the right plot in Figure 4.6 with the left plot in Figure 4.5, the estimate of $L(r) - r$ obtained with the parametric estimate of the intensity function is similar to the estimate obtained with the nonparametric estimate of the intensity function.

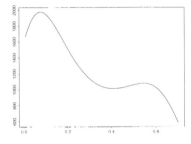
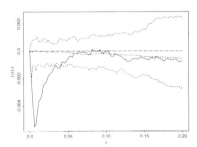

Figure 4.6 *Left: fitted parametric intensity as a function of the second spatial coordinate. Right: estimate of $(L(r) - r)$ using the parametric estimate of the intensity function (solid line), average and envelopes calculated from 39 simulations of an inhomogeneous Poisson process (dashed lines), and theoretical value of $L(r) - r$ for a Poisson process (long-dashed line).*

4.3.5 Nonparametric estimation of g

For estimation of the pair correlation function, we assume for convenience that $g(\xi,\eta) = g(\|\xi - \eta\|)$ is isotropic. We can express g in terms of the derivative K' of K, cf. (4.6), but since $\hat{K}$ is usually a step-function, it is not easy to estimate K' from $\hat{K}$.

Alternatively, an edge-corrected kernel estimate (Fiksel 1988, Stoyan & Stoyan 1994, Baddeley et al. 2000) is given by

$$\hat{g}(r) = \frac{1}{\sigma_d r^{d-1}|W|} \sum_{\xi,\eta \in x}^{\neq} \frac{k_b(r - \|\eta - \xi\|)}{\widehat{\rho(\xi)\rho(\eta)} |W \cap W_{\eta - \xi}|}. \qquad (4.21)$$

Here $k_b(u) = k(u/b)/b$, $u \in \mathbb{R}$, for a kernel $k(\cdot)$ and band width $b > 0$, and it is assumed that $|W \cap W_\xi| > 0$ for $\|\xi\| \le r$ (again we may in fact use the weaker condition (4.15) with $B = b(0, r)$). Proceeding as in the proof of Lemma 4.2,

$$\mathbb{E} \frac{1}{\sigma_d r^{d-1}|W|} \sum_{\xi,\eta \in X_W}^{\neq} \frac{k_b(r - \|\eta - \xi\|)}{\rho(\xi)\rho(\eta)|W \cap W_{\eta - \xi}|} = \int_0^\infty g(v) k_b(r - v)(v/r)^{d-1} dv.$$

Thus, for sufficiently small b and if $\widehat{\rho(\xi)\rho(\eta)}$ is close to $\rho(\xi)\rho(\eta)$, $\hat{g}(r)$ is expected to be close to $g(r)$.

The kernel $k(\cdot)$ may e.g. be the uniform kernel $k(u) = \mathbf{1}[|u| < 1/2]$ or the Epanečnikov kernel (4.13). Simulation studies and approximate calculations of the variance of the estimator (Stoyan & Stoyan 2000, Snethlage 2000) show that the smallest variance for a given b is obtained with the uniform kernel. In the sequel we use this kernel for (4.21).

The estimate (4.21) is sensitive to the choice of band width $b > 0$, and it is biased upwards at small distances r, see the discussion at page 186 in Stoyan & Stoyan (1994) and Example 4.7 below. In order to reduce the bias for small r, Snethlage (2000) suggests to use an adaptive band width by replacing $k_b(r - \|\eta - \xi\|)$ in (4.21) with

$$k_b^*(r, \|\eta - \xi\|) = k_{\min(b,r)}(r - \|\eta - \xi\|). \qquad (4.22)$$

Results obtained using k_b and k_b^* are compared in Example 4.7. For the planar stationary case, Stoyan & Stoyan (1994) recommend $b = c/\sqrt{\rho}$ with $0.1 \le c \le 0.2$. This is based on simulations and practical experience when $d = 2$ and $50 \le n(x) \le 300$ (D. Stoyan, 2002, personal communication).

EXAMPLE 4.5 (*g-function for Norwegian spruces*) The right plot in Figure 4.3 shows the estimated g-function for the Norwegian spruces, using the adaptive kernel (4.22) with $b = 5$ and estimating the squared intensity by (4.17). Also the g-function indicates repulsion.

NONPARAMETRIC ESTIMATION

EXAMPLE 4.6 (*g-function for mucous cell data*) For estimation of the pair correlation function for the mucous cell data, we use the adaptive kernel (4.22) with $b = 0.015$ in the numerator of (4.21) and $\bar{\rho}_{0.16}$ as the intensity function. The estimated g-function is shown in Figure 4.7 (left). The simulations for the envelopes are generated as for the L-function in Example 4.4. For comparison, the right plot shows the estimate obtained using instead $\hat{\rho}_{0.16}$. Considering the envelopes and the average of the estimates from the simulations, it is clear that more biased results are obtained with $\hat{\rho}_{0.16}$.

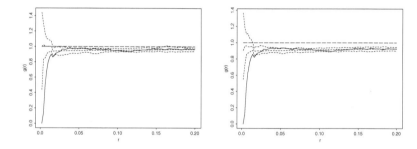

Figure 4.7 *Left: Estimate of g for mucous cell data (solid line), average and envelopes calculated from 39 simulations of an inhomogeneous Poisson process (dashed lines), and theoretical value of g for Poisson process (long-dashed line). Right: as left but using $\hat{\rho}_{0.16}$ instead of $\bar{\rho}_{0.16}$ in the estimation of g.*

EXAMPLE 4.7 (*g-function for weed data*) Estimates of the g-function for the *Veronica* spp. weed data in Figure 1.6 are shown in Figure 4.4 (right) and Figure 4.8 (left). The squared intensity is estimated by (4.17) where W is the union of the 45 observation frames. The adaptive kernel k_5^* is used in the right plot in Figure 4.4 and k_5 is used in the left plot in Figure 4.8. Both estimates of g clearly indicate aggregation. For both estimates, the envelopes are rather wide for 25cm $< r <$ 50cm where few interpoint distances are observed. The estimate obtained with k_5 is strongly biased under the Poisson model for small distances r. The estimate with k_5^* seems to be much less biased, but on the other hand exhibits large variability for small distances. This illustrates that one should be careful when interpreting estimates of $g(r)$ for small r.

EXAMPLE 4.8 (*g-function for hickory trees*) The right plot in Figure 4.8 shows the g-function for the data in Example 4.8 estimated using k_3^*. The clustering which is evident in Figure 1.1 is also clearly captured by the estimated g-function.

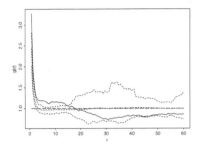 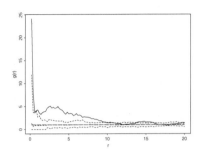

Figure 4.8 *Left: estimate of g using k_5 for Veronica spp. weed data (solid line), average and envelopes calculated from 39 simulations of an homogeneous Poisson process (dashed lines), and theoretical value of g for a Poisson process (long-dashed line). Right: estimate of g using k_3^* for the hickory trees (solid line), average and envelopes calculated from 39 simulations of an homogeneous Poisson process (dashed lines), and theoretical value of g for a Poisson process (long-dashed line).*

4.3.6 Nonparametric estimation of F, G, and J-functions

Reduced-sample estimators of F and G are easily derived using *minus sampling*: Let

$$d(\xi, B) = \inf\{\|\xi - \eta\| : \eta \in B\} \qquad (4.23)$$

be the shortest distance from a point $\xi \in \mathbb{R}^d$ to a set $B \subset \mathbb{R}^d$. Let $I \subset W$ denote a finite regular grid of points (chosen independently of X), and let $\#I_r$ denote the cardinality of the set $I_r = I \cap W_{\ominus r}$ for $r > 0$. Then we have the unbiased estimate

$$\hat{F}_{\mathrm{RS}}(r) = \sum_{\xi \in I_r} \mathbf{1}[d(\xi, x) \leq r]/\#I_r \qquad (4.24)$$

for $\#I_r > 0$, and the ratio-unbiased estimate

$$\hat{G}_{\mathrm{RS}}(r) = \sum_{\xi \in x \cap W_{\ominus r}} \mathbf{1}[d(\xi, x \setminus \xi) \leq r]/(\hat{\rho}|W_{\ominus r}|) \qquad (4.25)$$

for $|W_{\ominus r}| > 0$.

More efficient so-called *Kaplan-Meier estimates* of F and G have been proposed by Baddeley & Gill (1997):

$$\hat{F}_{\mathrm{KM}}(r) = 1 - \prod_{s \leq r}\left(1 - \frac{\#\{\xi \in I : d(\xi, x) = s, d(\xi, x) \leq d(\xi, \partial W)\}}{\#\{\xi \in I : d(\xi, x) \geq s, d(\xi, \partial W) \geq s\}}\right) \qquad (4.26)$$

MULTIVARIATE POINT PROCESSES

and

$$\hat{G}_{\mathrm{KM}}(r) = 1 - \prod_{s \leq r}\left(1 - \frac{\#\{\xi \in x : d(\xi, x \setminus \xi) = s, d(\xi, x \setminus \xi) \leq d(\xi, \partial W)\}}{\#\{\xi \in x : d(\xi, x \setminus \xi) \geq s, d(\xi, \partial W) \geq s\}}\right)$$
(4.27)

for $r > 0$, setting $0/0 = 0$. Baddeley & Gill (1997) also consider a Kaplan-Meier estimate for $K(r)$, and conclude that it is better than the reduced-sample estimate, but that estimates using edge correction factors (Section 4.3.3) are even better.

Given estimates $\hat{F}(r)$ and $\hat{G}(r)$ we obtain $\hat{J}(r) = (1 - \hat{G}(r))/(1 - \hat{F}(r))$ for $\hat{F}(r) < 1$. As illustrated in the following example, the variance of $\hat{J}(r)$ increases considerably as r increases.

EXAMPLE 4.9 (F, G, and J-functions for Norwegian spruces) The Kaplan-Meier estimates of F, G, and J for the Norwegian spruces are shown in Figure 4.9. All three summary statistics indicate that the point pattern is repulsive, which is in line with the conclusion in Examples 4.2 and 4.5 where L and g are used.

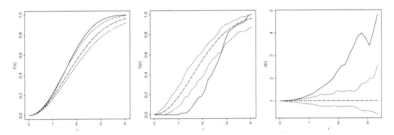

Figure 4.9 *Left to right: estimated F, G, and J-functions for the Norwegian spruces (solid line) and envelopes calculated from 39 simulations of a Poisson process (dashed line). The long-dashed curves show the theoretical values of F, G, and J for a Poisson process.*

EXAMPLE 4.10 (F, G, and J-functions for amacrine cell data) Figure 4.10 shows Kaplan-Meier estimates of F, G, and J for each type of points in Example 1.2. All three summary statistics indicate that both types of point patterns are repulsive.

4.4 Summary statistics for multivariate point processes

This section surveys summary statistics for a multivariate point process $X = (X_1, \ldots, X_k)$, cf. Section 2.2. Further material can be found in Lotwick & Silverman (1982), Diggle (1983), van Lieshout & Baddeley (1999), and the references therein.

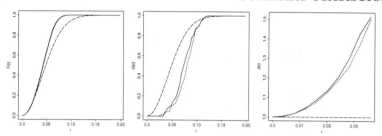

Figure 4.10 *Left to right: estimated F, G, and J-functions for on cells (solid line) and off cells (dashed line) in the amacrine cell data (F for on and off cells almost coincide). The long dashed line shows the theoretical values of F, G, J for a Poisson process.*

In the sequel we assume that each X_i is a point process on $\mathbb{R}^d$ with intensity function ρ_i and count function N_i. Summary statistics such as the K, L, g, F, G, J-functions for X_i are denoted $K_i, L_i, g_i, F_i, G_i, J_i$, and they can be estimated by the nonparametric methods in Section 4.3.

4.4.1 Definitions and properties

The correlation structure of the count functions can be expressed by the following *cross summary statistics*.

DEFINITION 4.8 *Let $i, j \in \{1, \ldots, k\}$ be different, and set $a/0 = 0$ for $a \geq 0$.*
(i) *Define the* cross moment measure *for points of types i and j by*

$$\alpha_{ij}(C) = \mathbb{E} \sum_{\xi \in X_i, \eta \in X_j} \mathbf{1}[(\xi, \eta) \in C], \quad C \subseteq \mathbb{R}^d \times \mathbb{R}^d. \tag{4.28}$$

If α_{ij} can be written as

$$\alpha_{ij}(C) = \int \int \mathbf{1}[(\xi, \eta) \in C] \rho_{ij}(\xi, \eta) \mathrm{d}\xi \mathrm{d}\eta, \quad C \subseteq \mathbb{R}^d \times \mathbb{R}^d, \tag{4.29}$$

where ρ_{ij} is a nonnegative function, then the cross pair correlation function *for points of types i and j is defined by*

$$g_{ij}(\xi, \eta) = \frac{\rho_{ij}(\xi, \eta)}{\rho_i(\xi) \rho_j(\eta)}.$$

(ii) *Suppose that the measure*

$$\mathcal{K}_{ij}(B) = \frac{1}{|A|} \mathbb{E} \sum_{\xi \in X_i, \eta \in X_j} \frac{\mathbf{1}[\xi \in A, \eta - \xi \in B]}{\rho_i(\xi) \rho_j(\eta)}, \quad B \subseteq \mathbb{R}^d, \tag{4.30}$$

MULTIVARIATE POINT PROCESSES

does not depend on the choice of $A \subset \mathbb{R}^d$ with $0 < |A| < \infty$. Then (X_i, X_j) is said to be *cross second order intensity reweighted stationary*, and we define
$$K_{ij}(r) = \mathcal{K}_{ij}(b(0,r)), \quad r > 0,$$
and
$$L_{ij}(r) = (K_{ij}(r)/\omega_d)^{1/d}, \quad r > 0.$$
Moreover, if $d = 2$, the *directional cross K-function* is defined by
$$K_{ij}(\varphi, \psi, r) = \mathcal{K}_{ij}(B(\varphi, \psi, r)), \tag{4.31}$$
cf. (4.7).

The cross statistics have interpretations analogous to the case of a univariate process. Cross second order intensity reweighted stationarity of (X_i, X_j) is satisfied if (X_i, X_j) is *stationary* (i.e. the distribution of (X_i, X_j) is invariant under translations in $\mathbb{R}^d$), or if just $g_{ij}(\xi, \eta) = g_{ij}(\xi - \eta)$ is invariant under translations. Note that the definitions of $\alpha_{ij}, g_{ij}, \mathcal{K}_{ij}$ are not symmetric in i and j. However, $K_{ij}(r) = K_{ji}(r)$ and $K_{ij}(\varphi, \psi, r) = K_{ji}(\varphi, \psi, r)$ are symmetric in i and j, since $\mathcal{K}_{ij}(B) = \mathcal{K}_{ji}(-B)$. If g_{ij} is isotropic, then $\alpha_{ij} = \alpha_{ji}$, $g_{ij} = g_{ji}$, and
$$K_{ij}(r) = \sigma_d \int_0^r t^{d-1} g_{ij}(t) \mathrm{d}t.$$
If (X_i, X_j) is isotropic (i.e. the distribution of (X_i, X_j) is invariant under rotations about the origin in $\mathbb{R}^d$), then $K_{ij}(\varphi, \psi, r) = K_{ij}(r)(\varphi - \psi)/\pi$.

PROPOSITION 4.4 If X_i and X_j are independent, then (X_i, X_j) is cross second order reweighted stationary, $\mathcal{K}_{ij}(B) = |B|$, $L_{ij}(r) = r$, and $g_{ij} = 1$.

Proof. Consider first $\mathcal{K}_{ij}$: Since X_i and X_j are independent, by (4.2) and (4.30),
$$\mathcal{K}_{ij}(B) = \frac{1}{|A|} \mathbb{E}\left[\sum_{\xi \in X_i} \frac{\mathbf{1}[\xi \in A]}{\rho_i(\xi)}\right] \mathbb{E}\left[\sum_{\eta \in X_j} \frac{\mathbf{1}[\eta - \xi \in B]}{\rho_j(\eta)}\right] = |B|$$
does not depend on the choice of $A \subset \mathbb{R}^d$ with $0 < |A| < \infty$. From this we immediately obtain that $L_{ij}(r) = r$. By (4.2), (4.28)–(4.29), and the independence between X_i and X_j, $\rho_{ij}(\xi, \eta) = \rho_i(\xi)\rho_j(\eta)$ exists, so $g_{ij}(r) = 1$. $\square$

Both $\mathcal{K}_{ij}$ and g_{ij} are invariant under *independent thinning* of the points in X_i and X_j, where the retention probabilities are allowed to depend on the type of points. This follows along similar lines as in the proof of Proposition 4.2.

Lotwick & Silverman (1982) and Diggle (1983) discuss nonparametric tests for the hypotheses of *independence* and *random labelling* of the k point processes. The hypothesis of random labelling can be investigated by conditioning on both the superposition $X_1 \cup \ldots \cup X_k = y$, say, and the counts $N_1(W) = n_1, \ldots, N_k(W) = n_k$, say, and then calculate test or summary statistics after independent random selection of $n_1, \ldots, n_k$ type $1, \ldots, k$ pseudo-points from y. Concerning independence Lotwick & Silverman (1982) propose to condition on the interpoint arrangement of each of the types of point patterns, and then generate simulations by randomly translating one of the point patterns on a torus; here it is assumed that $W \subset \mathbb{R}^d$ is rectangular and $X_1, \ldots, X_k$ are stationary: see also Example 4.11 below..

4.4.2 The stationary case

We assume now that (X_i, X_j) is stationary with intensities $0 < \rho_i < \infty$ and $0 < \rho_j < \infty$.

DEFINITION 4.9 The *nearest-neighbour function* G_{ij} is the distribution function for the distance from a typical type i point to its nearest type j point, i.e.

$$G_{ij}(r) = \frac{1}{\rho_i |A|} \mathbb{E} \sum_{\xi \in X_i \cap A} \mathbf{1}[X_j \cap b(\xi, r) \neq \emptyset] \qquad (4.32)$$

for an arbitrary set $A \subset \mathbb{R}^d$ with $0 < |A| < \infty$. Moreover, define

$$J_{ij}(r) = (1 - G_{ij}(r))/(1 - F_j(r)) \quad \text{for } F_j(r) < 1.$$

The cross statistics G_{ij} and J_{ij} are not symmetric in i and j. By stationarity of (X_i, X_j), the definition of G_{ij} does not depend on the choice of the set A in (4.32). The definition of J_{ij} is due to van Lieshout & Baddeley (1999). Both definitions are in line with the definitions of the K and G-functions in the univariate case.

Under the hypothesis of independence between X_i and X_j we have the following relations between the different kind of F, G, and J-functions, where we let $F_{i \cup j}, G_{i \cup j}$, and $J_{i \cup j}$ denote the F, G, and J-functions for the superposition $X_i \cup X_j$.

PROPOSITION 4.5 Assume that X_i and X_j are stationary and independent. Then

$$G_{ij}(r) = F_j(r), \quad 1 - F_{i \cup j}(r) = (1 - F_i(r))(1 - F_j(r)), \qquad (4.33)$$

$$J_{ij}(r) = 1 \text{ (where defined)}, \qquad (4.34)$$

MULTIVARIATE POINT PROCESSES

and

$$J_{i\cup j}(r) = \frac{\rho_i}{\rho_i + \rho_j} J_i(r) + \frac{\rho_j}{\rho_i + \rho_j} J_j(r) \text{ (where defined).} \quad (4.35)$$

Proof. Note that with probability one, X_i and X_j have no points in common when they are independent and stationary. We have that

$$\mathbb{E} \sum_{\xi \in X_i \cap A} \mathbf{1}[X_j \cap b(\xi, r) \neq \emptyset] = \mathbb{E} \sum_{\xi \in X_i \cap A} P(X_j \cap b(\xi, r) \neq \emptyset | X_i)$$

$$= \rho_i |A| F_j(r)$$

since X_i and X_j are independent and X_j is stationary, so $G_{ij}(r) = F_j(r)$. Thus the first result in (4.33) and (4.34) are verified. The other result in (4.33) also follows from the independence between X_i and X_j. It can easily be shown that

$$1 - G_{i \cup j}(r) = \frac{\rho_i}{\rho_i + \rho_j}(1 - G_i(r))(1 - F_j(r)) + \frac{\rho_j}{\rho_i + \rho_j}(1 - G_j(r))(1 - F_i(r))$$

which combined with (4.33) gives (4.35). □

Note that $J_{i\cup j}$ is a linear combination of J_i and J_j; for extensions of Proposition 4.5 to the situation where all the k processes are independent, see van Lieshout & Baddeley (1996, 1999). The similar relation for e.g. the K-functions of $X_i, X_j, X_i \cup X_j$ is more complicated; see again van Lieshout & Baddeley (1996, 1999).

4.4.3 Nonparametric estimation

Nonparametric estimation of summary statistics associated to each single point process X_i and to the superposition $X_i \cup X_j$ are given by the methods in Section 4.3. For the cross summary statistics similar methods apply, so we shall only consider a few cases below.

LEMMA 4.3 *If (X_i, X_j) is cross second order intensity reweighted stationary, then for $B \subseteq \mathbb{R}^d$ with $|W \cap W_\xi| > 0$ for all $\xi \in B$,*

$$\sum_{\xi \in X_i \cap W, \eta \in X_j \cap W} \frac{\mathbf{1}[\eta - \xi \in B]}{\rho_i(\xi)\rho_j(\eta)|W \cap W_{\eta-\xi}|} \quad (4.36)$$

is an unbiased estimator of $\mathcal{K}_{ij}(B)$.

Proof. The proof is similar to that of Lemma 4.2. □

The condition on B can be weakened as commented after Lemma 4.2. In (4.36) we can substitute $\rho_i(\xi)\rho_j(\eta)$ with a nonparametric estimator $\widehat{\rho_i(\xi)\rho_j(\eta)}$ (which in the stationary case does not depend on ξ and

η, cf. Section 4.3.1), and thereby obtain nonparametric estimators of $K_{ij}, L_{ij}, \ldots$. If g_{ij} is isotropic, a kernel estimator for the cross pair correlation similar to that in (4.21) is given by

$$\hat{g}_{ij}(r) = \frac{1}{\sigma_d r^{d-1}|W|} \sum_{\xi \in X_i \cap W, \eta \in X_j \cap W} \frac{e_b(r - \|\eta - \xi\|)}{\widehat{\rho_i(\xi)\rho_j(\eta)}|W \cap W_{\eta - \xi}|}.$$

For stationary (X_i, X_j) the reduced-sample estimator of $G_{ij}(r)$ is

$$\sum_{\xi \in X_i \cap W_{\ominus r}} \mathbf{1}[d(\xi, X_j \cap W) \leq r]/(\hat{\rho}_i |W_{\ominus r}|)$$

and the Kaplan-Meier estimator is 1 minus

$$\prod_{s \leq r} \left(1 - \frac{\#\{\xi \in X_i \cap W : d(\xi, X_j \cap W) = s, d(\xi, X_j \cap W) \leq d(\xi, \partial W)\}}{\#\{\xi \in X_i \cap W : d(\xi, X_j \cap W) \geq s, d(\xi, \partial W) \geq s\}}\right),$$

compare with (4.25) and (4.27). By substitution of the estimated F_j and G_{ij} functions, we obtain an estimator of J_{ij}.

EXAMPLE 4.11 (*Cross J and L-functions for amacrine cell data*) Figure 4.11 shows estimates of the J_{12}, J_{21}, and $(L_{12}(r) - r)$-functions for the data in Example 1.2, assuming stationarity and where 1 and 2 correspond to on and off cells, respectively. As suggested in Lotwick & Silverman (1982), the envelopes are obtained from simulations obtained by wrapping the observation window on a torus and randomly translating the point pattern of on-cells. The cross J-functions do not provide evidence against the hypothesis of independence. The cross L-function, however, lies above the upper envelope for a range of r-values.

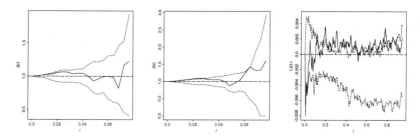

Figure 4.11 *The estimated J_{12}-function (left), J_{21}-function (middle), and $(L_{12}(r) - r)$-function (right) for the amacrine cell data (solid line). Dashed curves are envelopes calculated from 39 simulations where the pattern of on-cells is randomly translated on a torus. The long-dashed curves show the theoretical values of the summary statistics under the hypothesis of independence.*

SUMMARY STATISTICS FOR MARKED POINT PROCESSES

EXAMPLE 4.12 (*Cross L-function for mucous cell data*) An estimate of the $(L_{12}(r) - r)$-function for the data in Example 1.3 is shown in Figure 4.12, where 1 and 2 correspond to ECL and other cells, respectively. The intensity functions for the two types of cells are obtained by multiplying the parametric estimate in Example 4.4 with the proportions of ECL cells and other cells, respectively. This corresponds to maximum likelihood estimation in a model where each type of points are realisations of independent inhomogeneous Poisson processes with intensity functions proportional to the parametric intensity model used in Example 4.4. The $(L_{12}(r) - r)$-function indicates repulsion between the two types of points. The large deviations of $(L_{12}(r) - r)$ from zero for the larger values of r could indicate that the intensity functions for the two types of points are not proportional.

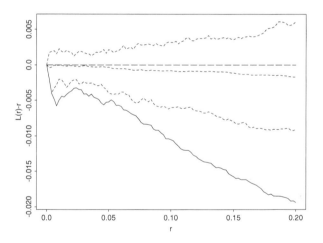

Figure 4.12 *Estimated $L_{12}(r) - r$-function for the mucous cell data. Dashed curves are envelopes calculated from 39 simulations of the two independent inhomogeneous Poisson processes estimated by maximum likelihood. The long-dashed curve is the theoretical value of $L_{12}(r) - r$ for a Poisson process.*

4.5 Summary statistics for marked point processes

Summary statistics for marked point processes other than multitype point processes are studied in Stoyan & Stoyan (1994), Schlather (2001), and the references therein. In this section we describe briefly what is meant by the mark distribution and how this can be estimated. Other and more complicated summary statistics are treated in the above-mentioned references.

Consider a marked point process $X = \{(\xi, m_\xi) : \xi \in Y\}$ with points in $\mathbb{R}^d$ and mark space M, cf. Sections 2.2 and B.2. Assume that Y has intensity function ρ, and define the intensity measure of X (viewed as a point process on $S = \mathbb{R}^d \times M$) by

$$\tilde\mu(C) = \mathbb{E} \sum_{\xi \in Y} \mathbf{1}[(\xi, m_\xi) \in C], \quad C \subseteq M \times \mathbb{R}^d.$$

By the Radon-Nikodym theorem (see e.g. p. 422 in Billingsley 1995) we have

$$\tilde\mu(B \times A) = \int_B Q_\xi(A)\rho(\xi)\mathrm{d}\xi \quad \text{for } B \subseteq \mathbb{R}^d,\ A \subseteq M,$$

where Q_ξ is a probability measure on M.

DEFINITION 4.10 *If $Q_\xi = Q$ does not depend on $\xi \in \mathbb{R}^d$, then Q is called the* mark distribution.

If the mark distribution Q exists, then $Q(A) = \tilde\mu(B \times A)/\mu(B)$ for an arbitrary set $B \subseteq \mathbb{R}^d$ with $0 < \mu(B) < \infty$, where μ is the intensity measure for Y.

Suppose that X is stationary, i.e. its distribution is invariant under translation of the points Y in $\mathbb{R}^d$, and so ρ is constant. If $0 < \rho < \infty$, the mark distribution exists and is given by

$$Q(A) = \mathbb{E} \sum_{\xi \in Y} \mathbf{1}[\xi \in B, m_\xi \in A]/(\rho|B|) \tag{4.37}$$

for an arbitrary set $B \subseteq \mathbb{R}^d$ with $0 < |B| < \infty$. Thus we can interpret Q as the distribution of a mark at a *typical point* of Y. For a stationary multitype Poisson point process (Section 3.3), or more generally under the hypothesis of *random labelling* (i.e. when the marks are i.i.d. and independent of Y), Q is simply the marginal distribution of any mark m_ξ.

Suppose that X is a stationary disc process with mark space $M = [0, \infty)$. If we can observe all the radii of the discs with centres in a bounded window W, (4.37) suggests the estimator of $Q([0, r])$ given by

$$\sum_{\xi \in Y_W} \mathbf{1}[m_\xi \leq r]/(\hat\rho|W|)$$

where $\hat\rho$ is an estimate of ρ. This is just the empirical distribution function of the marks when we take $\hat\rho|W|$ equal to $n(Y_W)$, cf. (4.11). We conclude this chapter with a simple example of such a distribution.

SUMMARY STATISTICS FOR MARKED POINT PROCESSES   55

EXAMPLE 4.13 (*Spruces*) The histogram for the disc radii in Example 1.4 is shown in Figure 4.13.

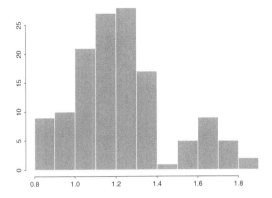

Figure 4.13 *Histogram for the disc radii in Example 1.4.*

CHAPTER 5

Cox processes

The class of Poisson processes is usually a too simplistic model class for real data, but it can be successfully used for constructing more flexible model classes. One important case is Cox processes; another is Markov point processes (Chapter 6). Cox processes are models for aggregated or clustered point patterns caused by e.g. an environmental random heterogeneity. Certain Cox processes are obtained by clustering of the points around another point process, and are hence special cases of so-called cluster processes.

Sections 5.1–5.2 consider some simple examples and basic properties of Cox processes. Section 5.3 considers when Neyman-Scott processes (Neyman & Scott 1958) become Cox processes. Sections 5.4–5.5 deal with shot noise Cox processes, including the processes from Section 5.3, Poisson/gamma processes (Wolpert & Ickstadt 1998), and shot noise G Cox processes (Brix 1999). Sections 5.6–5.7 deal with log Gaussian Cox processes (Coles & Jones 1991, Møller, Syversveen & Waagepetersen 1998). Section 5.8 concerns multivariate Cox processes.

Monte Carlo methods and simulation-based inference for Cox processes are covered by the material in Chapters 8 and 10, and perfect simulation for shot noise Cox processes is discussed in Appendix D. Further material on Cox and cluster processes can be found in Grandell (1976), Diggle (1983), Daley & Vere-Jones (1988), Stoyan et al. (1995), and the recent review by Møller & Waagepetersen (2002).

5.1 Definition and simple examples

A Cox process is a natural extension of a Poisson process, obtained by considering the intensity function of the Poisson process as a realisation of a random field. Such processes were studied in a seminal paper by Cox (1955) under the name *doubly stochastic Poisson processes*, but are today usually called Cox processes.

In accordance with the exposition of Poisson processes in Chapter 3, we let $S \subseteq \mathbb{R}^d$ in this section and Section 5.2.

DEFINITION 5.1 Suppose that $Z = \{Z(\xi) : \xi \in S\}$ is a nonnegative random field so that with probability one, $\xi \to Z(\xi)$ is a locally integrable function. If the conditional distribution of X given Z is a Poisson process on S with intensity function Z, then X is said to be a *Cox process driven by Z*.

REMARK 5.1 That Z is a random field means that $Z(\xi)$ is a random variable for all $\xi \in S$ (for an introduction to random fields, see e.g. Adler 1981). If $\rho(\xi) = \mathbb{E}Z(\xi)$ exists and is locally integrable, then with probability one, $Z(\xi)$ is a locally integrable function (in fact, ρ is the intensity function, cf. Section 5.2, and for the specific models of SNCPs in this book, ρ exists and is locally integrable). The intensity measure of the Poisson process $X|Z$ is

$$M(B) = \int_B Z(\xi)\mathrm{d}\xi, \quad B \subseteq S. \tag{5.1}$$

This is a random measure, and we can define the Cox process in terms of M instead of Z. For readers familiar with measure theory, it should be obvious how many of the definitions and results in this chapter extend to a general metric space S as in Appendix B and a random locally finite diffuse measure M on S.

In the special case where Z is deterministic, X simply becomes a Poisson process with intensity function $\rho = Z$. Below two other simple examples are given; further specific constructions are found in the succeeding sections.

EXAMPLE 5.1 (*Mixed Poisson processes*) A *mixed Poisson process* is a Cox process where $Z(\xi) = Z_0$ is given by a common positive random variable for all locations $\xi \in S$, i.e. $X|Z_0$ follows a homogeneous Poisson process with intensity Z_0. In general, apart from the Poisson case where Z_0 is almost surely constant, $N(A)$ and $N(B)$ are positive correlated for disjoint bounded sets $A, B \subseteq S$. A special tractable case of a mixed Poisson process occurs when Z_0 is gamma distributed; then $N(A)$ follows a negative binomial distribution for bounded $A \subset S$.

EXAMPLE 5.2 (*Thinning of Cox processes*) Random independent thinning in a Cox process results in a new Cox process. Specifically, suppose that X is a Cox process driven by Z, and $\Pi = \{\Pi(\xi) : \xi \in S\} \subseteq [0, 1]$ is a random field which is independent of (X, Z). Conditional on Π, let X_{thin} denote the point process obtained by independent thinning of the points in X with retention probabilities Π. Then X_{thin} is a Cox process driven by $Z_{\mathrm{thin}}(\xi) = \Pi(\xi)Z(\xi)$. This follows by combining Proposition 3.7 and Definition 5.1.

An example is shown in Figure 5.1 (left) where X is a homogeneous Poisson process and Π follows a *logistic process*, i.e. $\log(\Pi/(1 - \Pi))$ is

DEFINITION AND SIMPLE EXAMPLES

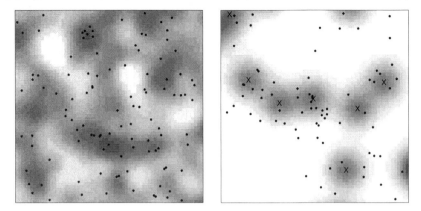

Figure 5.1 *Simulations of a logistic process (left) and a Thomas process (right, see Example 5.3 in Section 5.3) and the associated random intensity functions (in grey scale). The crosses in the right plot show the cluster centres for the Thomas process.*

a Gaussian process with zero mean and isotropic covariance function $c(r) = \exp(-(10r)^2)$ (for details, see Section 5.6).

Usually in applications Z is unobserved, and so we cannot distinguish a Cox process X from its corresponding Poisson process $X|Z$ when only one realisation of X_W is available (where W denotes the observation window). Which of the two models might be most appropriate, i.e. whether Z should be random or "systematic"/deterministic, depends on

- prior knowledge: in a Bayesian setting, incorporating prior knowledge of the intensity function of a Poisson process leads to a Cox process model; nonparametric Bayesian smoothing for the intensity function is treated in Heikkinen & Arjas (1998);

- the scientific questions to be investigated: if e.g. we want to investigate the dependence of certain covariates associated to Z, these may be treated as systematic terms, while unobserved effects may be treated as random terms (for an example, see Benes, Bodlak, Møller & Waagepetersen 2001);

- the particular application: if it seems difficult to model an aggregated point pattern with a parametric class of inhomogeneous Poisson processes (e.g. a class of polynomial intensity functions), Cox process models such as shot noise Cox processes (Section 5.4) and log Gaussian Cox processes (Section 5.6) may allow more flexibility and/or a more parsimonious parametrisation.

5.2 Basic properties

Distributional properties of a Cox process X follow immediately by conditioning on Z and exploiting the properties of the Poisson process $X|Z$. Thereby we easily obtain the following results.

The void probabilities are given by

$$v(B) = \mathbb{E}\exp\left(-\int_B Z(\xi)\mathrm{d}\xi\right) \quad \text{for bounded } B \subseteq S,$$

and the generating functional by

$$G_X(u) = \mathbb{E}\exp\left(-\int_S (1-u(\xi))Z(\xi)\mathrm{d}\xi\right), \tag{5.2}$$

for functions $u : S \to [0,1]$, cf. Proposition 3.3.

The intensity function is

$$\rho(\xi) = \mathbb{E}Z(\xi) \tag{5.3}$$

and the pair correlation function is given by

$$g(\xi,\eta) = \mathbb{E}[Z(\xi)Z(\eta)]/[\rho(\xi)\rho(\eta)] \tag{5.4}$$

provided $Z(\xi)$ has finite variance for all $\xi \in S$. This illustrates that the moment measure and the reduced factorial moment measure (and more generally the reduced moment measures considered in Section C.1.1) can easily be expressed in terms of moments of Z. For most specific models, $g \geq 1$, but as shown in Section 5.6.2 there exist some exceptions to this. As noted in Section 4.1.2 second order intensity reweighted stationarity is ensured by translation invariance of $g(\xi,\eta) = g(\xi - \eta)$, in which case $\mathcal{K}$ or K can be calculated from (4.5).

For $A, B \subseteq S$ with $\mathrm{Var}N(A) < \infty$ and $\mathrm{Var}N(B) < \infty$,

$$\mathbb{C}\mathrm{ov}(N(A), N(B)) = \int_A \int_B \mathbb{C}\mathrm{ov}(Z(\xi), Z(\eta))\mathrm{d}\xi\mathrm{d}\eta + \mu(A \cap B)$$

where as usual $\mu(B) = \mathbb{E}N(B)$ is the intensity measure, cf. (4.1) and (5.3)–(5.4). Note that $\mathrm{Var}N(A) \geq \mathbb{E}N(A)$ with equality only when X is a Poisson process. In other words, a Cox process exhibits over-dispersion when compared to a Poisson process.

If $S = \mathbb{R}^d$ and Z is stationary and/or isotropic (meaning that the distribution of Z is invariant under translations/rotations), then X is stationary and/or isotropic. Explicit expressions of the F, G, and J-functions for a stationary Cox process are in general difficult to obtain, but certain relations can be established for particular cases, cf. Sections 5.3–5.4.

Consider a Cox process X restricted to a set $B \subseteq S$ with $|B| < \infty$. By (3.11), the density of X_B with respect to the standard Poisson process is given by

$$f(x) = \mathbb{E}\left[\exp\left(|B| - \int_B Z(\xi)\mathrm{d}\xi\right) \prod_{\xi \in x} Z(\xi)\right] \qquad (5.5)$$

for finite point configurations $x \subset B$. An explicit expression of the mean in (5.5) is usually unknown, and the integral $\int_B Z(\xi)\mathrm{d}\xi$ can be difficult to calculate.

5.3 Neyman-Scott processes as Cox processes

In this section we consider those Neyman-Scott processes (Neyman & Scott 1958) which are also Cox processes. These processes are further special cases of the shot noise Cox processes studied in Sections 5.4–5.5.

Let C be a stationary Poisson process on $\mathbb{R}^d$ with intensity $\kappa > 0$. Conditional on C, let $X_c, c \in C$, be independent Poisson processes on $\mathbb{R}^d$ where X_c has intensity function

$$\rho_c(\xi) = \alpha k(\xi - c) \qquad (5.6)$$

where $\alpha > 0$ is a parameter and k is a *kernel* (i.e. for all $c \in \mathbb{R}^d$, $\xi \to k(\xi - c)$ is a density function). Then $X = \cup_{c \in C} X_c$ is a special case of a *Neyman-Scott process* with cluster centres C and clustres $X_c, c \in C$ (in the general definition of a Neyman-Scott process, $n(X_c)$ given C is not restricted to be a Poisson variate, see e.g. Stoyan et al. 1995). By Proposition 3.6, X is also a Cox process on $\mathbb{R}^d$ driven by

$$Z(\xi) = \sum_{c \in C} \alpha k(\xi - c). \qquad (5.7)$$

Clearly, Z in (5.7) is stationary and locally integrable, and it is also isotropic if $k(\xi) = k(\|\xi\|)$ is isotropic. The intensity is $\rho = \alpha\kappa$, and the pair correlation function is given by

$$g(\xi) = 1 + h(\xi)/\kappa, \qquad (5.8)$$

where

$$h(\xi) = \int k(\eta)k(\xi + \eta)\mathrm{d}\eta$$

is the density for the difference between two independent points which each have density k. This is verified later in a more general context in Proposition 5.1. Furthermore,

$$J(r) = \int k(\xi)\exp\left(-\alpha \int_{\|\eta\| \leq r} k(\xi + \eta)\mathrm{d}\eta\right)\mathrm{d}\xi,$$

see Bartlett (1975), van Lieshout & Baddeley (1996), and Møller (2003b); this is also a special case of Proposition 5.2 in Section 5.4.4. Thus $J(r)$

is nonincreasing for $r > 0$ with range $(\exp(-\alpha), 1)$. So $F(r) < G(r)$ for $r > 0$.

Simulation of a Neyman-Scott process is in principle easy using its construction as a Poisson cluster process. Section 5.5 and Appendix D describe in the more general context of shot noise Cox processes how simulation can be performed.

EXAMPLE 5.3 (*Thomas and Matérn cluster processes*) Two mathematically tractable models are given by

(I) the *Matérn cluster process* (Matérn 1960, 1986) where
$$k(\xi) = \mathbf{1}[\|\xi\| \le r]/(\omega_d r^d)$$
is the uniform density on the ball $b(0, r)$;

(II) the *Thomas process* (Thomas 1949) where
$$k(\xi) = \exp(-\|\xi\|^2/(2\omega^2))/(2\pi\omega^2)^{d/2}$$
is the density for $N_d(0, \omega^2 I_d)$, i.e. for d independent normally distributed variables with mean 0 and variance $\omega^2 > 0$.

In both cases k is isotropic. We refer to the kernels in (I) and (II) several times in the sequel.

The Thomas process is isotropic with
$$g(\xi) = 1 + \left(4\pi\omega^2\right)^{-d/2} \exp\left(-\|\xi\|^2/\left(4\omega^2\right)\right)/\kappa. \tag{5.9}$$

The K-function can be expressed in terms of the cumulative distribution function for a χ^2-distribution with d degrees of freedom, and for $d = 2$ we simply obtain
$$K(r) = \pi r^2 + [1 - \exp(-r^2/(4\omega^2))]/\kappa. \tag{5.10}$$

Figure 5.1 (right) shows a simulation of a Thomas process with $\kappa = 10$, $\alpha = 10$, and $\omega^2 = 0.1$.

The Matérn cluster process is studied in e.g. Santaló (1976) and Stoyan et al. (1995). Expressions for its summary statistics are more complicated than for the Thomas process.

5.4 Shot noise Cox processes

We can obviously modify the definition of a Neyman-Scott process in many ways. Following Brix & Kendall (2002), Møller & Waagepetersen (2002), and Møller (2003b) we consider now such extensions. In the sequel, by a *kernel* we mean a function $k(\cdot, \cdot)$ on $\mathbb{R}^d \times \mathbb{R}^d$ where $k(c, \cdot)$ is a density function for all $c \in \mathbb{R}^d$.

SHOT NOISE COX PROCESSES

DEFINITION 5.2 Let X be a Cox process on $\mathbb{R}^d$ driven by

$$Z(\xi) = \sum_{(c,\gamma)\in\Phi} \gamma k(c,\xi) \qquad (5.11)$$

where $k(\cdot,\cdot)$ is a kernel and Φ is a Poisson point process on $\mathbb{R}^d \times (0,\infty)$ with a locally integrable intensity function ζ. Then X is called a *shot noise Cox process (SNCP)*.

Section 5.8.3 discusses how to extend the definition of an SNCP to the multivariate case. Recall that we want Z to be locally integrable with probability one, cf. Definition 5.1. By Remark 5.1, this is satisfied if the function defined by

$$\rho(\xi) = \int \gamma k(c,\xi)\zeta(c,\gamma)\mathrm{d}c\mathrm{d}\gamma \qquad (5.12)$$

is finite and locally integrable. From (5.3), (5.11), and Theorem 3.2 follows immediately that (5.12) is the intensity function (provided $\rho(\cdot) < \infty$).

Combining (5.2) and (5.11) we obtain the generating functional

$$G_X(u) = \mathbb{E} \prod_{(c,\gamma)\in\Phi} \exp\left(\gamma \int u(\xi) k(c,\xi)\mathrm{d}\xi - \gamma\right)$$

for functions $u : \mathbb{R}^d \to [0,1]$. Hence, applying (3.6) for the generating functional of the Poisson process Φ,

$$G_X(u) = \exp\left\{-\int\int\left[1 - \exp\left(\gamma \int u(\xi) k(c,\xi)\mathrm{d}\xi - \gamma\right)\right]\zeta(c,\gamma)\mathrm{d}c\mathrm{d}\gamma\right\}. \qquad (5.13)$$

5.4.1 Shot noise Cox processes as cluster processes

By Proposition 3.6, we can view $X|\Phi$ as the superposition of independent Poisson processes $X_{(c,\gamma)}$, $(c,\gamma) \in \Phi$, where $X_{(c,\gamma)}$ has intensity function $\gamma k(c,\cdot)$. Thus X is an example of a *Poisson cluster process* (see e.g. Daley & Vere-Jones, 1988). The set C of c's is countable but not necessarily locally finite (this is exemplified later), so C is not always a point process in the sense used in this book. We shall nevertheless refer to C as the *centre process* and to $X_{(c,\gamma)}$ as the *cluster* with centre c, intensity γ, and dispersion density $k(c,\cdot)$. The centre points are also called *parent* or *mother* points and the clusters for *offspring* or *daughter* points.

Viewing SNCPs as Poisson cluster processes, they seem natural when the aggregation of e.g. plants is caused by seed setting mechanisms (Brix & Chadoeuf 2000).

5.4.2 Relation to marked point processes

Suppose that $\tilde{\zeta}(c) = \int_0^\infty \zeta(c,\gamma) \mathrm{d}\gamma$ is finite for all $c \in \mathbb{R}^d$. Then Φ is a marked Poisson point process (Section 3.3) with points $C \sim \mathrm{Poisson}(\mathbb{R}^d, \tilde{\zeta})$ and marks which conditional on C are mutually independent with density $\zeta(c,\gamma)/\tilde{\zeta}(c)$ for a mark γ associated to a point $c \in C$ (taking $a/0 = 0$ for $a \geq 0$).

Suppose that $\zeta_A(\gamma) = \int_A \zeta(c,\gamma) \mathrm{d}c < \infty$ for some $A \subseteq \mathbb{R}^d$ and all $\gamma > 0$. Then $\Phi \cap (A \times (0, \infty))$ is a marked Poisson process with points $\{\gamma : (c,\gamma) \in \Phi \cap (A \times (0, \infty))\} \sim \mathrm{Poisson}((0,\infty), \zeta_A)$ and marks $\{c : (c,\gamma) \in \Phi \cap (A \times (0, \infty))\}$ which conditional on the points γ are mutually independent with density $\zeta(c,\gamma)/\zeta_A(\gamma)$ on A.

5.4.3 Examples

In many of our examples of specific SNCPs, ζ is of the form

$$\zeta(c,\gamma) = \chi(\gamma), \tag{5.14}$$

where χ is a nonnegative locally integrable function on $(0, \infty)$. This is equivalent to assuming that Φ is stationary under translations of C. By (5.12),

$$\rho(\xi) = \int k(c,\xi) \mathrm{d}c \int \gamma \chi(\gamma) \mathrm{d}\gamma \tag{5.15}$$

provided the integrals are finite. If furthermore $k(c,\xi) = k(\xi - c)$ is invariant under translations (where $k(\cdot)$ is a density), we have that Z and hence X is stationary. For short we refer to this as *the stationary case of an SNCP*. Then X has intensity

$$\rho = \int \gamma \chi(\gamma) \mathrm{d}\gamma$$

provided the integral is finite. Finally, if also $k(\xi) = k(\|\xi\|)$ is isotropic, then X is isotropic.

EXAMPLE 5.4 (*Neyman-Scott processes*) A Neyman-Scott process can be viewed as a special case of an SNCP if we allow the γ's to coincide: Suppose that ζ is of the form (5.14) but all the γ's are equal to a parameter $\alpha > 0$ so that $\chi(\alpha) > 0$. Then we interpret χ as a Dirac δ-function, i.e. we set $\kappa = \chi(\alpha)$ and $\int_0^\infty h(\gamma)\chi(\gamma) \mathrm{d}\gamma = \kappa h(\alpha)$ so that X is a Neyman-Scott process with intensity $\rho = \alpha\kappa$.

EXAMPLE 5.5 Suppose that ζ is of the form (5.14) where $\kappa = \int_0^\infty \chi(\gamma) \mathrm{d}\gamma < \infty$. Then as mentioned in Section 5.4.2, Φ is a marked Poisson process, where C is a stationary Poisson process with intensity κ and the marks γ given C are i.i.d. with density χ/κ.

EXAMPLE 5.6 (*Shot noise G Cox processes*) A *shot noise G Cox process* as introduced in Brix (1999) is an SNCP where ζ is of the form (5.14) with
$$\chi(\gamma) = \kappa \gamma^{-\alpha-1} \exp(-\tau\gamma)/\Gamma(1-\alpha) \qquad (5.16)$$
where $\kappa > 0$, $\alpha < 1$, and $\tau > 0$. These restrictions on the parameters κ, α, τ are equivalent to local integrability of χ; Brix (1999) includes also the case $\tau = 0$ and $0 < \alpha < 1$. Furthermore, "G" refers to that
$$\mu_G(B) = \sum_{(c,\gamma) \in \Phi} \gamma \mathbf{1}[c \in B], \quad B \subseteq \mathbb{R}^d,$$
is a so-called G-measure, see Brix (1999) or Møller & Waagepetersen (2002).

The intensity function is given by
$$\rho(\xi) = \kappa \tau^{\alpha-1} \int k(c,\xi) dc$$
which reduces to $\rho = \kappa \tau^{\alpha-1}$ in the stationary case.

The distribution of Φ depends on α as described below.

- The case $\alpha < 0$: Then Φ is a special case of the marked Poisson process in Example 5.5, where C is a stationary Poisson process with intensity $-\kappa \tau^\alpha/\alpha$, the γ's for $(c,\gamma) \in \Phi$ are i.i.d. and independent of C, and each γ is gamma distributed with shape parameter $-\alpha$ and inverse scale parameter τ.

- The case $0 \le \alpha < 1$: The situation is now less simple as the set C is not locally finite, since $\int_0^\infty \chi(\gamma) d\gamma = \infty$. As $\{(c,\gamma) \in \Phi : c \in A\}$ and $\{(c,\gamma) \in \Phi : c \in B\}$ are independent for disjoint sets $A, B \subset \mathbb{R}^d$, we can for simplicity assume that C is concentrated on a bounded set $B \subset \mathbb{R}^d$. Then Φ is a special case of the marked Poisson process in Section 5.4.2, where $\{\gamma : (c,\gamma) \in \Phi\}$ is an inhomogeneous Poisson process on $(0, \infty)$ with intensity function $|B|\chi$, and the c's are i.i.d. uniformly distributed marks on B which are independent of the γ's.

For $\alpha = 0$, we have a so-called *Poisson-gamma process* and μ_G is a so-called gamma-measure (Daley & Vere-Jones 1988, Wolpert & Ickstadt 1998). The left plots in Figure 5.2 show simulations of Poisson-gamma processes with k given by a bivariate normal density as for the Thomas process in Example 5.3 with $\omega^2 = 0.001$, and with (κ, τ) equal to $(15, 0.1)$ (upper left) and $(7.5, 0.05)$ (lower left), respectively. The simulations are on $B = [0,1]^2$ and are generated as described in Section 5.5, using the truncation $\epsilon = 0.0001$ and the extended window $B_{\text{ext}} = [-0.25, 1.25]^2$ (see Section 5.5 for details). The right plots in Figure 5.2 show the corresponding realisations of $\Phi \cap (B_{\text{ext}} \times (\epsilon, \infty))$. Note that the intensity $\rho = \kappa/\tau$ is the same for the two sets of parameter values. With the small

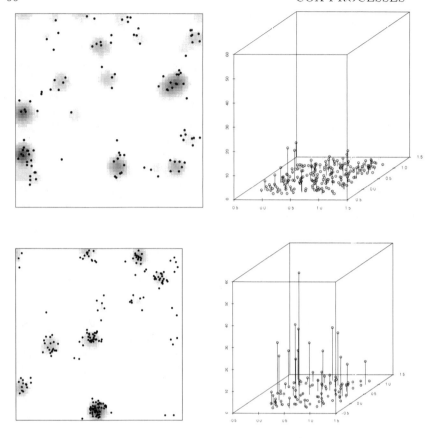

Figure 5.2 *Left: simulations of Poisson-gamma processes (see text for details). Right: corresponding realisations of $\Phi \cap (B_{ext} \times (\epsilon, \infty))$ (only points (c, γ) with $\gamma > 0.02$ are shown).*

values of κ and γ we obtain fewer and larger clusters than when the larger values are used.

5.4.4 Summary statistics

The form (5.11) of Z and the Slivnyak-Mecke theorem allow us to establish a number of useful results as shown below.

PROPOSITION 5.1 Let X be an SNCP. If
$$\beta(\xi, \eta) = \int \int \gamma^2 k(c, \xi) k(c, \eta) \zeta(c, \gamma) \mathrm{d}c \mathrm{d}\gamma$$

is finite for all $\xi, \eta \in \mathbb{R}^d$, then the pair correlation function is given by

$$g(\xi, \eta) = 1 + \beta(\xi, \eta)/(\rho(\xi)\rho(\eta)). \tag{5.17}$$

Proof. By (5.12) and Jensen's inequality, $\rho(\xi)^2 \leq \beta(\xi, \xi)$ is finite for all $\xi \in \mathbb{R}^d$. By (5.11) (5.12) and Theorem 3.3 (Section 3.2),

$$\mathbb{E}[Z(\xi)Z(\eta)]$$

$$= \mathbb{E} \sum_{(c,\gamma),(c',\gamma')\in\Phi}^{\neq} \gamma k(c,\xi)\gamma' k(c',\eta) + \mathbb{E} \sum_{(c,\gamma)\in\Phi} \gamma^2 k(c,\xi)k(c,\eta)$$

$$= \int\int\int\int \gamma k(c,\xi)\gamma' k(c',\eta)\zeta(c,\gamma)\zeta(c',\gamma')\mathrm{d}c\mathrm{d}\gamma\mathrm{d}c'\mathrm{d}\gamma'$$

$$+ \int\int \gamma^2 k(c,\xi)k(c,\eta)\zeta(c,\gamma)\mathrm{d}c\mathrm{d}\gamma$$

$$= \rho(\xi)\rho(\eta) + \beta(\xi,\eta)$$

which is finite, whereby (5.17) follows from (5.4). □

By (5.17), $g \geq 1$. In the stationary case, $g(\xi, \eta) = g(\xi - \eta)$ is of the same form as for a Neyman-Scott process, cf. (5.8). So $g(\xi - \eta) - 1$ depends up to a constant only on the choice of the kernel $k(\xi - \eta)$. This is exemplified below.

EXAMPLE 5.7 (*Shot noise G Cox processes*) Consider a stationary shot noise G Cox process as in Example 5.6 with $k(c, \xi) = k(\xi - c)$. Then the pair correlation function exists and is given by

$$g(\xi) = 1 + \frac{1-\alpha}{\kappa\tau^\alpha} \int k(\eta)k(\xi + \eta)\mathrm{d}\eta. \tag{5.18}$$

The class $\{(\rho, g) : \kappa > 0, \alpha > 0\}$ obtained from (5.8) and the class $\{(\rho, g) : \kappa > 0, \alpha < 1, \tau > 0\}$ obtained from (5.18) agree, so for any given kernel k we cannot distinguish between these classes. So consideration of nonparametric estimates $\hat{\rho}$ and $\hat{g}$ is not useful for choosing between a Neyman-Scott or a shot noise G Cox process model for a given point pattern.

Møller (2003b) establishes a very simple description of the so-called reduced Palm distributions (Appendix C.2) of an SNCP which makes it possible to verify the following result.

PROPOSITION 5.2 For a stationary SNCP with intensity $\rho > 0$,

$$J(r) = \frac{1}{\rho}\int\int \gamma k(c)\exp\left(-\int_{\|\eta\|\leq r} \gamma k(\eta + c)\mathrm{d}\eta\right)\chi(\gamma)\mathrm{d}c\mathrm{d}\gamma, \quad r > 0.$$

It follows that $J(r)$ is nonincreasing for $r > 0$, with $\lim_{r \to 0} J(r) = 1$ and

$$\lim_{r \to \infty} J(r) = \int \gamma \exp(-\gamma) \chi(\gamma) \mathrm{d}\gamma / \rho.$$

If the kernel has finite range R, i.e. $k(\xi) = 0$ for $\|\xi\| \geq R$, then $J(r)$ is constant for $r \geq 2R$.

5.5 Approximate simulation of SNCPs

Although SNCPs are tractable for mathematical analysis, at least when compared to many other types of point process models, simulation is needed for performing statistical inference (Wolpert & Ickstadt 1998, Brix 1999, Best, Ickstadt & Wolpert 2000, Brix & Chadoeuf 2000, Brix & Diggle 2001, Brix & Kendall 2002, Møller & Waagepetersen 2002). In this section we follow Møller (2003b) and discuss how to obtain approximate simulations of X, or more precisely X_B where $B \subset \mathbb{R}^d$ is bounded (whereby X_B is finite). The method is directly based on the construction of an SNCP and rather straightforward, but a truncation is needed to deal with the possibly infinite point process Φ entering in the definition of the random intensity function. A perfect (or exact) simulation algorithm is given in the more technical Appendix D. Furthermore, conditional simulation for Φ given X_B is treated in Section 10.2.2.

The random intensity function of X_B is

$$Z_B(\xi) = \mathbf{1}[\xi \in B] \sum_{(c,\gamma) \in \Phi} \gamma k(c, \xi) \qquad (5.19)$$

which for $\xi \in B$ depends on all centres $c \in C$ with $k(c, \xi) > 0$, so the sum can be infinite. We assume that we can choose $B_{\text{ext}} \subseteq \mathbb{R}^d$ and $\epsilon \geq 0$ so that $\int_{B_{\text{ext}}} \int_\epsilon^\infty \zeta(c,\gamma) \mathrm{d}c \mathrm{d}\gamma$ becomes finite, and approximate (5.19) by replacing Φ by the truncated finite Poisson process $\Phi \cap (B_{\text{ext}} \times (\epsilon, \infty))$. A positive ϵ is required in Example 5.6 with $\alpha \geq 0$ whereas $\epsilon = 0$ suffices in Examples 5.4–5.5 and Example 5.6 with $\alpha < 0$.

Now, an approximate simulation of X_B is obtained by simulating first the Poisson process $\Phi \cap (B_{\text{ext}} \times (\epsilon, \infty))$, and next the associated independent Poisson processes $X_{(c,\gamma)} \cap B$ with intensity functions $\xi \to \gamma k(c,\xi) \mathbf{1}[\xi \in B]$, $(c,\gamma) \in \Phi \cap (B_{\text{ext}} \times (\epsilon, \infty))$. Note that in applications edge effects may enter, as B_{ext} is typically a bounded region so that $B \subset B_{\text{ext}}$; another kind of error occurs when $\epsilon > 0$. Below we quantify the error of such approximate simulations.

Let q_B be the probability that some cluster with centre $c \notin B_{\text{ext}}$ or intensity $\gamma \leq \epsilon$ has a point in B. Conditional on Φ, the probability that a cluster $X_{(c,\gamma)}$ has at least one point in B is given by

$$p_B(c,\gamma) = 1 - \exp\left(-\gamma \int_B k(c,\xi)\mathrm{d}\xi\right).$$

Hence

$$q_B = 1 - \mathbb{E} \prod_{(c,\gamma)\in\Phi} (1-p_B(c,\gamma))^{\mathbf{1}[c\notin B_{\text{ext}} \text{ or } \gamma\leq\epsilon]}$$

and by analogy with (5.13) this reduces to

$$q_B = 1 - \exp\left(-\int\int \mathbf{1}[c\notin B_{\text{ext}} \text{ or } \gamma\leq\epsilon]p_B(c,\gamma)\zeta(c,\gamma)\mathrm{d}c\mathrm{d}\gamma\right).$$

Though $p_B(c,\gamma)$ for $c\notin B_{\text{ext}}$ or $\gamma\leq\epsilon$ might be negligible, possibly q_B is not.

An alternative quantity is given by

$$M_B = \sum_{(c,\gamma)\in\Phi} \mathbf{1}[c\notin B_{\text{ext}} \text{ or } \gamma\leq\epsilon]n(X_{(c,\gamma)}\cap B),$$

i.e. M_B is the number of missing points when we make an approximate simulation of X_B by ignoring those clusters $X_{(c,\gamma)}$ with $c\notin B_{\text{ext}}$ or $\gamma\leq\epsilon$. By Theorem 3.2 (Section 3.2),

$$\mathbb{E}M_B = \mathbb{E}\mathbb{E}[M_B|\Phi] = \mathbb{E}\sum_{(c,\gamma)\in\Phi}\mathbf{1}[c\notin B_{\text{ext}} \text{ or } \gamma\leq\epsilon]\int_B \gamma k(c,\xi)\mathrm{d}\xi$$

$$= \int\int \mathbf{1}[c\notin B_{\text{ext}} \text{ or } \gamma\leq\epsilon]\int_B \gamma k(c,\xi)\zeta(c,\gamma)\mathrm{d}\xi\mathrm{d}c\mathrm{d}\gamma.$$

In order to evaluate q_B and M_B it is convenient to consider a function $k_B^{\text{dom}} : \mathbb{R}^d \times \mathbb{R}^d \to (0,\infty)$ so that

(C1) $k_B^{\text{dom}}(c,\xi) \geq k(c,\xi)$ if $\xi\in B$, and $k_B^{\text{dom}}(c,\xi) = 0$ if $\xi\notin B$;

(C2) we can easily calculate the integral

$$a_B^{\text{dom}}(c) = \int_B k_B^{\text{dom}}(c,\xi)\mathrm{d}\xi, \quad c\in\mathbb{R}^d.$$

We illustrate this in Example 5.8 below, using the following proposition.

PROPOSITION 5.3 We have that

$$q_B \leq 1 - \exp\left(-\int\int \mathbf{1}[c\notin B_{\text{ext}} \text{ or } \gamma\leq\epsilon]\right.$$
$$\left.(1-\exp(-\gamma a_B^{\text{dom}}(c)))\zeta(c,\gamma)\mathrm{d}c\mathrm{d}\gamma\right)$$

and

$$\mathbb{E}M_B \leq \int\int \mathbf{1}[c\notin B_{\text{ext}} \text{ or } \gamma\leq\epsilon]\gamma a_B^{\text{dom}}(c)\zeta(c,\gamma)\mathrm{d}c\mathrm{d}\gamma.$$

Proof. Follows immediately from the expressions of q_B and M_B above, using (C1). □

For specific models of ζ the upper bounds in Proposition 5.3 may be evaluated by numerical methods as demonstrated in the following example.

EXAMPLE 5.8 Let $B = b(0, R)$ and $B_{\text{ext}} = b(0, R+r)$, and consider the stationary case (5.14) of an SNCP where the kernel $k(c, \xi) = k(\xi - c)$ is given by either the uniform kernel (I) with band width r or the Gaussian kernel (II) in Example 5.3. In case (I) $X_{(c,\gamma)} \cap B = \emptyset$ whenever $c \notin B_{\text{ext}}$, so the error is only due to the truncation of the γ when $\epsilon > 0$. Setting $k_B^{\text{dom}}(c, \xi) = \sup_{\eta \in B} k(c, \eta)$ for $\xi \in B$, (C1)–(C2) are satisfied: for $s = \|c\| \geq 0$, in case (I)

$$a_I(s) \equiv a_B^{\text{dom}}(c) = (R/r)^d \mathbf{1}[s \leq R+r],$$

while in case (II)

$$a_{II}(s) \equiv a_B^{\text{dom}}(c) = \left[\omega_d R^d / (2\pi\omega^2)^{d/2}\right] \exp\left[-\mathbf{1}[s > R](s-R)^2/(2\omega^2)\right].$$

Making a shift to polar coordinates and noting that $\sigma_d = d\omega_d$, we obtain in case (I)

$$q_B \leq 1 - \exp\left(-\omega_d(R+r)^d \int_0^\epsilon (1 - \exp(-(R/r)^d \gamma))\chi(\gamma)\mathrm{d}\gamma\right)$$

and

$$\mathbb{E}M_B \leq \omega_d(R(R+r)/r)^d \int_0^\epsilon \gamma\chi(\gamma)\mathrm{d}\gamma,$$

while in case (II)

$$q_B \leq 1 - \exp\left(-\sigma_d \int\int \mathbf{1}[s > R+r \text{ or } \gamma \leq \epsilon]s^{d-1}\right.$$
$$\left.(1 - \exp(-\gamma a_{II}(s)))\chi(\gamma)\mathrm{d}s\mathrm{d}\gamma\right)$$

and

$$\mathbb{E}M_B \leq \sigma_d \int\int \mathbf{1}[s > R+r \text{ or } \gamma \leq \epsilon]s^{d-1}\gamma a_{II}(s)\chi(\gamma)\mathrm{d}s\mathrm{d}\gamma$$
$$= \sigma_d \int_0^\infty s^{d-1}a_{II}(s)\mathrm{d}s \int_0^\infty \gamma\chi(\gamma)\mathrm{d}\gamma - \sigma_d \int_0^{R+r} s^{d-1}a_{II}(s)\mathrm{d}s \int_\epsilon^\infty \gamma\chi(\gamma)\mathrm{d}\gamma.$$

These integrals can easily be determined by numerical methods for Neyman-Scott and shot noise G Cox processes.

For example, consider a Poisson-gamma process with $\chi(\gamma) = \exp(-\tau\gamma) \times \kappa/\gamma$. In case (I), Table 5.1 contains values of the upper bounds for q_B

and $\mathbb{E}M_B$ for $\kappa = 100$, $\tau = 0.5$, $R = \sqrt{0.5}$, and varying values of ϵ and r. The exact values of $\mathbb{E}M_B = \kappa(1 - \exp(-\tau\epsilon))/\tau$ can easily be calculated in this case and are also shown in the table. The expected numbers of points $|B_{\text{ext}}| \int_\epsilon^\infty \chi(\gamma) d\gamma$ in the truncated process are given in the column $\mathbb{E}\Phi^{\text{trunc}}$. From the exact values of $\mathbb{E}M_B$ it seems that $\epsilon = 0.001$ is sufficient, since the expected number of missing points is then only 0.05 % of $\mathbb{E}n(X_B) = \pi R^2 \kappa/\tau = \pi 100$. We then, depending on the value of r, need to simulate between 1473 and 2238 mother points on average. The upper bounds for $\mathbb{E}M_B$ and q_B are rather conservative but improve with increasing values of r and decreasing values of ϵ.

ϵ	r	$q_B \leq$	$\mathbb{E}M_B \leq$	$\mathbb{E}M_B$	%	$\mathbb{E}\Phi^{\text{trunc}}$
0.10	0.10	1.00	499.05	15.32	0.05	505
0.10	0.20	1.00	157.60	15.32	0.05	637
0.10	0.30	1.00	86.33	15.32	0.05	786
e−3	0.10	1.00	5.11	0.16	0.05	1437
e−3	0.20	0.96	1.62	0.16	0.05	1815
e−3	0.30	0.83	0.88	0.16	0.05	2238
e−5	0.10	9.73e−2	5.12e−2	1.57e−3	5e−4	2379
e−5	0.20	3.18e−2	1.62e−2	1.57e−3	5e−4	3006
e−5	0.30	1.75e−2	8.85e−3	1.57e−3	5e−4	3705
e−7	0.10	1.02e−3	5.12e−4	1.57e−5	5e−6	3322
e−7	0.20	3.23e−4	1.62e−4	1.57e−5	5e−6	4196
e−7	0.30	1.77e−4	8.85e−5	1.57e−5	5e−6	5172

Table 5.1 *Errors for approximate simulation of a Poisson-gamma process in case (I). The columns $q_B \leq$, $\mathbb{E}M_B \leq$, and $\mathbb{E}M_B$ contain the upper bounds for q_B and M_B and the exact value of $\mathbb{E}M_B$, respectively (and e.g. 5e−4 means 5×10^{-4}). The sixth colum is $\mathbb{E}M_B/\mathbb{E}n(X_B)$ in %.*

In case (II) we restrict attention to the upper bound for $\mathbb{E}M_B$ which can be computed using one-dimensional numerical integration. Upper bounds for $\mathbb{E}M_B$ are given in Table 5.2 with (R, κ, τ) as before, $\omega = 0.05$, and varying values of ϵ and r. The values $\epsilon = 10^{-5}$ and $r = 0.25$ may be sufficient since we then on average lose at most 0.19 points (again $\mathbb{E}n(X_B) = \pi 100$).

ϵ	10^{-3}		10^{-5}		10^{-7}	
r	0.15	0.25	0.15	0.25	0.15	0.25
$\mathbb{E}M_B \leq$	37.16	18.65	18.71	0.19	18.53	0.006

Table 5.2 *Upper bounds for $\mathbb{E}M_B$ in case (II).*

REMARK 5.2 (*Truncation with fixed number of points*) An alternative approach to truncation is considered in Wolpert & Ickstadt (1998) and Brix (1999). Assume that $B_{\text{ext}} \subseteq \mathbb{R}^d$ is chosen so that $\zeta_{B_{\text{ext}}}(\gamma)$ (defined in Section 5.4.2 with $A = B_{\text{ext}}$) is positive for all $\gamma > 0$, and that

$$q(t) = \int_t^\infty \zeta_{B_{\text{ext}}}(\gamma) d\gamma$$

is finite for all $t > 0$ (as before, in applications B_{ext} is typically a bounded region containing B). Let $y_1 < y_2 < \ldots$ be the points of a standard Poisson process on $(0, \infty)$. As in Remark 3.4, it follows that $\{\gamma_j : j = 1, 2, \ldots\}$ with $\gamma_j = q^{-1}(y_j)$ is Poisson$((0, \infty), \zeta_{B_{\text{ext}}})$. Wolpert & Ickstadt (1998) and Brix (1999) approximate Z_B by

$$Z_B(\xi) \approx \sum_{j=1}^{J} \gamma_j k(c_j, \xi)$$

where $J \in \mathbb{N}$ is fixed and specified by the user, and the c_j's are conditionally independent with density $\zeta(c_j, \gamma_j)/\zeta_{B_{\text{ext}}}(\gamma_j)$ given $\{\gamma_j : j = 1, 2, \ldots\}$. Methods for evaluating the tail sum

$$\sum_{j=J+1}^{\infty} \gamma_j k(c_j, \xi)$$

are discussed in Wolpert & Ickstadt (1998) and Brix (1999). An advantage of our approach, where we truncate the domain of the mother points instead of the number of mother points, is that we can compute bounds on errors related directly to X_B.

5.6 Log Gaussian Cox processes

We now consider the case where $Y = \log Z$ is a Gaussian field, i.e. for any integer $n > 0$, locations $\xi_1, \ldots, \xi_n \in \mathbb{R}^d$, and numbers $a_1, \ldots, a_n \in \mathbb{R}$, $\sum_{i=1}^n a_i Y(\xi_i)$ follows a normal distribution.

DEFINITION 5.3 Let X be a Cox process on $\mathbb{R}^d$ driven by $Z = \exp(Y)$ where Y is a Gaussian field. Then X is said to be a *log Gaussian Cox process (LGCP)*.

Such models have independently been introduced in astronomy by Coles & Jones (1991) and in statistics by Møller et al. (1998). The definition of an LGCP can easily be extended in a natural way to a multivariate LGCP (Møller et al. 1998; see Section 5.8.2) and to multivariate spatio-temporal LGCPs (Brix & Møller 2001, Brix & Diggle 2001). More generally, we could consider Cox processes where $X|Y$ has intensity function $h(Y)$ for some nonnegative function h, but as we shall see the exponential transformation $h(Y) = \exp(Y)$ is particularly convenient. Another tractable case is to let Z be a χ^2 field,

$$Z(\xi) = Y_1(\xi)^2 + \ldots + Y_m(\xi)^2 \qquad (5.20)$$

where $Y_1, \ldots, Y_m$ are independent Gaussian fields with zero mean.

The distribution of (X, Y) is completely determined by the mean and covariance function

$$m(\xi) = \mathbb{E}Y(\xi) \quad \text{and} \quad c(\xi, \eta) = \mathbb{C}\text{ov}(Y(\xi), Y(\eta)).$$

Because of the richness of possible mean and covariance functions, LGCPs are flexible models for aggregation as demonstrated in Møller et al. (1998). Brix & Møller (2001) consider a situation where the aggregation is due to different soil conditions. Brix & Møller (2001) and Møller & Waagepetersen (2003) consider situations where m is a linear or polynomial function. Benes, Bodlak, Møller & Waagepetersen (2002) consider a case where m depends on covariates. Further examples of applications are given in Chapter 10.

During the last few years there has been some debate concerning which of the two model classes of SNCPs and LGCPs is most appropriate, see Wolpert & Ickstadt (1998), Richardson (2003), Møller (2003a), and Møller & Waagepetersen (2002).

5.6.1 Conditions on the covariance function

For simplicity we assume in this section that $c(\xi, \eta) = c(\xi - \eta)$ is translation invariant and of the form

$$c(\xi) = \sigma^2 r(\xi/\alpha) \qquad (5.21)$$

where $\sigma^2 > 0$ is the variance, r is a known *correlation function*, and $\alpha > 0$ is a correlation parameter. In our examples of applications we have isotropy, i.e. $r(\xi) = r(\|\xi\|)$, whereby the covariance between $Y(\xi)$ and $Y(\eta)$ only depends on the distance $\|\xi - \eta\|$.

A given function $r : \mathbb{R}^d \to [-1, 1]$ with $r(0) = 1$ for all $\xi \in \mathbb{R}^d$ is a correlation function for a Gaussian field if and only if r is positive semi-definite, i.e. $\sum_{i,j} a_i a_j r(\xi_i, \xi_j) \geq 0$ for all $a_1, \ldots, a_n \in \mathbb{R}$, $\xi_1, \ldots, \xi_n \in \mathbb{R}^d$, $n \in \mathbb{N}$. Whether a given function is positive semi-definite may be best answered through a spectral analysis, see e.g. Christakos (1984), Wackernagel (1995), and Schlater (1999).

It is necessary to impose weak conditions on m and r in order to get a well-defined and finite integral $\int_B Z(\xi) d\xi$ for bounded $B \subset \mathbb{R}^d$. For example, we may require that $\xi \to Y(\xi)$ is continuous with probability one. This is satisfied for the models of m and r considered in this book.

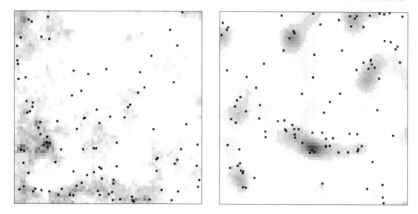

Figure 5.3 *Simulations of LGCPs with exponential (left) and Gaussian (right) correlation functions. The associated simulated random intensity functions are shown in grey scale.*

In fact if m is continuous and there exist $\epsilon > 0$ and $\delta > 0$ such that
$$1 - r(\xi) < \delta/(-\log(\|\xi\|))^{1+\epsilon}$$
for all ξ in an open ball with centre 0, then $\xi \to Y(\xi)$ is continuous with probability one (Theorem 3.4.1 in Adler 1981). A stronger condition on r, which is often easier to check, is that there exist $\tilde{\epsilon} > 0$ and $\tilde{\delta} > 0$ such that
$$1 - r(\xi) < \tilde{\delta}\|\xi\|^{\tilde{\epsilon}}$$
for all ξ in an open ball with centre 0 (Møller et al. 1998).

The *power exponential family* consists of correlation functions of the form
$$r(\xi) = \exp\left(-\|\xi\|^{\delta}\right), \quad 0 \leq \delta \leq 2. \tag{5.22}$$
The parameter δ controls the smoothness of the realizations of the Gaussian field and is typically fixed at a value chosen from a priori knowledge. The special case $\delta = 1$ is the *exponential correlation function*, $\delta = 1/2$ the *stable correlation function*, and $\delta = 2$ the *Gaussian correlation function*. Figure 5.3 shows simulations of LGCPs on $[0,1]^2$ with $m = 4.10$, $\sigma = 1$, and exponential ($\alpha = 0.14$) or Gaussian ($\alpha = 0.1$) correlation functions. The associated simulated intensity functions are also shown in Figure 5.3. Note that the intensity function is smoother for the Gaussian correlation function (Adler 1981). Further examples of simulated LGCPs are given in Møller et al. (1998).

For the applications we have considered, it has sufficed to use power correlation functions or to let
$$Y(\xi) = Y_1(\xi) + \ldots + Y_m(\xi) \tag{5.23}$$

be a sum of independent Gaussian fields $Y_1, \ldots, Y_m$ each with correlation function of the form (5.22) (in which case the covariance function of Y is a linear combination of power exponential correlation functions). Other types of correlation functions are considered in Møller et al. (1998) and in the survey by Schlater (1999).

5.6.2 Summary statistics

LGCPs are tractable for mathematical analysis, cf. Møller et al. (1998). Particularly, the intensity and pair correlation functions possess some nice properties.

PROPOSITION 5.4 The intensity and pair correlation function of an LGCP are given by

$$\rho(\xi) = \exp(m(\xi) + c(\xi,\xi)/2), \quad g(\xi,\eta) = \exp(c(\xi,\eta)). \qquad (5.24)$$

Proof. The normal distribution with mean ζ and variance σ^2 has Laplace transform $\exp(\zeta + \sigma^2 t/2)$, $t \in \mathbb{R}$. Hence

$$\rho(\xi) = \mathbb{E}\exp(Z(\xi)) = \exp(m(\xi) + c(\xi,\xi)/2).$$

Furthermore,

$$\mathbb{E}\exp(Z(\xi) + Z(\eta)) = \exp(m(\xi) + m(\eta) + c(\xi,\xi)/2 + c(\eta,\eta)/2 + c(\xi,\eta))$$
$$= \rho(\xi)\rho(\eta)\exp(c(\xi,\eta))$$

so

$$g(\xi,\eta) = \mathbb{E}\exp(Z(\xi) + Z(\eta))/(\rho(\xi)\rho(\eta)) = \exp(c(\xi,\eta)).$$

$\square$

Similarly, the densities of higher order reduced moment measures can be found, cf. Theorem 1 in Møller & Waagepetersen (2003). For models of the type (5.23), g is a log linear combination of correlation functions,

$$\log g(\xi,\eta) = \sum_1^m c_i(.,\xi,\eta)$$

where c_i is the covariance function of Y_i, $i = 1, \ldots, m$.

By (5.24), $g \geq 1$ if and only if $c \geq 0$. If c is of the form (5.21), a pair correlation function with values smaller than one is obtained with the cardinal sine correlation function $r(\xi) = \sin(\|\xi\|)/\|\xi\|$. Isotropy of $r(\xi) = r(\|\xi\|)$ implies the restriction $g \geq \exp(-1/d)$ (Matérn 1960, Adler 1981).

The simple relationships in (5.24) indicate that in many respects LGCPs are more tractable for mathematical analysis than SNCPs, cf. the discussion in Møller et al. (1998). Particularly, there is a one-to-one correspondence between (m,c) and (ρ,g), and so the distribution of

(X, Y) is uniquely determined by (ρ, g). This relationship can be used for selecting an appropriate class of parametric models for (m, c) when considering a nonparametric estimate of (ρ, g). Under the assumption that the covariance function is translation invariant, second order reweighted stationarity is automatically satisfied for an LGCP no matter whether ρ is homogeneous or not. Moreover, stationarity respective isotropy of an LGCP is equivalent to stationarity respective isotropy of (m, c) or equivalently of (ρ, g).

We conclude by establishing a result similar to Proposition 5.4 but for a χ^2 Cox process.

PROPOSITION 5.5 *For a χ^2 Cox process with Z of the form (5.20) and where c_i is the covariance function of Y_i, the intensity and pair correlation function are given by*

$$\rho(\xi) = \sum_{i=1}^{m} c_i(\xi, \xi), \quad g(\xi, \eta) = 1 + 2 \sum_{i=1}^{m} c_i(\xi, \eta)^2 / (\rho(\xi)\rho(\eta)).$$

Proof. Clearly,

$$\rho(\xi) = \sum_{i=1}^{m} \mathbb{E} Y_i(\xi)^2 = \sum_{i=1}^{m} c_i(\xi, \xi).$$

Furthermore, since $\mathrm{Cov}(Y_i(\xi)^2, Y_i(\eta)^2) = 2c_i(\xi, \eta)^2$ (see e.g. p. 169 in Adler 1981),

$$\mathbb{E}(Z(\xi)Z(\eta)) = \sum_{i=1}^{m} \mathbb{E}\left(Y_i(\xi)^2 Y_i(\eta)^2\right) + \sum_{i \neq j} \mathbb{E} Y_i(\xi)^2 \mathbb{E} Y_j(\eta)^2$$

$$= 2 \sum_{i=1}^{m} c_i(\xi, \eta)^2 + \sum_{i=1}^{m} c_i(\xi, \xi) c_i(\eta, \eta) + \sum_{i \neq j} c_i(\xi, \xi) c_j(\eta, \eta)$$

whereby the expression for g is easily obtained. □

5.7 Simulation of Gaussian fields and LGCPs

We now turn to simulation of an LGCP. In contrast to the case of Neyman-Scott processes and SNCPs, we do not have problems with boundary effects, since the distribution of an LGCP restricted to a bounded region B depends only on the distribution of

$$Y_B = \{Y(\xi) : \xi \in B\}.$$

As Y_B does not in general have a finite representation in a computer, we approximate Y_B by a random step function Y_B^{step} with constant value $Y(\eta)$ within disjoint cells C_η centred around a finite set I of locations $\eta \in B$ so that $B = \cup_{\eta \in I} C_\eta$. Given a simulation of Y_B^{step} we can use the methods in Section 3.2.3 for simulating the Poisson process with intensity function $\exp(Y_B^{\text{step}})$. So the basic problem is to simulate the Gaussian random field $\tilde{Y} = (Y(\eta))_{\eta \in I}$ of the step heights.

A review of methods for simulation of Gaussian random fields is given in Schlater (1999). In the following we briefly describe two methods that we find particularly useful.

Consider the case $d = 2$. When $c(\xi, \eta) = c(\xi - \eta)$ is invariant under translations and B is rectangular, an efficient way of simulating $\tilde{Y}$ is obtained using *circulant embedding* and the *fast Fourier transform*, cf. Wood & Chan (1994), Dietrich & Newsam (1993) or Section 6.1 in Møller et al. (1998). Briefly, this works as follows (a more detailed account is given in Appendix E). Let $I \subset B$ denote a rectangular grid which is embedded in a rectangular grid I_{ext} wrapped on a torus. Then a block circulant matrix $C = \{C_{\xi\eta}\}_{\xi,\eta \in I_{\text{ext}}}$ can be constructed so that $\{C_{\xi\eta}\}_{\xi,\eta \in I}$ is the covariance matrix of $\tilde{Y}$. Since C is block circulant, it can easily be diagonalised by means of the two-dimensional discrete Fourier transform with associated matrix $F_\otimes$ given in Appendix E. Suppose that C has nonnegative eigenvalues. Then we can extend $\tilde{Y}$ to a larger Gaussian field $\tilde{Y}_{\text{ext}} = (\tilde{Y}(\xi))_{\xi \in I_{\text{ext}}}$ with covariance matrix C: set

$$\tilde{Y}_{\text{ext}} = \Gamma Q + \mu_{\text{ext}} \tag{5.25}$$

where Γ follows a standard multivariate normal distribution, the restriction of μ_{ext} to I agrees with the mean of $\tilde{Y}$, and

$$Q = \bar{F}_\otimes \Lambda^{1/2} F_\otimes,$$

where $\bar{F}_\otimes$ is the complex conjugate of $F_\otimes$ and Λ is the diagonal matrix of eigenvalues for C. Using the two-dimensional fast Fourier transform, a very fast simulation algorithm for $\tilde{Y}_{\text{ext}}$ and hence $\tilde{Y}$ is obtained even when the dimension of $\tilde{Y}$ is of order several thousands.

We use this method for the simulations in Figure 5.3 and in connection with the MCMC algorithm considered in Section 10.2.3. The circulant embedding method is e.g. implemented in the R package RandomFields by Martin Schlater.

REMARK 5.3 If the covariance matrix of $\tilde{Y}$ is positive definite, an alternative to circulant embedding is to use the *Choleski decomposition* LL^T of the covariance matrix where L is a lower triangular matrix (see e.g. Ripley 1981). This may be advantageous if the covariance function is not translation invariant or B is far from being rectangular. Then we still have an identity of the form (5.25) but with $\tilde{Y}_{\text{ext}} = \tilde{Y}$, $Q = L^\mathsf{T}$, and μ_{ext} equal to the mean of $\tilde{Y}$. The Choleski decomposition is, on the other hand, only practically applicable if the dimension of $\tilde{Y}$ is moderate. It is used in e.g. Møller & Waagepetersen (2003).

5.8 Multivariate Cox processes

Definition 5.1 immediately extends to the multivariate case as follows.

DEFINITION 5.4 Suppose that $Z_i = \{Z_i(\xi) : \xi \in S\}$, $i = 1, \ldots, k$, are nonnegative random fields so that for $i = 1, \ldots, k$, $\xi \to Z_i(\xi)$ is a locally integrable function with probability one. Conditional on $Z = (Z_1, \ldots, Z_k)$, suppose that $X_1, \ldots, X_k$ are independent Poisson processes with intensity functions $Z_1, \ldots, Z_k$, respectively. Then $X = (X_1, \ldots, X_k)$ is said to be a *multivariate Cox process driven by Z*.

Note that dependence between the fields $Z_1, \ldots, Z_k$ implies dependence between the components $X_1, \ldots, X_k$.

5.8.1 Summary statistics

As in the univariate case, we immediately obtain the cross pair correlation function for X_i and X_j,

$$g_{ij}(\xi, \eta) = \mathbb{E}[Z_i(\xi) Z_j(\eta)] / [\rho_i(\xi) \rho_j(\eta)]$$

where

$$\rho_i(\xi) = \mathbb{E} Z_i(\xi)$$

is the intensity function for X_i, while summary statistics based on nearest neighbour distances are rarely tractable.

Diggle (1983) considers the following three general classes of models for a bivariate Cox process (X_1, X_2):

(i) A *linked Cox process*, i.e. when Z_1 and Z_2 are proportional.

(ii) A *balanced Cox process*, i.e. when $Z_1 + Z_2$ is constant.

(iii) $Z_i = \Pi_i Z_0$, $i = 1, 2$, where Π_1, Π_2, Z_0 are nonnegative random fields so that $\Pi_1 + \Pi_2 = 1$, Π_1 and Z_0 are independent, and for $i = 0, 1, 2$, $\xi \to Z_i(\xi)$ is a locally integrable function.

There is a positive dependence between X_1 and X_2 for a linked Cox process and a negative dependence for a balanced Cox process. The model class (iii) covers both (i) and (ii), as we obtain a linked Cox process if Π_1 is constant, and a balanced Cox process if Z_0 is constant.

Consider (X_1, X_2) given by (iii) and suppose that $Y = (Y_1, Y_2)$ is a linked Cox process driven by (Z_0, Z_0). We can then view (X_1, X_2) as the result of a random independent thinning of each of the components Y_1 and Y_2 with retention probabilities Π_1 and Π_2, respectively, provided (Y, Z_0) and Π_1 are independent. We obtain the relationships

$$\frac{\rho_1(\xi)\rho_1(\eta)g_1(\xi,\eta)}{\mathbb{E}[\Pi_1(\xi)\Pi_1(\eta)]} = \frac{\rho_2(\xi)\rho_2(\eta)g_2(\xi,\eta)}{\mathbb{E}[\Pi_2(\xi)\Pi_2(\eta)]} = \frac{\rho_1(\xi)\rho_2(\eta)g_{12}(\xi,\eta)}{\mathbb{E}[\Pi_1(\xi)\Pi_2(\eta)]}$$
$$= \rho_0(\xi)\rho_0(\eta)g_0(\xi,\eta)$$

MULTIVARIATE COX PROCESSES

where $\rho_0 = \rho_1 + \rho_2$ is the intensity function and g_0 is the pair correlation function for Y_i, $i = 1, 2$. For a linked Cox process (Π_1 constant) this reduces to
$$g_0 = g_1 = g_2 = g_{12},$$
while for a balanced Cox process (Z_0 constant) ρ_0 is constant and $g_0 = 1$.

5.8.2 Multivariate log Gaussian Cox processes

Definition 5.3 can naturally be extended as follows (Møller et al. 1998).

DEFINITION 5.5 Let $X = (X_1, \ldots, X_k)$ be a multivariate Cox process driven by $Z = (\exp(Y_1), \ldots, \exp(Y_k))$ where $Y = (Y_1, \ldots, Y_k)$ is a k-dimensional Gaussian field. Then X is called a *multivariate log Gaussian Cox process (multivariate LGCP)*.

Each X_i is then an LGCP. The distribution of (X, Y) is determined by the mean and covariance functions
$$m_i(\xi) = \mathbb{E}Y_i(\xi) \quad c_{ij}(\xi, \eta) = \mathbb{C}\mathrm{ov}(Y_i(\xi), Y_j(\eta)), \quad 1 \leq i \leq j \leq k.$$
As in Proposition 5.4 we obtain
$$\rho_i(\xi) = \exp(m_i(\xi) + c_{ii}(\xi, \xi)/2), \quad g_{ij}(\xi, \eta) = \exp(c_{ij}(\xi, \eta)).$$

Møller et al. (1998) and Brix & Møller (2001) consider an easy construction obtained by linear combinations of independent Gaussian fields Z_j, $j = 1, \ldots, l$, with zero mean and variance 1:
$$Y_i(\xi) = m_i(\xi) + \sum_{j=1}^{l} a_{ij} Z_j(\xi), \ i = 1, \ldots, k, \tag{5.26}$$
where the a_{ij} are real parameters. This provides a large family of multivariate LGCPs with flexibility for modelling positive and negative dependence between the X_i, and the simulation techniques in Section 5.7 immediately extend to such models. Figure 5.4 shows simulations of bivariate LGCPs where $m_1 = m_2 = 4.10$, $l = 1$, and (a_{11}, a_{21}) is either $(1, 1)$ or $(1, -1)$, whereby either a linked or a balanced Cox process is obtained. A positive respective negative dependence between X_1 and X_2 is clearly visible in Figure 5.4 for the linked respective balanced Cox process.

EXAMPLE 5.9 (*Weed*) Brix & Møller (2001) use the construction (5.26) to model the bivariate point pattern of weed plants in Example 1.5 with $k = 2$, $l = 3$ and with $a_{12} = a_{21} = 0$. The random field Z_3 models environmental effects common for the weed plants, while Z_1 and Z_2

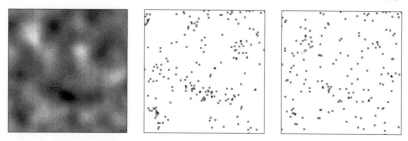

Figure 5.4 *Left: simulation on $[0,1]^2$ of a zero mean Gaussian field Z_1 with Gaussian correlation function. Middle: a bivariate LGCP with $Y_1 = Y_2 = 4.10 + Z_1$. Right: a bivariate LGCP with $Y_1 = 4.1 + Z_1$ and $Y_2 = 4.1 - Z_1$.*

model effects particular for each weed type (due e.g. to the seed banks of the plants).

5.8.3 Multivariate shot noise Cox processes

Definition 5.2 can be extended as follows.

DEFINITION 5.6 Let $X = (X_1, \ldots, X_k)$ be a multivariate Cox process driven by $Z = (Z_1, \ldots, Z_k)$ where

$$Z_i(\xi) = \sum_{(c,\gamma) \in \Phi_i} \gamma k_i(c, \xi), \ i = 1, \ldots, k,$$

where for each i, $k_i(\cdot, \cdot)$ is a kernel and Φ_i is a Poisson point process on $R^d \times (0, \infty)$ with a locally integrable intensity function ζ_i. Then X is called a *multivariate shot noise Cox process (multivariate SNCP)*.

Each X_i is then an SNCP. However, for the joint distribution of the X_i it remains to specify the joint distribution of the Φ_i. If the Φ_i are independent we just obtain the trivial case where the X_i are independent. There are different possibilities for constructing dependence between the Φ_i, e.g. the examples of multivariate Poisson processes given in Section 3.3.2, leading to different types of multivariate SNCPs which need to be considered separately. However, we do not pursue this further.

CHAPTER 6

Markov point processes

Markov point processes are models for point processes with interacting points. Such models are constructed by considering a density for a point process with respect to a Poisson process and imposing certain conditions ensuring a Markov property. Often a repulsive behaviour is modelled, though there exist possibilities for modelling attraction as well. Markov (or Gibbs) point processes arose in statistical physics for the description of large interacting particle systems, see e.g. Ruelle (1969), Preston (1976), and Georgii (1988). Markov point processes were introduced in spatial statistics by Ripley & Kelly (1977). Van Lieshout (2000) provides a recent account of the state of the art of Markov point processes in spatial statistics; see also the reviews in Ripley (1977), Baddeley & Møller (1989), and Stoyan et al. (1995).

Section 6.1 describes the setting and some other background material used in this chapter. Section 6.2 concerns pairwise interaction point processes, which provide the simplest and most used class of finite Markov point processes. Section 6.3 deals with general finite Markov point process. Section 6.4 surveys how to extend a finite Markov point process to an infinite Markov or Gibbs point process defined on $\mathbb{R}^d$ and establishes some results for the summary statistics of a (finite or infinite) Markov point process. Section 6.5 concerns inhomogeneous Markov point processes. Finally, Section 6.6 discusses how everything in the previous sections can be extended to multivariate and marked point processes.

Monte Carlo methods and simulation algorithms for Markov point processes are covered by the material given in Chapters 7, 8, and 11. Simulation-based inference for Markov point processes is treated in Chapter 9. Appendix F concerns so-called nearest-neighbour Markov point processes, a class of models that generalises Markov point processes.

6.1 Finite point processes with a density

Consider a point process X on $S \subseteq \mathbb{R}^d$ with density f with respect to the standard Poisson process $\text{Poisson}(S, 1)$. For similar reasons as in Section 3.2.4, in order to obtain a well defined density we need in general that $|S| < \infty$, so in this and the following sections we assume that $|S| < \infty$ unless otherwise stated. The density is then concentrated on the set of *finite point configurations* contained in S,

$$N_\mathrm{f} = \{x \subset S : n(x) < \infty\},$$

and by Proposition 3.1 (Section 3.1.2), for $F \subseteq N_\mathrm{f}$,

$$P(X \in F) = \sum_{n=0}^{\infty} \frac{\exp(-|S|)}{n!} \int_S \cdots \int_S \mathbf{1}[\{x_1, \ldots, x_n\} \in F]$$
$$f(\{x_1, \ldots, x_n\}) \mathrm{d}x_1 \cdots \mathrm{d}x_n \qquad (6.1)$$

where the term for $n = 0$ is read as $\exp(-|S|)\mathbf{1}[\emptyset \in F]f(\emptyset)$. If $|S| = 0$ then $P(X = \emptyset) = 1$, so in applications of interest, $|S| > 0$.

An example is the density for $X \sim \mathrm{Poisson}(S, \rho)$ with $\mu(S) = \int_S \rho(\xi)\mathrm{d}\xi < \infty$, where

$$f(x) = \exp(|S| - \mu(S)) \prod_{\xi \in x} \rho(\xi), \qquad (6.2)$$

cf. (3.11). In this chapter we shall construct much more interesting models exhibiting interaction between the points.

In most cases f is only specified up to proportionality $f \propto h$ where $h : N_\mathrm{f} \to [0, \infty)$ is a known function. The *normalising constant*

$$c = \sum_{n=0}^{\infty} \frac{\exp(-|S|)}{n!} \int_S \cdots \int_S h(\{x_1, \ldots, x_n\}) \mathrm{d}x_1 \cdots \mathrm{d}x_n$$

is unknown for the models considered in this chapter except in the special case where X is a Poisson process, see Proposition 3.8 (Section 3.2.4). In the terminology of statistical physics, c is the *partition function*.

In Section 9.1.2 we discuss likelihood inference for X conditional on $n(X) = n$ where n is a fixed integer (e.g. the number of points in a given point pattern). We then consider the conditional density $f_n \propto f$ so that

$$P(X \in F | n(X) = n) = \int_S \cdots \int_S \mathbf{1}[\{x_1, \ldots, x_n\} \in F]$$
$$f_n(\{x_1, \ldots, x_n\}) \mathrm{d}x_1 \cdots \mathrm{d}x_n. \qquad (6.3)$$

If we specify $f_n \propto h_n$ where h_n is a known function, the normalising constant

$$c_n = \int_S \cdots \int_S h_n(\{x_1, \ldots, x_n\}) \mathrm{d}x_1 \cdots \mathrm{d}x_n \qquad (6.4)$$

is in general unknown except in the binomial case.

REMARK 6.1 For readers familiar with measure theory it should be obvious how most of the definitions and results in this chapter extend to densities f with respect to a Poisson process with a finite intensity measure μ defined on a general space S, cf. Remark 3.1 (Section 3.1.1). A few

FINITE POINT PROCESSES WITH A DENSITY

times in this book we consider densities f with respect to Poisson(S, ρ), where $S \subseteq \mathbb{R}^d$ and $\mu(S) = \int_S \rho(\xi) \mathrm{d}\xi < \infty$, i.e.

$$P(X \in F) = \sum_{n=0}^{\infty} \frac{\exp(-\mu(S))}{n!} \int_S \cdots \int_S \mathbf{1}[\{x_1, \ldots, x_n\} \in F]$$
$$f(\{x_1, \ldots, x_n\}) \rho(x_1) \cdots \rho(x_n) \mathrm{d}x_1 \cdots \mathrm{d}x_n. \quad (6.5)$$

6.1.1 Papangelou conditional intensity and stability conditions

We now define a fundamental characteristic which will play a key role in the rest of this book.

DEFINITION 6.1 The *Papangelou conditional intensity* for a point process X with density f is defined by

$$\lambda^*(x, \xi) = f(x \cup \xi)/f(x), \quad x \in N_\mathrm{f}, \ \xi \in S \setminus x, \quad (6.6)$$

taking $a/0 = 0$ for $a \geq 0$ (Kallenberg 1984).

Note that λ^* does not depend on the normalising constant of f. For a Poisson process with intensity function ρ, by (3.11), $\lambda^*(x, \xi) = \rho(\xi)$ does not depend on x. Heuristically, $\lambda^*(x, \xi) \mathrm{d}\xi$ can be interpreted as the conditional probability of X having a point in an infinitesimal region containing ξ and of size $\mathrm{d}\xi$ given the rest of X is x. We say that X (or f) is *attractive* if

$$\lambda^*(x, \xi) \leq \lambda^*(y, \xi) \quad \text{whenever } x \subset y \quad (6.7)$$

and *repulsive* if

$$\lambda^*(x, \xi) \geq \lambda^*(y, \xi) \quad \text{whenever } x \subset y. \quad (6.8)$$

Intuitively, given that $X \setminus \xi = x$, attractivity means that the chance that $\xi \in X$ is an increasing function of x, while repulsivity means the opposite.

We shall often consider functions $h : N_\mathrm{f} \to [0, \infty)$ which are *hereditary*, that is,

$$h(x) > 0 \Rightarrow h(y) > 0 \quad \text{for } y \subset x. \quad (6.9)$$

If f is hereditary, then there is a one-to-one correspondence between f and λ^*.

We now introduce some useful stability conditions.

DEFINITION 6.2 Suppose that $\phi^* : S \to [0, \infty)$ is a function so that $c^* = \int_S \phi^*(\xi) \mathrm{d}\xi$ is finite. For a given function $h : N_\mathrm{f} \to [0, \infty)$, *local stability* (or ϕ^*-*local stability*) means that

$$h(x \cup \xi) \leq \phi^*(\xi) h(x) \quad (6.10)$$

for all $x \in N_f$ and $\xi \in S \setminus x$. *Ruelle stability* (or ϕ^*-Ruelle stability; Ruelle 1969) means that $h(x) \leq \alpha \prod_{\xi \in x} \phi^*(\xi)$ for some positive constant α and all $x \in N_f$.

Ruelle stability means that h is dominated by an unnormalised Poisson density, and it implies integrability of h with respect to Poisson$(S,1)$. Local stability implies Ruelle stability and that h is hereditary. If $f \propto h$ is ϕ^*-locally stable, then $\lambda^*(x,\xi) \leq \phi^*(\xi)$. Local stability is satisfied by most point process models (Geyer 1999, Kendall & Møller 2000). One example, where Ruelle but not local stability is satisfied, is a Lennard-Jones process (Example 6.4 below). As shown in Chapters 7 and 11, local stability plays an important role in simulation algorithms.

6.2 Pairwise interaction point processes

Let the situation be as in the beginning of Section 6.1.

6.2.1 Definitions and properties

In many applications we have a *pairwise interaction point process*,

$$f(x) \propto \prod_{\xi \in x} \phi(\xi) \prod_{\{\xi,\eta\} \subseteq x} \phi(\{\xi,\eta\}) \tag{6.11}$$

where ϕ is an *interaction function*, i.e. a nonnegative function for which the right hand side in (6.11) is integrable with respect to Poisson$(S,1)$. The *range of interaction* is defined by

$$R = \inf\{r > 0 : \text{for all } \{\xi,\eta\} \subset S, \phi(\{\xi,\eta\}) = 1 \text{ if } \|\xi - \eta\| > r\}.$$

In the Poisson case, $\phi(\xi)$ is the intensity function and $\phi(\{\xi,\eta\}) = 1$, so $R = 0$. Apart from the Poisson case, pairwise interaction processes are analytically intractable because of the unknown normalising constant.

The density (6.11) is clearly hereditary. For $f(x) > 0$ and $\xi \notin x$,

$$\lambda^*(x,\xi) = \phi(\xi) \prod_{\eta \in x} \phi(\{\xi,\eta\}).$$

So f is repulsive if and only if $\phi(\{\xi,\eta\}) \leq 1$, in which case the process is locally stable if $\int_S \phi(\xi) \mathrm{d}\xi < \infty$ (taking $\phi^*(\xi) = \phi(\xi)$ in Definition 6.2). Most examples of pairwise interaction point processes are repulsive, and the attractive case ($\phi(\{\xi,\eta\}) \geq 1$ for all distinct $\xi,\eta \in S$) is in general not well defined.

A pairwise interaction point process is said to be *homogeneous* if the first order term $\phi(\xi)$ is constant and the second order interaction term is invariant under motions, i.e. $\phi(\{\xi,\eta\}) = \phi_2(\|\xi - \eta\|)$ where $\phi_2 : (0,\infty) \to [0,\infty)$. Most examples of pairwise interaction point process (and more generally Markov point processes) in this chapter will be homogeneous. Inhomogeneous pairwise interaction point processes (and inhomogeneous Markov point processes) are treated in Section 6.5.

For a given $n \in \mathbb{N}$, we say that $X|n(X) = n$ follows a *conditional pairwise interaction point process* if the conditional density is of the form

$$f_n(x) = \frac{1}{c_n} \prod_{\xi \in x} \phi(\xi) \prod_{\{\xi,\eta\} \subseteq x} \phi(\{\xi,\eta\}) \qquad (6.12)$$

for $n(x) = n$. Here the normalising constant given by (4.4) may be finite (so that f_n is well defined) even if $\phi(\{\xi,\eta\})$ is not less or equal to one for all distinct $\xi, \eta \in S$, see Example 6.1 below.

6.2.2 Examples of pairwise interaction point processes

In the examples below we consider the homogeneous case where $\phi(\xi)$ is constant and $\phi(\{\xi,\eta\}) = \phi_2(\|\xi - \eta\|)$.

EXAMPLE 6.1 (*Strauss process*) The simplest nontrivial pairwise interaction process is the *Strauss process* (Strauss 1975), with

$$\phi_2(r) = \gamma^{1[r \leq R]} \qquad (6.13)$$

setting $0^0 = 1$. Here $0 \leq \gamma \leq 1$ is an interaction parameter and $R > 0$ is the range of interaction. The process is repulsive and hence locally stable. The class of Strauss processes is characterised by the property that $\lambda^*(x,\xi)$ depends only on x through $n(x \cap b(\xi, R))$ (Kelly & Ripley 1976).

The density is usually written as

$$f(x) \propto \beta^{n(x)} \gamma^{s_R(x)} \qquad (6.14)$$

where $\beta > 0$ is a parameter and

$$s_R(x) = \sum_{\{\xi,\eta\} \subseteq x} \mathbf{1}[\|\xi - \eta\| \leq R]$$

is the number of R-close pairs of points in x. Strauss (1975) incorrectly assumed $\gamma > 1$ in order to obtain a model for clustering, but then (6.14) is not integrable (Kelly & Ripley 1976): If S contains a set A so that $|A| > 0$ and the diameter of A is less than R, then

$$\sum_{n=0}^{\infty} \frac{\beta^n}{n!} \int_S \cdots \int_S \gamma^{s_R(x)} \mathrm{d}x_1 \cdots \mathrm{d}x_n \geq \sum_{n=0}^{\infty} \frac{(\beta|A|)^n}{n!} \gamma^{n(n-1)/2} = \infty$$

for $\gamma > 1$.

If $\gamma = 1$ we obtain $X \sim \text{Poisson}(S, \beta)$, while for $\gamma < 1$ there is repulsion between R-close pairs of points in X. The special case where $\gamma = 0$

is called a *hard core process* with *hard core* R as the points are prohibited from being closer than distance R apart. The hard core process can be thought of as a conditional Poisson process of rate β where we have conditioned on the hard core condition. Simulated realisations of a Strauss process look more and more regular as γ decreases, cf. Figure 6.1. For small values of γ and large values of β or R, the points of the process tend to cluster around the points in a regular grid, cf. Møller (1999a).

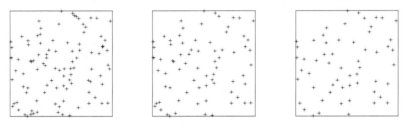

Figure 6.1 *Realisations of Strauss processes defined on the unit square $S = [0,1]^2$, when $\beta = 100$, $R = 0.05$, and $\gamma = 1, 0.5, 0$ (from left to right). The realisations are obtained by perfect simulation (for details, see Algorithm 11.7 in Section 11.2.6).*

The *conditional Strauss process* is defined by

$$f_n(x) \propto \gamma^{s_R(x)} \qquad (6.15)$$

(for $n(x) = n$). This is well defined for any value of $\gamma \geq 0$ and has been studied particularly for $\gamma > 1$ (Kelly & Ripley 1976, Ripley 1977, Gates & Westcott 1986, Geyer & Thompson 1995, Geyer 1999, Møller 1999a). However, in the case $\gamma > 1$, the unconditional process X obtained by imposing some distribution on $N = n(X)$ cannot be a pairwise interaction process: Let $a_n = P(N = n)$, then the density of X is

$$f(x) \propto (a_{n(x)}/c_{n(x)})\gamma^{s_R(x)}.$$

So to obtain a pairwise interaction density, we need $a_n/c_n \propto \beta^n$ for some $\beta > 0$. But as shown above, $f(x) \propto \beta^{n(x)}\gamma^{s_R(x)}$ is in general not integrable if $\gamma > 1$.

For small and modest values of R, simulated realisations of the conditional Strauss process with $\gamma > 1$ look much like realisations from either a binomial process or a "one-clump" process (Geyer & Thompson 1995, Geyer 1999, Møller 1999a). Thus the conditional Strauss process is not of much interest for spatial applications when $\gamma > 1$.

EXAMPLE 6.2 (*Multiscale process*) The Strauss process can be extended to a *multiscale process* (Penttinen 1984) where

$$\phi_2(r) = \gamma_i \quad \text{if } R_{i-1} < r \leq R_i \qquad (6.16)$$

with change points $0 = R_0 < R_1 < \ldots < R_k < R_{k+1} = \infty$. Here $k \in \mathbb{N}$, $\gamma_i \geq 0$, $i = 1, \ldots, k$, and $\gamma_{k+1} = 1$, so the process has range of interaction $R = R_k$. As exemplified below further restrictions on the parameters γ_i are needed to ensure integrability. The process can be used for approximating a general interaction function ϕ_2 (Ripley 1989, Heikkinen & Penttinen 1999, Berthelsen & Møller 2003).

The density is usually written as

$$f(x) \propto \beta^{n(x)} \prod_{i=1}^{k} \gamma_i^{s_i(x)} \qquad (6.17)$$

where $\beta > 0$ and

$$s_i(x) = \sum_{\{\xi,\eta\} \subseteq x} \mathbf{1}[R_{i-1} < \|\xi - \eta\| \leq R_i].$$

For $k = 1$, we have a Strauss process, assuming that $0 \leq \gamma_1 \leq 1$. For $k \geq 2$, integrability is ensured if either

(i) $0 < \gamma_1 \leq 1$ and $0 \leq \gamma_2 \leq 1, \ldots, 0 \leq \gamma_k \leq 1$, or

(ii) $\gamma_1 = 0$ and $\gamma_2 \geq 0, \ldots, \gamma_k \geq 0$.

In case (i) the process is repulsive and hence locally stable. In case (ii) γ_1 is a hard core parameter and different types of interaction can be modelled at different scales. The process is also locally stable in case (ii), since the hard core condition imposed by $\gamma_1 = 0$ implies the existence of an $n_0 \in \mathbb{N}$ so that $f(x) = 0$ whenever $n(x) > n_0$, and hence $\lambda^* \leq \beta \prod_{i=2}^{k} \max\{1, \gamma_i\}^{n_0 - 1}$.

EXAMPLE 6.3 (*Other examples with finite range of interaction*) A simple alternative to the Strauss process is to replace its step function ϕ_2 by a linearly decreasing function

$$\phi_2(r) = \mathbf{1}[r \leq R] r/R \qquad (6.18)$$

(Diggle, Gates & Stibbard, 1987; similarly we can modify the multiscale point process). This function depends only on the parameter R, but is not differentiable at $r = R$. A smoother function is

$$\phi_2(r) = \mathbf{1}[r \leq R](1 - (1 - r^2/R^2)) \qquad (6.19)$$

(Diggle, Fiksel, Grabarnik, Ogata, Stoyan & Tanemura 1994).

Thinking of the points of a point process as centres for circular territories (or influence zones) of diameter R, Penttinen (1984) proposed an "*overlap area process*" where

$$\phi_2(r) = \gamma^{|b(\xi, R/2) \cap b(\eta, R/2)|} \qquad (6.20)$$

depends only on $r = \|\xi - \eta\|$ and where $0 \leq \gamma \leq 1$. For instance, if $d = 2$

and $r \leq R$,
$$|b(\xi, R/2) \cap b(\eta, R/2)| = R^2/2 \arccos(r/R) - r(R^2 - r^2)^{1/2}/2.$$

For each model (6.18)–(6.20), the process is repulsive and hence locally stable, and R specifies the range of interaction.

EXAMPLE 6.4 (*Some examples with infinite range of interaction*) A *very-soft-core process* has
$$\phi_2(r) = 1 - \exp(-(r/\theta)^2)$$
where $\theta > 0$ (see e.g. Ogata & Tanemura 1984), and a *Lennard-Jones process* has
$$\phi_2(r) = \exp\left(\alpha_1(\sigma/r)^6 - \alpha_2(\sigma/r)^{12}\right)$$
where $\alpha_1 \geq 0$, $\alpha_2 > 0$, $\sigma > 0$ (see e.g. Ruelle 1969). Note that both processes have *infinite range of interaction* in the sense that $R \to \infty$ as S tends to $\mathbb{R}^d$ (see also Remark 6.4 below).

The very-soft-core process is repulsive and hence locally stable. The Lennard-Jones process is repulsive for $\alpha_1 = 0$, and neither repulsive nor attractive for $\alpha_1 > 0$ as
$$\phi_2(r) \lesseqgtr 1 \quad \text{for } r \lesseqgtr (\alpha_2/\alpha_1)^{1/6}\sigma.$$

The Lennard-Jones process is clearly not locally stable for $\alpha_1 > 0$, but it can be shown to be Ruelle stable.

As illustrated in Ripley (1989), functions ϕ_2 for an overlap area process and a very-soft-core process can be very close.

6.3 Markov point processes

Let the situation be as in the beginning of Section 6.1.

6.3.1 Definition and characterisation

The role of the Papangelou conditional intensity is similar to that of the local characteristics of a Markov random field (e.g. Besag 1974) when defining Markov properties. For instance, for a pairwise interaction process with range of interaction R, for $f(x) > 0$,
$$\lambda^*(x, \xi) = \phi(\xi) \prod_{\eta \in x : \|\xi - \eta\| \leq R} \phi(\{\xi, \eta\})$$
depends only on x through $x \cap b(\xi, R)$. In other words, if we say that ξ and η are *R-close-neighbours* whenever $\|\xi - \eta\| \leq R$, $\lambda^*(x, \xi)$ depends only on those $\eta \in x$ which are R-close-neighbours to ξ. This extends as follows.

MARKOV POINT PROCESSES

Let $\sim$ be a reflexive and symmetric relation on S, i.e. for all $\xi, \eta \in S$, $\xi \sim \xi$, and $\xi \sim \eta$ implies $\eta \sim \xi$.[†] We say that ξ and η are *neighbours* if $\xi \sim \eta$, and define the *neighbourhood* of ξ by $N_\xi = \{\eta \in S : \eta \sim \xi\}$.

One example is given by $\xi \sim \eta$ if and only if $A_\xi \cap A_\eta \neq \emptyset$, where $A_\xi \subseteq S$. The most commonly used relation is the *R-close-neighbourhood relation* defined by $A_\xi = b(\xi, R/2)$, i.e. $N_\xi = b(\xi, R)$. We can think of A_ξ as a territory and N_ξ as an influence zone associated to ξ.

DEFINITION 6.3 Let $h : N_f \to [0, \infty)$ denote any hereditary function. If for all $x \in N_f$ with $h(x) > 0$ and all $\xi \in S \setminus x$,

$$h(x \cup \xi)/h(x) \text{ depends on } x \text{ only through } x \cap N_\xi \quad (6.21)$$

then h is said to be a *Markov function* (with respect to $\sim$). Particularly, a density (with respect to Poisson$(S, 1)$) which is a Markov function is called a *Markov density function*, and a point process with a Markov density function is called a *Markov point process*.

The hereditary condition corresponds to a relaxation of the positivity condition in the Hammersley-Clifford theorem for random fields defined on a lattice, see Besag (1974) and Clifford (1990). In the sequel it is convenient to define $\lambda^*(x, \xi)$ in terms of h, though h need not be a (normalised or unnormalised) density, i.e. for $x \in N_f$ and $\xi \in S \setminus x$, $\lambda^*(x, \xi) = h(x \cup \xi)/h(x)$ if $h(x) > 0$, and $\lambda^*(x, \xi) = 0$ otherwise.

REMARK 6.2 From a computational and simulation point of view, the important property of a Markov point process is the *local Markov property* (6.21), namely that $\lambda^*(x, \xi)$ depends only on "local information". This property is exploited in the computations for the examples of applications in Chapter 9.

The class of Markov functions is characterised by the following so-called *Hammersley-Clifford-Ripley-Kelly theorem* due to Ripley & Kelly (1977) (see also Clifford 1990 for a historical account on this an other types of Hammersley-Clifford type theorems). We say that a function $\phi : N_f \to [0, \infty)$ is an *interaction function* if $\phi(x) = 1$ whenever there exist $\xi, \eta \in x$ with $\xi \not\sim \eta$.

THEOREM 6.1 A function $h : N_f \to [0, \infty)$ is a Markov function if and only if there is an interaction function ϕ such that

$$h(x) = \prod_{y \subseteq x} \phi(y), \quad x \in N_f. \quad (6.22)$$

[†] We also use the symbol $\sim$ to specify a distribution for a stochastic quantity, but it will always be clear from the context how $\sim$ should be interpreted.

Then, for $h(x) > 0$,

$$\lambda^*(x,\xi) = \prod_{y \subseteq x} \phi(y \cup \xi), \quad x \in N_{\mathrm{f}}, \ \xi \in S \setminus x. \tag{6.23}$$

Proof. Clearly, (6.22) implies (6.23). Hence h given by (6.22) is obviously hereditary and locally Markov.

Suppose that h is Markov. Define ϕ inductively by $\phi(\emptyset) = h(\emptyset)$, $\phi(x) = 1$ if there exist $\xi, \eta \in x$ with $\xi \not\sim \eta$, and

$$\phi(x) = h(x) / \prod_{y \subset x} \phi(y) \tag{6.24}$$

otherwise, taking $0/0 = 1$. Then ϕ is an interaction function. To show that $h(x)$ is of the form (6.22) we consider three cases below.

1) The case $h(x) = 0$ and $\prod_{y \subset x} \phi(y) = 0$: Then $h(x) = 0 = \prod_{y \subseteq x} \phi(y)$.

2) The case $h(x) = 0$ and $\prod_{y \subset x} \phi(y) > 0$: By definition of λ^* and ϕ, $\lambda^*(x \setminus \kappa, \kappa) = 0$ for all $\kappa \in x$ and $h(y) > 0$ for all $y \subset x$, so $\lambda^*(x \setminus \{\kappa, \zeta\}, \zeta) > 0$ for all $\{\kappa, \zeta\} \subseteq x$. Suppose there exist $\xi, \eta \in x$ with $\xi \not\sim \eta$. Then by (6.21) we obtain a contradiction:

$$0 = \lambda^*(x \setminus \eta, \eta) = \lambda^*(x \setminus \{\xi, \eta\}, \eta) > 0.$$

Thus by definition $\phi(x) = 0$, so $h(x) = 0 = \prod_{y \subseteq x} \phi(y)$.

3) The case $h(x) > 0$: Then, since h is hereditary, $h(y) > 0$ for all $y \subseteq x$. We use induction on $n = n(x)$ to show that $h(x)$ is of the required form. If either $n = 0$ or $\xi \sim \eta$ for all $\xi, \eta \in x$, then by definition of ϕ and the induction hypothesis, $\phi(y) > 0$ for all $y \subset x$, and $h(x) = \prod_{y \subseteq x} \phi(y)$. Suppose that $x = z \cup \{\xi, \eta\}$ where $\xi \not\sim \eta$ and $z \cap \{\xi, \eta\} = \emptyset$. By (6.21),

$$h(x) = \lambda^*(z \cup \xi, \eta) h(z \cup \xi) = \lambda^*(z, \eta) h(z \cup \xi) = \frac{h(z \cup \eta)}{h(z)} h(z \cup \xi).$$

Combining this with the induction hypothesis,

$$h(x) = \frac{\prod_{y \subseteq z \cup \eta} \phi(y)}{\prod_{y \subseteq z} \phi(y)} \prod_{y \subseteq z \cup \xi} \phi(y) = \prod_{y \subseteq z} \phi(y \cup \eta) \prod_{y \subseteq z \cup \xi} \phi(y) = \prod_{y \subseteq x} \phi(y)$$

since $\phi(y) = 1$ whenever $\{\xi, \eta\} \subseteq y$. $\square$

REMARK 6.3 It follows from (6.24) that ϕ in (6.22) is uniquely determined in terms of h. It also follows that no special structure of the space S is required: Theorem 6.1 holds for any nonnegative function h defined on any space S equipped with a reflexive symmetric relation.

By (6.22), a Markov point process X has density of the form

$$f(x) = \prod_{y \subseteq x} \phi(y), \tag{6.25}$$

MARKOV POINT PROCESSES

where $c = 1/\phi(\emptyset)$ is the normalising constant. In the terminology of statistical physics, X is a *finite Gibbs point process* with *energy*

$$U(x) = -\sum_{y \subseteq x: y \neq \emptyset} \log \phi(y)$$

(taking $\log 0 = -\infty$) and *partition function* c, the distribution of X is a "*finite Gibbs state*", and f is a "*grand canonical ensemble*". Furthermore, a conditional density of the form

$$f_n(x) \propto \exp(-U(x)) \text{ for } n(x) = n$$

is called a "*canonical ensemble*".

DEFINITION 6.4 For a given interaction function ϕ, define the *range of interaction* by

$$R = \inf\{r > 0 : \text{for all } x \in N_f, \phi(x) = 1 \text{ if } \|\xi - \eta\| > r \text{ for some } \xi, \eta \in x\}. \tag{6.26}$$

REMARK 6.4 By Definition 6.4, $\|\xi - \eta\| > R$ implies $\phi(z \cup \{\xi, \eta\}) = 1$ for any $z \in N_f$, and so by (6.23), $\lambda^*(x, \xi) = \lambda^*(x \cap b(\xi, R), \xi)$. Suppose that $\phi(x)$ is defined for all locally finite $x \subset \mathbb{R}^d$. Then we may extend Definition 6.4 and define the range of interaction R by (6.26) where now x is any locally finite subset of $\mathbb{R}^d$. Usually, we assume that R defined in this way is finite (Example 6.4 provides some exceptions), and then we say that ϕ has *finite range of interaction*.

6.3.2 Examples

Pairwise interaction processes provide the simplest examples of Markov point processes, and they are mainly models for a repulsive behaviour, cf. Section 6.2. As illustrated in the following examples, attractive Markov point processes will usually be of *infinite order of interaction*, that is, $\phi(y)$ can be different from one no matter how large $n(y)$ is.

EXAMPLE 6.5 (*Widom-Rowlinson model/area-interaction point process*) The *Widom-Rowlinson penetrable spheres model* (Widom & Rowlinson 1970) is given by

$$f(x) \propto \beta^{n(x)} \gamma^{-|U_{x,R}|} \tag{6.27}$$

where $U_{x,R} = \cup_{\xi \in x} b(\xi, R)$ and $\beta > 0$, $R > 0$, $\gamma > 0$. This process has been studied in great detail in Baddeley & van Lieshout (1995) under the name *area-interaction point process*.

The process is Markov with respect to the $2R$-neighbourhood relation, since

$$\lambda^*(x, \xi) = \beta \gamma^{-|b(\xi, R) \setminus \cup_{\eta \in x: \|\xi - \eta\| \leq 2R} b(\eta, R)|}.$$

It is attractive for $\gamma \geq 1$ and repulsive for $0 < \gamma \leq 1$. In both cases it is locally stable, with $\lambda^* \leq \beta$ if $\gamma \geq 1$, and $\lambda^* \leq \beta\gamma^{-\omega_d R^d}$ if $0 < \gamma \leq 1$. Simulated realisations for the attractive case are given by the marginal point patterns in Figure 6.2 (see Example 6.8 for details).

By the inclusion-exclusion formula,

$$|U_{x,R}| = \sum_{y \subseteq x : y \neq \emptyset} (-1)^{n(y)+1} |\cap_{\xi \in y} b(\xi, R)|$$

so the interaction function is given by

$$\phi(y) = \beta\gamma^{-\omega_d R^d} \quad \text{for } n(y) = 1$$

and

$$\phi(y) = \gamma^{(-1)^{n(y)}|\cap_{\xi \in y} b(\xi, R)|} \quad \text{for } n(y) \geq 2.$$

Hence there is an infinite order of interaction.

REMARK 6.5 The complicated form of the interaction function for the area-interaction process illustrates that it is not always useful to decompose the density as in (6.25).

EXAMPLE 6.6 (*Geyer's triplet process*) Geyer (1999) introduced two extensions/modifications of the Strauss process, mainly to illustrate how easy it is to invent new models and do inference for them. One is the *triplet process* with density

$$f(x) \propto \beta^{n(x)} \gamma^{s_R(x)} \delta^{t_R(x)}.$$

Here $s_R(x)$ is defined as in the Strauss process,

$$t_R(x) = \sum_{\{\xi,\eta,\kappa\} \subseteq x} \mathbf{1}[\|\xi - \eta\| \leq R, \|\xi - \kappa\| \leq R, \|\eta - \kappa\| \leq R],$$

$\beta > 0$, and either

(i) $0 \leq \gamma \leq 1$ and $0 \leq \delta \leq 1$, or

(ii) $\gamma > 1$ and $0 < \delta < 1$.

The process is locally stable (see Geyer 1999), and it is clearly Markov with respect to the R-close-neighbourhood relation, with interaction function

$$\phi(\xi) = \beta, \quad \phi(\{\xi, \eta\}) = \gamma^{\mathbf{1}[\|\xi-\eta\| \leq R]},$$

$$\phi(\{\xi, \eta, \kappa\}) = \delta^{\mathbf{1}[\|\xi-\eta\| \leq R, \|\xi-\kappa\| \leq R, \|\eta-\kappa\| \leq R]},$$

and $\phi(y) = 1$ for $n(y) \geq 4$. The process is repulsive in case (i), but neither attractive nor repulsive in case (ii).

MARKOV POINT PROCESSES   93

6.3.3 A spatial Markov property

We now establish a *spatial Markov property* which among other things is useful for handling edge effects and for the theory in Section 6.4.1. Define the *neighbourhood* of $B \subseteq S$ by

$$N_B = \cup_{\xi \in B} N_\xi = \{\xi \in S : \xi \sim \eta \text{ for some } \eta \in B\},$$

and the *boundary* by $\partial B = N_B \setminus B$.

PROPOSITION 6.1 Let X be a Markov point process with density f and interaction function ϕ.

(i) If $A, B \subset S$ so that $A \cap N_B = \emptyset$, then X_A and X_B are conditionally independent given X_C where $C = S \setminus (A \cup B)$.

(ii) For $B \subset S$, we have that $X_B | X_{S \setminus B} \sim X_B | X_{\partial B}$ and conditional on $X_{\partial B} = x_{\partial B}$ with $f(x_{\partial B}) > 0$, the point process X_B is Markov with density

$$f_B(x_B | x_{\partial B}) = \frac{1}{c(x_{\partial B})} \prod_{y \subseteq x_B \cup x_{\partial B} : y_B \neq \emptyset} \phi(y) \qquad (6.28)$$

with respect to Poisson$(B, 1)$, where $c(x_{\partial B})$ is a normalising constant, and where the product is set to one if $x_B = \emptyset$.

Proof. The standard Poisson process on S is equivalent to $Y_A \cup Y_B \cup Y_C$ where $Y_A, Y_B,$ and Y_C are independent standard Poisson processes on $A, B,$ and C, respectively. Thereby

$$P(X_A \in F_1, X_B \in F_2, X_C \in F_3) =$$
$$\mathbb{E}\big(\mathbf{1}[Y_A \in F_1, Y_B \in F_2, Y_C \in F_3] f(Y_A \cup Y_B \cup Y_C)\big)$$

for $F_1, F_2, F_3 \subseteq N_f$. Hence, by (6.25) and since $\phi(y) = 1$ if $y_A \neq \emptyset$ and $y_B \neq \emptyset$, (X_A, X_B, X_C) has joint density

$$g(x_A, x_B, x_C) = \prod_{y \subseteq x_C} \phi(y) \prod_{y \subseteq x_A \cup x_C : y_A \neq \emptyset} \phi(y) \prod_{y \subseteq x_B \cup x_C : y_B \neq \emptyset} \phi(y)$$

with respect to (Y_A, Y_B, Y_C) where $x_A, x_B,$ and x_C denote finite point configurations in $A, B,$ and C, respectively. So the marginal density of X_C is given by

$$g_C(x_C) = \mathbb{E} g(Y_A, Y_B, x_C) = f(x_C) \mathbb{E} g_{A|C}(Y_A, x_C) \mathbb{E} g_{B|C}(Y_B, x_C)$$

where

$$g_{A|C}(Y_A, x_C) = \prod_{y \subseteq Y_A \cup x_C : y_A \neq \emptyset} \phi(y)$$

and

$$g_{B|C}(Y_B, x_C) = \prod_{y \subseteq Y_B \cup x_C : y_B \neq \emptyset} \phi(y).$$

Since $\mathbb{E}g_{A|C}(Y_A, x_C) \geq P(Y_A = \emptyset) > 0$ and similarly $\mathbb{E}g_{B|C}(Y_B, x_C) > 0$, we have that $g_C(x_C) > 0$ if and only if $f(x_C) > 0$. Thus for $f(x_C) > 0$, the conditional density of (X_A, X_B) given $X_C = x_C$ is

$$g(x_A, x_B, x_C)/g_C(x_C) \propto g_{A|C}(x_A, x_C)g_{B|C}(x_B, x_C).$$

Thereby (i) follows. Letting $C = \partial B$ and $A = S \setminus (B \cup \partial B)$, we obtain (ii), since the conditional density of X_B given $X_C = x_C$ is proportional to $g_{B|C}(x_B, x_C)$. □

REMARK 6.6 The spatial Markov property offers a way of handling *edge effects* as follows. Suppose we only observe X_W where $W \subset S$ is a proper subset of S, and that $W \setminus \partial(S \setminus W)$ has positive volume. With the *conditional approach*, we choose a subset $B \subseteq W \setminus \partial(S \setminus W)$ of positive volume and base statistical inference on the conditional distribution of X_B given $X_{S \setminus B}$, which by Proposition 6.1 coincides with the conditional distribution of $X_B | X_{\partial B}$. Upon observing a realisation x of X_W, the important point is that since $B \cup \partial B \subseteq W$, we can (at least up to a normalising constant) evaluate the conditional density of $X_B = x_B$ given $X_{\partial B} = x_{\partial B}$. Other approaches for handling edge effects in statistical inference for Markov point processes are discussed in Section 9.1.1.

6.4 Extensions of Markov point processes to $\mathbb{R}^d$

Extensions of finite Markov (or Gibbs) point processes to $S = \mathbb{R}^d$ is especially of interest in statistical physics, where very large particle systems are modelled, often with the aim of studying phase transition behaviour as briefly explained below. In spatial statistics such extensions are mainly of interest in order to obtain stationarity under translations (see Section 6.4.2) and to study edge effects (see Section 9.1).

6.4.1 Infinite Gibbs point processes

As the theory is rather technical we give only a review on infinite extensions. More comprehensive treatments can be found in Ruelle (1969), Preston (1976), and the other references mentioned in the following.

Suppose we are given a symmetric and reflexive relation $\sim$ on $\mathbb{R}^d$. We then extend the definition of an *interaction function* from Section 6.3.1 by replacing N_f with the space N_{lf} of locally finite point configurations in $\mathbb{R}^d$. That is, a function $\phi : N_{lf} \to [0, \infty)$ is an interaction function if $\phi(y) = 1$ whenever there are two points $\xi, \eta \in y$ with $\xi \nsim \eta$. For the sake of simplicity we make the following assumptions:

(i) There exists $R < \infty$ so that $\xi \nsim \eta$ whenever $\|\xi - \eta\| \leq R$ (hence ϕ is of finite range of interaction $\leq R$, see Remark 6.4);

EXTENSIONS OF MARKOV POINT PROCESSES TO $\mathbb{R}^D$

(ii) *Ruelle stability*, i.e. there is a constant $\alpha > 0$, and a locally integrable function $\phi^* : \mathbb{R}^d \to [0, \infty)$, so that for all finite $x \subset \mathbb{R}^d$, $\prod_{y \subseteq x} \phi(y) \leq \alpha \prod_{\xi \in x} \phi^*(\xi)$.

Finally, we extend the definition of the *Papangelou conditional intensity* to the locally finite case by defining

$$\lambda^*(x, \xi) = \prod_{y \subseteq x \setminus \xi} \phi(y \cup \xi), \quad x \in N_{\text{lf}}, \; \xi \in \mathbb{R}^d, \tag{6.29}$$

where the product is finite by (i). The case $\xi \in x$ in (6.29) is only included for convenience (for example in connection with (6.30) below). For a finite Gibbs point process with interaction function ϕ, (6.29) agrees with (6.23) if $\xi \notin x$ and $n(x) < \infty$.

There are three equivalent characterisations of a *Gibbs point process* on $\mathbb{R}^d$ (see Georgii 1976, Nguyen & Zessin 1979): X is a Gibbs point process with interaction function ϕ if either 1), 2), or 3) is satisfied:

1) *Local specification*: for each bounded $B \subset \mathbb{R}^d$, $X_B | X_{\mathbb{R}^d \setminus B} \sim X_B | X_{\partial B}$ has conditional density f_B given by (6.28).

2) *Integral characterisation*: for functions $h : \mathbb{R}^d \times N_{\text{lf}} \to [0, \infty)$,

$$\mathbb{E} \sum_{\xi \in X} h(\xi, X \setminus \xi) = \int \mathbb{E}[h(\xi, X) \lambda^*(X, \xi)] \, d\xi. \tag{6.30}$$

3) *Differential characterisation*: this is a certain relation between λ^* and the so-called reduced Palm distributions; the details are given in Example C.1 in Section C.2.1.

Using a local specification is the most common approach, and this is equivalent to the so-called *DLR-equations* (where DLR refers to the pioneering work on Gibbs distributions by Dobrushin, Liggett and Ruelle). Clearly, if X is contained in a bounded set $S \subset \mathbb{R}^d$, then 1) with $B = S$ states that X is Markov with respect to the R-close-neighbourhood relation. Furthermore, 1) is useful in connection to handling edge effects (see Remark 6.6 and Section 9.1.1). For the purposes in Section 6.4.2, 2) is most useful. Note that (6.30) extends the Slivnyak-Mecke theorem (Theorem 3.2, Section 3.2.1), and it is easily verified to hold for a finite Markov point process (by arguments as in Example C.1 in Section C.2.1).

Conditions for *existence of Gibbs distributions* can be found in Ruelle (1969) and Preston (1976). For example, for *translation invariant pairwise interaction functions*, i.e. when $\phi(\xi) = \beta > 0$ is constant for $\xi \in \mathbb{R}^d$, $\phi(\{\xi, \eta\}) = \phi_2(\xi - \eta)$ for distinct $\xi, \eta \in \mathbb{R}^d$, and $\phi(y) = 1$ whenever $n(y) \geq 3$, there exists a stationary Gibbs point process under fairly weak conditions (see e.g. Georgii 1976): Suppose that $\phi_2 \leq c$ for some constant $c > 0$, and that for all $\xi \in \mathbb{R}^d \setminus \{0\}$, $\phi_2(\xi) = 0$ if and only if

$\|\xi\| < \delta$ for some constant $\delta \geq 0$, and

$$\int_{\|\xi\| \geq a} \min\{\log \phi_2(\xi), 0\} \mathrm{d}\xi > -\infty \quad \text{for all } a > \delta.$$

Then a stationary Gibbs point process exists in each of the following cases.

- *Repulsive case*: $\delta = 0$ and $\phi_2 \leq 1$.
- *Divergent case*: $\delta = 0$ and there is an increasing function ψ on $(0, R]$ so that $\log \phi_2(\xi) \leq \psi(\|\xi\|)$ whenever $0 < \|\xi\| \leq R$, and $\psi(r)r^d \to -\infty$ as $r \to 0$.
- *Hard core case*: $\delta > 0$.

A stationary Gibbs distribution exists also for the *Widom-Rowlinson penetrable spheres model* where ϕ is defined as in Example 6.5.

For a given interaction function ϕ, more than one Gibbs point process may exist. Even if ϕ is invariant under translations in $\mathbb{R}^d$, the uniqueness of a stationary Gibbs distribution is not guaranteed. *Phase transition* means existence but nonuniqueness of a Gibbs distribution. Such cases are of great interest in statistical physics (other definitions of phase transition are also used by physicists), where one is particularly interested in finding so-called *critical values* which divide parametric models for interaction functions into those where phase transitions happens or not (roughly speaking, phase transitions happens if the interaction is sufficiently strong). For a few models there exist some results on existence of critical values, but their exact values are unknown. Even for the simplest type of a *hard core Gibbs process*, i.e. when $\phi_2(\xi) = \mathbf{1}[\|\xi\| > \delta]$, it has yet not been verified if phase transition happens for sufficiently large values of $\beta > 0$. See the above-mentioned references, Meester & Roy (1996), Mase, Møller, Stoyan, Waagepetersen & Döge (2001), and the references therein. Phase transition in the Widom-Rowlinson model is discussed later in Example 6.8.

6.4.2 Summary statistics

Only a few properties are known for the summary statistics of a (finite or infinite) Gibbs/Markov point process. In this section we establish some theoretical results based on the integral characterisation of a Gibbs point process. Simulation (Section 7.1) is needed when dealing with nonparametric estimates of summary statistics (Section 4.3).

PROPOSITION 6.2 *A point process X with Papangelou conditional intensity λ^* has intensity function*

$$\rho(\xi) = \mathbb{E}\lambda^*(X, \xi) \tag{6.31}$$

and second order product density

$$\rho^{(2)}(\xi,\eta) = \mathbb{E}[\lambda^*(X,\xi)\lambda^*(X \cup \xi,\eta)]. \tag{6.32}$$

Proof. Both equations follow immediately from Definitions 4.1–4.3 using (6.30) once for (6.31) and twice for (6.32). □

Similar expressions for densities of higher order reduced moment measures (Section C.1.1) are easily derived, but apart from the Poisson case where $\lambda^*(x,\xi) = \rho(\xi)$ is the intensity function, closed form expressions are unknown. Stoyan & Grabarnik (1991) study the marked point process $\{(\xi, 1/\lambda^*(X,\xi) : \xi \in X\}$ for which (6.30) leads to certain interesting relations involving the intensity function and the pair correlation function.

PROPOSITION 6.3 Let X be a stationary Gibbs point process with intensity $\rho > 0$, Papangelou conditional intensity λ^*, and translation invariant interaction function ϕ. Then

$$G(r) = \mathbb{E}\big[\lambda^*(X,0)\mathbf{1}[X \cap b(0,r) \neq \emptyset]\big]/\rho, \tag{6.33}$$

and for $F(r) < 1$,

$$J(r) = \mathbb{E}\big[\lambda^*(X,0) \,\big|\, X \cap b(0,r) = \emptyset\big]/\rho. \tag{6.34}$$

When ϕ is of finite range of interaction R and $r \geq R$ with $F(r) < 1$,

$$J(r) = \beta/\rho \tag{6.35}$$

is constant where $\beta = \lambda^*(\emptyset, 0) = \phi(0)$.

Proof. For an arbitrary set $A \subset \mathbb{R}^d$ with $0 < |A| < \infty$, by (4.9) and (6.30),

$$G(r) = \frac{1}{\rho|A|} \int_A \mathbb{E}\big[\lambda^*(X,\xi)\mathbf{1}[X \cap b(\xi,r) \neq \emptyset]\big] \mathrm{d}\xi$$

whereby (6.33) follows by stationarity of X and since $\lambda^*(x,\xi) = \lambda(x - \xi, 0)$ because of (6.29) and translation invariance of ϕ. Combining (4.8), (4.10), and (6.33), we obtain

$$J(r) = \frac{1 - G(r)}{1 - F(r)} = \frac{1 - \mathbb{E}\lambda^*(X,0)/\rho + \mathbb{E}\big[\lambda^*(X,0)\mathbf{1}[X \cap b(0,r) = \emptyset]\big]/\rho}{P(X \cap b(0,r) = \emptyset)}.$$

Hence, using (6.31), (6.34) is obtained. Finally, (6.29) and (6.34) imply (6.35). □

6.5 Inhomogeneous Markov point processes

A Gibbs point process on $\mathbb{R}^d$ with a translation invariant interaction function ϕ is said to be *homogeneous*. Similarly, a finite Markov point process on $S \subset \mathbb{R}^d$ with $|S| < \infty$ is said to be homogeneous if its interaction function can be extended to a translation invariant interaction function $\phi(x)$ defined for all $x \in N_{\mathrm{lf}}$. This section reviews three types of *inhomogeneous Gibbs/Markov point processes*; see also the review in Jensen & Nielsen (2001).

6.5.1 First order inhomogeneity

Suppose that $\phi(x)$ is not constant for $n(x) = 1$, but $\phi(x)$ is translation invariant for $n(x) \geq 2$. Inhomogeneous Markov point processes of this type have been used in e.g. Ogata & Tanemura (1986) and Stoyan & Stoyan (1998). Such a model may be natural if the intensity function is nonconstant but the interaction between neighbouring points is translation invariant. One example is a hard core process given by $\lambda^*(x, \xi) = \phi(\xi) \prod_{\eta \in x \setminus \xi} \mathbf{1}[\|\xi - \eta\| > R]$. In the finite case, this is obtained by conditioning in an inhomogeneous Poisson process with intensity function $\phi(\xi)$ on the event that no pair of points are R-close.

6.5.2 Thinning of homogeneous Markov point processes

Baddeley et al. (2000) consider an independent thinning of a stationary Gibbs point process X where the retention probabilities $p(\xi)$, $\xi \in \mathbb{R}^d$, are nonconstant and unknown. Hence the thinned point process Y, say, is nonstationary with an intensity function which is proportional to $p(\cdot)$, and the summary statistics K, L, g are the same for X and Y, cf. Propositions 4.2–4.3 (Sections 4.1.1–4.1.2). These properties are exploited in the semiparametric statistical analysis considered in Baddeley et al. (2000). Clearly Y is not Markov, but this is less important as the local Markov property of X can be exploited when using MCMC methods, see Example 8.3 (Section 8.6.1) and Baddeley et al. (2000) for details.

6.5.3 Transformation of homogeneous Markov point processes

Consider a homogeneous Gibbs/Markov point process X with interaction function ϕ. Let $h : S \to T$ be a bijective differentiable mapping where $S, T \subseteq \mathbb{R}^d$ and the Jacobian $J_h(\xi) = |\det \mathrm{d}h(\xi)^\mathsf{T}/\mathrm{d}\xi|$ is nonzero for all $\xi \in S$. The point process $Y = h(X)$ is then a Gibbs/Markov point process on T with interaction function given by $\phi_h(\emptyset) = \phi(\emptyset)$, $\phi_h(\xi) = \phi(h^{-1}(\xi))/J_h(\xi)$ for $\xi \in S$, and $\phi_h(x) = \phi(h^{-1}(x))$ for $2 \leq n(x) < \infty$. Thus Y is homogeneous if and only if J_h is constant.

Such transformation models were introduced in Jensen & Nielsen (2000) when S is bounded and X and Y are finite Markov point processes. As explained above, they can also be defined in the Gibbs case

MARKED AND MULTIVARIATE MARKOV POINT PROCESSES 99

where $S = \mathbb{R}^d$. Jensen & Nielsen (2000) and Nielsen & Jensen (2003) study the inferential aspects in detail, and an application is given in Nielsen (2000) for a data set similar to the mucous cell data (Example 1.3 in Chapter 1; see also Example 9.3 in Section 9.1.4). If h is known, statistical methods for analysing X apply directly for analysing Y because of the one-to-one correspondence.

6.6 Marked and multivariate Markov point processes

Most of the definitions and results for Markov point processes (and their extensions to nearest-neighbour Markov point processes as considered in Appendix F) immediately extend to the case of a general state space S, and hence to the case of marked point processes, including multivariate point processes. In the following we survey such extensions.

6.6.1 Finite marked and multivariate point processes with a density

Using a notation as in Sections 2.2 and 3.3, let Poisson$(T \times M, \rho)$ be a marked Poisson process defined on $S = T \times M$, where $T \subset \mathbb{R}^d$, $|T| < \infty$, $M \subseteq \mathbb{R}^p$, and $\rho(\xi, m) = p(m)$, where p is a discrete or continuous density on M. In words, under Poisson$(T \times M, \rho)$ the points and marks are independent, the points follow a standard Poisson process on T, and the marks are independent and identically distributed with mark density p. In this and the following sections we consider marked point processes $X = \{(\xi, m_\xi) : \xi \in Y\}$ which are absolutely continuous with respect to Poisson$(T \times M, \rho)$.

The density f of X is concentrated on N_f where now

$$N_f = \{\{(\xi_1, m_1), \ldots, (\xi_n, m_n)\} \subset T \times M : n < \infty\}$$

is the set of finite marked point configurations in $S = T \times M$. For $F \subseteq N_f$, in the continuous case of p,

$$P(X \in F) =$$

$$\sum_{n=0}^{\infty} \frac{\exp(-|T|)}{n!} \int_T \int_M \cdots \int_T \int_M \mathbf{1}[\{(\xi_1, m_1), \ldots, (\xi_n, m_n)\} \in F]$$
$$f(\{(\xi_1, m_1), \ldots, (\xi_n, m_n)\}) p(m_1) \cdots p(m_n) d\xi_1 dm_1 \cdots d\xi_n dm_n$$

where the term for $n = 0$ is read as $\exp(-|T|)\mathbf{1}[\emptyset \in F]f(\emptyset)$. This is just a special case of (6.5), since we can view X as a finite point process with a density proportional to

$$f(\{(\xi_1, m_1), \ldots, (\xi_n, m_n)\}) p(m_1) \cdots p(m_n)$$

with respect to Poisson$(S, 1)$. In the discrete case of p,

$$P(X \in F) =$$
$$\sum_{n=0}^{\infty} \frac{\exp(-|T|)}{n!} \int_T \sum_{m_1 \in M} \cdots \int_T \sum_{m_n \in M} \mathbf{1}[\{(\xi_1, m_1), \ldots, (\xi_n, m_n)\} \in F]$$
$$f(\{(\xi_1, m_1), \ldots, (\xi_n, m_n)\}) p(m_1) \cdots p(m_n) \mathrm{d}\xi_1 \cdots \mathrm{d}\xi_n$$

where again the term for $n = 0$ is read as $\exp(-|T|)\mathbf{1}[\emptyset \in F]f(\emptyset)$. Also in this case X can be viewed as an ordinary finite point process with density with respect to a standard Poisson process on S, cf. Remark 3.1 in Section 3.1.1. In both cases, the *Papangelou conditional intensity* is defined as in (6.6) by

$$\lambda^*(x, (\xi, m)) = f(x \cup (\xi, m))/f(x), \quad x \in N_\mathrm{f}, \ (\xi, m) \in (T \times M) \setminus x.$$

It should be obvious how properties like attractivity, repulsivity, local stability, etc. can be defined in a similar way as in Section 6.1.

In the special case $M = \{1, \ldots, k\}$, X is a multitype process which as usual is identified with a multivariate point process $(X_1, \ldots, X_k)$, cf. Section 2.2. It is then natural to consider the density of $(X_1, \ldots, X_k)$ with respect to k independent standard Poisson processes on T. It follows straightforwardly from the discrete case of p considered above that $(X_1, \ldots, X_k)$ has a density $f(x_1, \ldots, x_k)$ which is related to the density $f(\{(\xi_1, m_1), \ldots, (\xi_n, m_n)\})$ by

$$f(x_1, \ldots, x_k) = \exp(|T|(k-1)) f(\cup_{i=1}^k \{(\xi, i) : \xi \in x_i\}) \prod_{i=1}^k p(i)^{n(x_i)}.$$

Defining

$$\lambda_i^*(x_1, \ldots, x_n, \xi) = f(x_1, \ldots, x_i \cup \xi, \ldots, x_k)/f(x_1, \ldots, x_k), \quad \xi \in T \setminus x_i,$$

for the different types $i = 1, \ldots, k$, then

$$\lambda_i^*(x_1, \ldots, x_n, \xi) = p(i)\lambda^*(\cup_{j=1}^k \{(\eta, j) : \eta \in x_j\}, (\xi, i)).$$

6.6.2 Definition and characterisation of marked and multivariate Markov point processes

All definitions and almost all results for Markov point processes (and nearest-neighbour Markov point processes) immediately extend to the case of a general state space S, and hence to the case of marked point processes, including multivariate point processes. For instance, the neighbourhood relation $\sim$ is now defined on $T \times M$, and as noted in Remarks 6.3 and F.1, the Hammersley-Clifford type Theorems 6.1 and F.1 immediately carry over (see Sections 6.3.1 and F.1).

MARKED AND MULTIVARIATE MARKOV POINT PROCESSES

A special case occurs if the relation is essentially independent of the marks, i.e. when $(\xi_1, m_1) \sim (\xi_2, m_2)$ if and only if $\xi_1 \sim_T \xi_2$ where $\sim_T$ is a symmetric and reflexive relation on T. Then the marks given the points are a *Markov random field* on the neighbouring-graph defined by the points (Baddeley & Møller 1989). In fact one approach of marked point processes is to consider the points as fixed and model the marks conditional on the points by a Markov random field. We do not pursue this in the present book.

6.6.3 Examples of marked and multivariate Markov point processes

EXAMPLE 6.7 (*Marked pairwise interaction point processes*) A *marked pairwise interaction point process* has a density of the form

$$f(x) \propto \prod_{\xi \in y} \phi((\xi, m_\xi)) \prod_{\{\xi,\eta\} \subseteq y} \phi(\{(\xi, m_\xi), (\eta, m_\eta)\})$$

for $x = \{(\xi, m_\xi) : \xi \in y\} \in N_f$, where $\phi(\cdot) \geq 0$ is chosen so that the right hand side above is integrable.

For example, a *Strauss type disc process* with $M \subseteq (0, \infty)$ is given by

$$\phi((\xi, m_\xi)) = \beta, \quad \phi(\{(\xi, m_\xi), (\eta, m_\eta)\}) = \gamma^{\mathbf{1}[\|\xi-\eta\| \leq m_\xi + m_\eta]},$$

(taking $0^0 = 1$) where $\beta > 0$ and $0 \leq \gamma \leq 1$ are parameters. This is Markov with respect to the *overlapping disc relation* defined by $(\xi, m_\xi) \sim (\eta, m_\eta)$ if and only if $\|\xi - \eta\| \leq m_\xi + m_\eta$. For $\gamma = 1$, a Boolean model is obtained (Section 3.3); for $\gamma < 1$, there is repulsion between overlapping discs; if $\gamma = 0$, a conditional Boolean model is obtained, where we have conditioned on that the discs are not allowed to overlap.

In Example 9.2 (Section 9.1.5) we consider another disc process model with

$$\phi((\xi, m_\xi)) = \beta(m_\xi), \quad \phi(\{(\xi, m_\xi), (\eta, m_\eta)\}) = \gamma^{|b(\xi, m_\xi) \cap b(\eta, m_\eta)|},$$

(taking $0^0 = 1$) where $\beta(m_\xi) > 0$ takes a finite number of values depending on m_ξ and $0 \leq \gamma \leq 1$. Other examples of marked pairwise interaction point processes are given in Baddeley & Møller (1989) and van Lieshout (2000).

EXAMPLE 6.8 (*Widom-Rowlinson penetrable spheres mixture model*) Consider a bivariate point process (X_1, X_2) with density

$$f(x_1, x_2) \propto \beta_1^{n(x_1)} \beta_2^{n(x_2)} \mathbf{1}[d(x_1, x_2) > R]$$

where $\beta_1, \beta_2, R > 0$ are parameters and $d(x_1, x_2)$ is the shortest distance between a point in x_1 and a point in x_2. This is called a *Widom-Rowlinson penetrable spheres mixture model* (Widom & Rowlinson 1970).

It can be obtained by conditioning in two independent Poisson processes $X_1 \sim \text{Poisson}(T, \beta_1)$ and $X_2 \sim \text{Poisson}(T, \beta_2)$, where we condition on the event that $d(X_1, X_2) > R$.

The model is of great interest in statistical physics, since in the symmetric case $\beta_1 = \beta_2 = \beta$ it provides the simplest case of phase transition in the continuum, i.e. when $T \subset \mathbb{R}^d$ with $d \geq 2$ (Ruelle 1971, Chayes, Chayes & Kotecky 1995, Häggström, van Lieshout & Møller 1999). More precisely, it can be verified that phase transition occurs when β is sufficiently large, while no phase transition occurs when β is sufficiently small, but it is yet an open problem to verify if there exists a critical value of β (see the discussion in Häggström et al. 1999). Note that increasing T to infinity and keeping (β, R) fixed correspond to both keeping T fixed, increasing β to infinity, and decreasing R to 0: more precisely, for $k \to \infty$, considering two independent $\text{Poisson}([-k, k]^2, \beta)$-processes with R fixed corresponds to considering two independent $\text{Poisson}([-1, 1]^2, k\beta)$-processes and replacing R by R/k. Figure 6.2 shows realisations of the symmetric Widom-Rowlinson model within a unit square and for different values of β and R. The effect of phase transition is visible, as the patterns in Figure 6.2 are dominated by one type of points when $R = 0.2$ and β increases (the top left and right plots), and similarly when $R = 0.1$ (the bottom left and right plots); see also the discussion in Häggström et al. (1999) on critical values for this model.

The marginal density for each component X_i is a kind of modified attractive area-interaction process. For example, X_1 has density

$$f(x_1) \propto \sum_{n=0}^{\infty} \beta_1^{n(x_1)} \beta_2^n \exp(-|T|)/n!$$

$$\int_T \cdots \int_T \mathbf{1}[d(x_1, \{\xi_1, \ldots, \xi_n\}) > R] \mathrm{d}\xi_1 \cdots \mathrm{d}\xi_n$$

$$\propto \beta_1^{n(x_1)} \sum_{n=0}^{\infty} \left[\beta_2 | T \setminus U_{x_1, R}|\right]^n / n!$$

$$\propto \beta_1^{n(x_1)} \exp\left(-\beta_2 |U_{x_1, R} \cap T|\right)$$

where $U_{x_1, R}$ is defined as in Example 6.5. Except for the intersection with T, $f(x_1)$ is of a similar form as the density of an area-interaction process (6.27) with $\beta = \beta_1 > 0$ and $\gamma = \exp(-\beta_2) > 1$. By symmetry, X_2 has a density proportional to $\beta_2^{n(x_2)} \exp\left(-\beta_1 |U_{x_2, R} \cap T|\right)$.

In the case $\beta_1 = \beta_2 = \beta$, the marginal density for the superposition $Z = X_1 \cup X_2$ is a repulsive continuum random cluster model, see Remark F.2 in Section F.3.

The conditional distribution of one component given the other com-

MARKED AND MULTIVARIATE MARKOV POINT PROCESSES 103

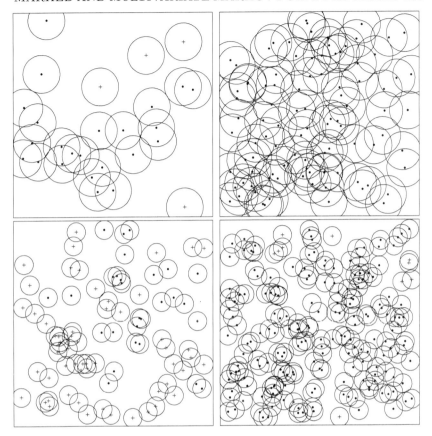

Figure 6.2 *Realisations of the Widom-Rowlinson model using perfect simulation (see Section 11.2.4) with $\beta = 30, R = .2$ (top left), $\beta = 80, R = .2$ (top right), $\beta = 100, R = .1$ (bottom left) and $\beta = 200, R = .1$ (bottom right).*

ponent is a homogeneous Poisson process:

$$X_1 | X_2 \sim \text{Poisson}(T \setminus U_{X_2,R}, \beta_1)$$

and

$$X_2 | X_1 \sim \text{Poisson}(T \setminus U_{X_1,R}, \beta_2).$$

Some of the results above extend to the case of a k-type Widom-Rowlinson penetrable spheres mixture model $(X_1, \ldots, X_k)$ with $k \geq 2$ and density

$$f(x_1, \ldots, x_k) \propto \prod_{i=1}^{k} \beta_i^{n(x_i)} \mathbf{1}[d(x_i, x_j) > R, \ 1 \leq i < j \leq k].$$

For $i = 1, \ldots, k$,
$$X_i | (X_1, \ldots, X_{i-1}, X_{i+1}, \ldots, X_k) \sim \text{Poisson}(T \setminus \cup_{j:j \neq i} U_{X_j, R}, \beta_i).$$
If $\beta_1 = \ldots = \beta_k = \beta$, then $Z = \cup_{i=1}^k X_i$ is a repulsive continuum random cluster model with $\gamma = 1/k$ in (F.10). However, for $k \geq 3$, there is no simple closed form expression for the marginal density of X_i.

EXAMPLE 6.9 (*Continuum Ising models*) Georgii & Haggström (1996) consider a class of bivariate point processes with a density of the form

$$f(x_1, x_2) \propto \beta_1^{n(x_1)} \beta_2^{n(x_2)} \prod_{\xi \in x_1} \prod_{\eta \in x_2} \phi(\xi, \eta) \tag{6.36}$$

where $0 \leq \phi(\cdot, \cdot) \leq 1$. Such processes are called *continuum Ising models* because of their analogy to the Ising model defined on a lattice. Georgii & Haggström (1996) consider also a further extension by adding an interaction term for pairs of points of the same type, and they study issues like uniqueness and phase transition for infinite extensions. See also Georgii (2000). The Widom-Rowlinson penetrable spheres mixture model is clearly a special case of a continuum Ising model.

Apart from the trivial case where $\phi(\cdot, \cdot) = 1$ and from the case of the Widom-Rowlinson penetrable spheres mixture model, the marginal distributions for X_1, X_2, and $X_1 \cup X_2$ are complicated. However, like in the Widom-Rowlinson penetrable spheres mixture model, the conditional distribution for one component given the other component is a Poisson process and hence tractable. For instance,

$$X_1 | X_2 \sim \text{Poisson}(T, \rho_{1, X_2})$$

where

$$\rho_{1, X_2}(\xi) = \beta_1 \prod_{\eta \in X_2} \phi(\xi, \eta).$$

An extension of the model to $k \geq 2$ components is considered in Section 7.1.3.

6.6.4 Summary statistics for multivariate Markov point processes

Let $(X_1, \ldots, X_k)$ be a multivariate point process so that $p(i) > 0$ for $i = 1, \ldots, k$ (we exclude the trivial case $p(i) = 0$ as this implies that $X_i = \emptyset$). Proposition 6.2 extends to the following results which can be verified along similar lines. The intensity function for X_i is

$$\rho_i(\xi) = \mathbb{E} \lambda_i^*(X_1, \ldots, X_k, \xi) / p(i)$$

and the second order cross product density for X_i and X_j is

$$\rho_{ij}^{(2)}(\xi, \eta) = \mathbb{E}[\lambda_i^*(X_1, \ldots, X_k, \xi) \lambda_j^*(X_1, \ldots, X_i \cup \xi, \ldots, X_k, \eta)] / (p(i) p(j)).$$

Proposition 6.3 extends also; see van Lieshout & Baddeley (1999).

CHAPTER 7

Metropolis-Hastings algorithms for point processes with an unnormalised density

This chapter is concerned with Metropolis-Hastings algorithms for simulation of spatial point processes defined by an unnormalised density. The algorithms have found applications especially for simulation of Markov point process models restricted to a bounded region, cf. Chapters 6 and 9. The algorithms also apply for instance to conditional simulation of the mother points for a shot noise Cox process, cf. Chapters 5 and 10. Section 7.1 describes various Metropolis-Hastings algorithms; other kinds of simulation algorithms, particularly those based on spatial birth-death processes, are studied in Chapter 11. A minimum of Markov chain theory is used in Section 7.1, while Section 7.3 provides the theory explaining why the algorithms work, using the background material in Section 7.2 on Markov chain theory. Readers less interested in mathematical details may prefer to skip Sections 7.2–7.3. For a history of MCMC algorithms related to spatial point processes, see Section A.1.

7.1 Description of algorithms

To establish the notation in Sections 7.1.1–7.1.2, X denotes a point process with unnormalised density h with respect to Poisson$(B, 1)$, where $B \subset \mathbb{R}^d$ with $0 < |B| < \infty$ (since $|B| = 0$ implies $P(X = \emptyset) = 1$, we omit this trivial case). Note that the process may be defined on a larger space $S \supset B$, and we may have conditioned on the process outside B to obtain h, see e.g. Proposition 6.1 and Remark 6.6 in Section 6.3.3.

Apart from Poisson processes it is typically not easy to generate directly independent simulations of interesting examples of spatial point processes defined by an unnormalised density (perfect simulation, a more complicated topic introduced in Section 11.2, provides methods of generating i.i.d. simulations from an unnormalised density). Therefore *Markov chain Monte Carlo (MCMC)* is needed. An MCMC algorithm is a recipe for generation of a *Markov chain*, $Y_0, Y_1, \ldots$ that is, the conditional distribution of Y_{m+1} given $(Y_0, \ldots, Y_m)$ is identical to the conditional distribution of Y_{m+1} given Y_m. Usually, the chain forms a sample of approximate simulations of a specified target distribution — in our context the distribution of X — where "approximate simulations" means that the distribution of Y_m converges towards that of X as $m \to \infty$. Almost all of the constructions of MCMC chains we consider are *reversible* with respect to the target distribution, that is, if Y_m follows the target distribution, then (Y_m, Y_{m+1}) and (Y_{m+1}, Y_m) are identically distributed. Under weak conditions the MCMC algorithms in this section produces *irreducible* Markov chains. Briefly, irreducibility of a Markov chain means that the chain can reach any state no matter where it is started (see Section 7.2.1 for details). Irreducibility and reversibility with respect to the target distribution imply that if the chain has a limiting distribution, it must be the target distribution. Irreducibility is needed in order to establish convergence of the chain, and perhaps more important, irreducibility implies consistency of Monte Carlo estimates based on $Y_0, Y_1, \ldots$ (see Sections 7.2 and 8.1 for further details). When $Y_0, Y_1, \ldots$ is used for Monte Carlo purposes, it is in general desirable that the dependence between two states Y_m and Y_{m+j} dies out quickly as j increases, in which case the chain is said to be well *mixing* (see also Section 8.1.3).

Simulation of X is considered in Section 7.1.2, and simulation of X conditional on $n(X) = n$ in Section 7.1.1. The algorithms in Section 7.1.1 can be used in combination with those in Section 7.1.2. Finally, Section 7.1.3 deals with simulation of marked and multivariate point processes.

7.1.1 Metropolis-Hastings algorithms for the conditional case of point processes with a density

Suppose we have conditioned on $n(X) = n$ for an integer $n \geq 1$, and want to simulate from the unnormalised conditional density h_n, cf. Section 6.1. It is then convenient to order the n points $X_1, \ldots, X_n$ in X and consider the unnormalised density for $(X_1, \ldots, X_n)$ given by

$$\pi(x_1, \ldots, x_n) = h_n(\{x_1, \ldots, x_n\}). \tag{7.1}$$

For ease of presentation we assume that $\pi(x_1, \ldots, x_n)$ is defined for all $(x_1, \ldots, x_n) \in B^n$, i.e. some points may be equal. This has no practical importance, since with probability one, the points $X_1, \ldots, X_n$ will be pairwise different: We thus simulate $\bar{X} = (X_1, \ldots, X_n) \sim \pi$ and return $X = \{X_1, \ldots, X_n\} \sim h_n$ (where $X \sim h_n$ means that the distribution of X is specified by the unnormalised density h_n).

Some of the earliest MCMC algorithms are concerned with this simulation problem, cf. Section A.1. Today there exists a variety of Metropolis-Hastings type algorithms which may apply for a given multivariate den-

DESCRIPTION OF ALGORITHMS

sity such as π, see e.g. Dellaportas & Roberts (2003). Nevertheless most simulation algorithms in the point process literature are given by either the Metropolis algorithm or the Gibbs sampler, cf. the discussion in Møller (1999a). These algorithms are particular cases of Algorithm 7.1 below.

Algorithm 7.1 generates a Markov chain $Y_0, Y_1, \ldots$. For each $\bar{x} = (x_1, \ldots, x_n) \in B^n$ and each $i \in \{1, \ldots, n\}$, $q_i(\bar{x}, \cdot)$ is a given density function on B, and we define for $x = \{x_1, \ldots, x_n\}$ and $\xi \in B$,

$$r_i(\bar{x}, \xi) = \frac{h_n((x \setminus x_i) \cup \xi) q_i((x_1, \ldots, x_{i-1}, \xi, x_{i+1}, \ldots, x_n), x_i)}{h_n(x) q_i(\bar{x}, \xi)} \quad (7.2)$$

where we take $a/0 = 1$ for $a \geq 0$. Suppose the current state of the chain is $Y_m = \bar{x} = (x_1, \ldots, x_n)$. Then it is proposed to replace a uniformly selected point x_i in $\bar{x}$ by a point ξ generated from $q_i(\bar{x}, \cdot)$, and with probability

$$\alpha_i(\bar{x}, \xi) = \min\{1, r_i(\bar{x}, \xi)\} \quad (7.3)$$

we set $Y_{m+1} = (x_1, \ldots, x_{i-1}, \xi, x_{i+1}, \ldots, x_n)$; otherwise we retain $Y_{m+1} = \bar{x}$. We call $q_i(\bar{x}, \cdot)$ the *proposal density*, $\alpha_i(\bar{x}, \xi)$ the *acceptance probability*, and $r_i(\bar{x}, \xi)$ the *Hastings ratio*. If the unconditional process is Markov, it may be exploited that in the Hastings ratio,

$$h_n((x \setminus x_i) \cup \xi)/h_n(x) = \lambda^*(x \setminus x_i, \xi)/\lambda^*(x \setminus x_i, x_i)$$

depends only on local information.

The issue of how to generate Y_0 is usually problem specific. We may e.g. let Y_0 be a realisation of a binomial process or Y_0 may be some "feasible" point pattern so that $\pi(Y_0) > 0$. Formally, the algorithm proceeds as follows, where we assume that Y_0 is given; for example, Y_0 may be a realisation of a binomial point process or it may be chosen so that $\pi(Y_0) > 0$.

ALGORITHM 7.1 (*Fixed number of points Metropolis-Hastings algorithm*) For $m = 0, 1, \ldots$, given that $Y_m = \bar{x} = (x_1, \ldots, x_n) \in B^n$ and setting $x = \{x_1, \ldots, x_n\}$, generate Y_{m+1} as follows:

(i) draw $I_m \sim \text{Uniform}(\{1, \ldots, n\})$ and $R_m \sim \text{Uniform}([0, 1])$, and given $I_m = i$, draw $\xi_m \sim q_i(\bar{x}, \cdot)$;

(ii) given $I_m = i$, set

$$Y_{m+1} = \begin{cases} (x_1, \ldots, x_{i-1}, \xi_m, x_{i+1}, \ldots, x_n) & \text{if } R_m \leq r_i(\bar{x}, \xi_m) \\ \bar{x} & \text{otherwise.} \end{cases}$$

Here I_m, R_m, $m = 0, 1, \ldots$, are mutually independent, and given I_m and Y_m, ξ_m is conditionally independent of R_m and all previous variables used for generating $Y_0, \ldots, Y_m$.

REMARK 7.1 Note that in Algorithm 7.1 (and in all MCMC algorithms considered in the following for simulation from a given target density π),

$$P(\pi(Y_{m+1}) > 0 | \pi(Y_m) > 0) = 1. \qquad (7.4)$$

The natural state space of the chain is therefore given by

$$E_n = \{\bar{x} \in B^n : \pi(\bar{x}) > 0\}, \qquad (7.5)$$

i.e. it is natural to let $Y_0 \in E_n$ whereby all $Y_m \in E_n$ (with probability one). For ease of presentation we assume in Section 7.3.1 that $Y_0 \in E_n$. It is, however, not in general necessary that $Y_0 \in E_n$, since Y_m may reach E_n after some iterations.

Instead of the "random point update" I_m in (i), a *systematic updating scheme* might be used, for example, cycling "from one end to the other" ($I_0 = 1, \ldots, I_{n-1} = n, I_n = 1, \ldots, I_{2n-1} = n, \ldots$) or "from one end to the other and then back again" ($I_0 = 1, \ldots, I_{n-1} = n, I_n = n, \ldots, I_{2n-1} = 1, \ldots$). The properties of Algorithm 7.1 are studied in Section 7.3.1; here is just a summary of the results: By Proposition 7.11, the Markov chain Y_m is reversible with respect to π. Proposition 7.11 also gives weak conditions ensuring irreducibility, conditions ensuring convergence of the chain towards π, and a condition ensuring uniform ergodicity (a concept defined in Section 7.2.3) so that both a bound on the rate of convergence is obtainable and a central limit theorem holds (Proposition 7.10).

ALGORITHM 7.2 (*Metropolis algorithm*) This is the special case of Algorithm 7.1 where each $q_i(\bar{x}, \cdot) = q_i(x_i, \cdot)$ depends only on x_i and where $q_i(x_i, \xi) = q_i(\xi, x_i)$ is symmetric.

This is effectively the Metropolis, Rosenbluth, Rosenbluth, Teller & Teller (1953) algorithm. Since the q_i-terms cancel in the Hastings ratio (7.2) when $q_i(x_i, \xi) > 0$, a proposal to a more likely state is always accepted. Usually, $q(x_i, \xi) \propto \mathbf{1}[\xi \in N_{x_i}]$ is the uniform density on a neighbourhood $N_{x_i} \subset \mathbb{R}^d$ of x_i, where $|N_{x_i}| > 0$ and where $\xi \in N_{x_i}$ implies $x_i \in N_\xi$ (in order to obtain symmetry for q_i, N_{x_i} may have points outside B, so we define $\pi(\bar{y}) = 0$ if a proposed state $\bar{y}$ has a point $y_i \notin B$). For instance, N_{x_i} could be a square centred at x_i; Allen & Tildesley (1987, Section 4.4) discuss how to adjust the side length of the square. In the simplest version, $N_\xi = B$ for all $\xi \in B$.

In the following algorithm, we assume that for all $i = 1, \ldots, n$ and all $\bar{x} = (x_1, \ldots, x_n) \in E_n$, the marginal density of $\bar{X}_{-i} = (X_1, \ldots, X_{i-1}, X_{i+1}, \ldots, X_n)$ given by

$$\pi_{-i}(\bar{x}_{-i}) = \int_B \pi(x_1, \ldots, x_{i-1}, \xi, x_{i+1}, \ldots, x_n) \mathrm{d}\xi$$

DESCRIPTION OF ALGORITHMS

is strictly positive, and define the conditional density of X_i given the rest $\bar{X}_{-i} = x_{-i}$ by

$$\pi_i(x_i|\bar{x}_{-i}) = \pi(\bar{x})/\pi_{-i}(\bar{x}_{-i}).$$

For $\bar{x} \notin E_n$, we let $\pi_i(\cdot|\bar{x}_{-i})$ denote an arbitrary density for X_i.

ALGORITHM 7.3 (*Gibbs sampler*) This is the special case of Algorithm 7.1 where each $q_i(\bar{x}, \cdot) = \pi_i(\cdot|\bar{x}_{-i})$.

Gibbs sampling for point processes was introduced in Ripley (1977, 1979).[†] For $\bar{x} \in E_n$ fixed, $\pi_i(\xi|\bar{x}_{-i}) \propto h_n((x \backslash x_i) \cup \xi)$, so $r_i = 1$, and hence there is no need for generating the R_m in step (i) of Algorithm 7.1. Particularly, for a Markov point process, $\pi_i(\cdot|\bar{x}_{-i}) \propto \lambda^*(x \backslash x_i, \cdot)$ depends only on local information. Simulation from the conditional density $\pi_i(\cdot|\bar{x}_{-i})$ often involves *rejection sampling* as described in the following proposition, where Geo(p) denotes the geometric distribution with parameter $0 < p < 1$ and density $(1-p)^j p$, $j = 0, 1, \ldots$.

PROPOSITION 7.1 Suppose that ϕ^*-local stability is satisfied: $\lambda^*(x, \xi) \leq \psi^*(\xi)$ and $0 < c^* < \infty$ where $c^* = \int_B \phi^*(\xi) \mathrm{d}\xi$. Then for the Gibbs sampler, at the $(m+1)$th update in (i) of Algorithm 7.1, if $Y_m = \bar{x} \in E_n$ and $I_m = i$, we can generate ξ_m as follows:

for $j = 1, 2, \ldots$, generate $\xi_{jm} \sim \phi^*(\cdot)/c^*$ and $R_{jm} \sim$ Uniform$([0, 1])$ until the first time $R_{jm} \leq \lambda^*(x \backslash x_i, \xi_{jm})/\phi^*(\xi_{jm})$, and then return $\xi_m = \xi_{jm}$,

where ξ_{jm}, R_{jm}, $j = 1, 2, \ldots$, $m = 0, 1, \ldots$ are mutually independent. Furthermore, if

$$J = \inf\{j : R_{jm} \leq \lambda^*(x \backslash x_i, \xi_{jm})/\phi^*(\xi_{jm})\} - 1$$

denotes the number of steps before acceptance, then

$$J \sim \text{Geo}(a(x \backslash x_i)) \quad \text{with} \quad a(x \backslash x_i) = \int_B \lambda^*(x \backslash x_i, \xi) \mathrm{d}\xi / c^*.$$

Proof. For $A \subseteq B$,

$$P(\xi_{jm} \in A, R_{jm} \leq \lambda^*(x \backslash x_i, \xi_{jm})/\phi^*(\xi_{jm}))$$
$$= \int_A \frac{\phi^*(\xi)}{c^*} \frac{\lambda^*(x \backslash x_i, \xi)}{\phi^*(\xi)} \mathrm{d}\xi$$
$$= \int_A \lambda^*(x \backslash x_i, \xi) \mathrm{d}\xi / c^*.$$

[†] The name "Gibbs sampler" was coined later on by Geman & Geman (1984). See also Section A.1.

Hence the output ξ_m follows the correct distribution, since the conditional distribution of ξ_{jm} given that $R_{jm} \leq \lambda^*(x \setminus x_i, \xi_{jm})/\phi^*(\xi_{jm})$ has unnormalised density $\lambda^*(x \setminus x_i, \cdot)$. Setting $A = B$ we obtain that $J \sim \text{Geo}(a(x \setminus x_i))$ because the (ξ_{jm}, R_{jm}), $j = 1, 2, \ldots$, are i.i.d. □

Unless $\lambda^*(x \setminus x_i, \cdot)$ and ϕ^* are close enough, rejection sampling can be slow. For example, taking $\phi^*(\cdot) = \beta$ for a Strauss process (Example 6.1, Section 6.2.2),

$$a(x \setminus x_i) = \int_B \gamma^{\sum_{j:\, j \neq i} \mathbf{1}[\|\xi - x_j\| \leq R]} d\xi/|B|$$

can be very small for small values of γ or large values of R and n. Lotwick (1982) describes an efficient alternative to rejection sampling for such cases.

REMARK 7.2 Ripley (1977) also considers another type of Metropolis-Hastings algorithm: If $h_n(\cdot) > 0$ and $x = \{x_1, \ldots, x_n\}$ is the current state of the chain, generate $\xi \sim \text{Uniform}(B)$, select a point $\eta \in x \cup \xi$ with probability proportional to $h_n((x \cup \xi) \setminus \eta)$, and return $(x \cup \xi) \setminus \eta$ as the next state. Like in Gibbs sampling, the acceptance probability is always one (Møller 1999a). Ripley (1977) remarks that the algorithm can have poor mixing properties, since $\eta = \xi$ with a high probability unless h is close to a homogeneous Poisson process.

REMARK 7.3 If $h_n(\cdot)$ is highly multimodal (i.e. most of the probability mass is concentrated on a number of disjoint parts of E_n), e.g. in the inhomogeneous case or in the case with strong interaction, the Metropolis algorithm and the Gibbs sampler may be poorly mixing or even not irreducible (see Example 7.2 in Section 7.3.1). A *simulated tempering algorithm* may then be used instead (see Marinari & Parisi 1992, Geyer & Thompson 1995, Møller & Nicholls 1999, and Mase, Møller, Stoyan, Waagepetersen & Döge 2001).

7.1.2 Metropolis-Hastings algorithms for the unconditional case

We consider now a Metropolis-Hastings algorithm for simulation under the unnormalised density h with respect to Poisson$(B, 1)$. The algorithm attempts either to add or delete a point at each transition, and we extend it later to another Metropolis-Hastings algorithm by including a move step as in Algorithm 7.1. The algorithms are studied in detail in Geyer & Møller (1994), Geyer (1999), Møller (1999a), and in Section 7.3.2.

We give first an informal description. Let

$$N_f = \{x \subset B : n(x) < \infty\}.$$

DESCRIPTION OF ALGORITHMS

For $x \in N_f$, we let $p(x)$ be a given probability for "proposing a birth if x is the current state of the chain", and $q_b(x, \cdot)$ is a given density function on B for "the location of the point in a birth proposal". Furthermore, with probability $1 - p(x)$, if $x = \emptyset$ we "do nothing", while if $x \neq \emptyset$ we let $q_d(x, \cdot)$ be a given discrete density on x for "the selection of a point $\eta \in x$ in a death proposal". For a birth-proposal $x \to x \cup \xi$, the *acceptance probability* is given by

$$\alpha_b(x, \xi) = \min\{1, r_b(x, \xi)\},$$

with *Hastings ratio*

$$r_b(x, \xi) = \frac{h(x \cup \xi)(1 - p(x \cup \xi))q_d(x \cup \xi, \xi)}{h(x)p(x)q_b(x, \xi)} \quad (7.6)$$

taking $a/0 = 1$ for $a \geq 0$. For a death-proposal $x \to x \setminus \eta$, the acceptance probability is given by

$$\alpha_d(x, \eta) = \min\{1, r_d(x, \eta)\},$$

with Hastings ratio

$$r_d(x, \eta) = \frac{h(x \setminus \eta)p(x \setminus \eta)q_b(x \setminus \eta, \eta)}{h(x)(1 - p(x))q_d(x, \eta)} \quad (7.7)$$

taking again $a/0 = 1$ for $a \geq 0$.

Formally, the algorithm generates a Markov chain $Y_0, Y_1, \ldots$ as follows, where Y_0 is assumed to be given; for example, $Y_0 = \emptyset$ or Y_0 may be a realisation of a Poisson process.

ALGORITHM 7.4 (*Birth-death Metropolis-Hastings algorithm*) For $m = 0, 1, \ldots$, given $Y_m = x \in N_f$, generate Y_{m+1} as follows:
(i) draw $R'_m \sim$ Uniform$([0,1])$ and $R''_m \sim$ Uniform$([0,1])$;
(ii) if $R'_m \leq p(x)$ then generate $\xi_m \sim q_b(x, \cdot)$ and set

$$Y_{m+1} = \begin{cases} x \cup \xi_m & \text{if } R''_m \leq r_b(x, \xi_m) \\ x & \text{otherwise;} \end{cases}$$

(iii) if $R'_m > p(x)$ then
  (a) if $x = \emptyset$ then set $Y_{m+1} = x$
  (b) else generate $\eta_m \sim q_d(x, \cdot)$ and set

$$Y_{m+1} = \begin{cases} x \setminus \eta_m & \text{if } R''_m \leq r_d(x, \eta_m) \\ x & \text{otherwise.} \end{cases}$$

Here the random variables R'_m, R''_m, and ξ_m or η_m (used for generating Y_{m+1}) are mutually conditionally independent given the random variables used for the generation of $(Y_0, \ldots, Y_m)$.

REMARK 7.4 Like in Remark 7.1, since h is our unnormalised target density and (7.4) holds, the natural state space of the chain is

$$E = \{x \in N_\mathrm{f} : h(x) > 0\}. \tag{7.8}$$

Note that if $x, x \setminus \eta \in E$, then $r_d(x, \eta) = 1/r_b(x \setminus \eta, \eta)$. The Hasting ratios (7.6) and (7.7) depend only on h through the Papangelou conditional intensity, so if h is Markov, the Hastings ratios depend only on local information. For ease of presentation we assume in Section 7.3.2 that $Y_0 \in E$.

REMARK 7.5 Algorithm 7.4 provides a simple example of Green's reversible jump MCMC algorithm (Green 1995, Waagepetersen & Sorensen 2001). The algorithm easily extends to the slightly more general case in Remark 6.1 (Section 6.1) where f is a density with respect to Poisson(S, ρ) with $S \subseteq \mathbb{R}^d$ and $\int_S \rho(\xi) \mathrm{d}\xi < \infty$: if in (iii)(b) $q_b(x, \cdot)$ is still the density for a birth proposal, then simply replace $q_b(x, \xi)$ by $q_b(x, \xi)/\rho(\xi)$ in the Hastings ratios (7.6) and (7.7).

REMARK 7.6 We shall often consider the case where h is ϕ^*-locally stable, $p(\cdot) = p$ is given by a constant with $0 < p < 1$, $q_b(x, \xi) = \phi^*(\xi)/c^*$, and $q_d(x, \eta) = \mathbf{1}[\eta \in x]/n(x)$. We refer to this as the *locally stable version of the birth-death Metropolis-Hastings algorithm*. The Hastings ratio for a birth is then given by

$$r_b(x, \xi) = \frac{\lambda^*(x, \xi)(1 - p)c^*}{\phi^*(\xi)p(n(x) + 1)} \tag{7.9}$$

if $x \in E$, and $r_b(x, \xi) = 1$ otherwise. It may seem tempting instead to let $q_b(x, \xi) = \lambda^*(x, \xi)/c^*(x)$ where $c^*(x) = \int_B \lambda^*(x, \xi) \mathrm{d}\xi$ is a normalising constant. Then the Hastings ratio for a birth is given by replacing $c^*/\phi^*(\xi)$ in (7.9) by $c^*(x)/\lambda^*(x, \xi)$, and so for the different states $Y_m = x$ of the chain with $x \in E$ we need to calculate $c^*(x)$. In contrast the normalising constant c^* in (7.9) can be calculated once and for all.

The properties of Algorithm 7.4 are studied in Section 7.3.2; here is a summary of the results: Proposition 7.12 gives that the chain Y_m is reversible with respect to h. The chain is also irreducible and converges towards the distribution of X if the following weak conditions are satisfied: $Y_0 \in E$,

$$p(\emptyset) < 1, \tag{7.10}$$

and for all $x \in E$ with $x \neq \emptyset$ exists $\eta \in x$ such that

$$(1 - p(x))q_d(x, \eta) > 0 \quad \text{and} \quad h(x \setminus \eta)p(x \setminus \eta)q_b(x \setminus \eta, \eta) > 0, \tag{7.11}$$

cf. Proposition 7.13. Condition (7.10) means that the chain can remain

DESCRIPTION OF ALGORITHMS 115

in $\emptyset$ when $x = \emptyset$, and (7.11) means that the chain can move from x to $\emptyset$ in $n(x)$ steps when $x \neq \emptyset$. Furthermore, for a locally stable version of the birth-death Metropolis-Hastings algorithm, the Hastings ratio (7.9) for a birth is bounded by $(1-p)c^*/(p(n(x)+1))$, which is a decreasing function of $n(x)$. This is exploited in Proposition 7.14 to verify that the chain is geometrically ergodic or in fact V-uniformly ergodic, and sometimes it is even uniformly ergodic (these concepts are defined in Section 7.2.3). Geometric ergodicity implies that a central limit theorem (Proposition 7.10) applies for MCMC estimates.

Algorithm 7.4 may be combined with the fixed number of points Metropolis-Hastings Algorithm 7.1 as follows.

ALGORITHM 7.5 (*Birth-death-move Metropolis-Hastings algorithm*) Let $0 \leq q < 1$, and let $R_m''' \sim \text{Uniform}([0,1])$, $m = 0, 1, \ldots$, be mutually independent, and let each R_m''' be independent of $(Y_0, \ldots, Y_m)$. Then
for $m = 0, 1, \ldots$, given Y_m, if $R_m''' \leq q$,
then generate Y_{m+1} by a move step as in Algorithm 7.1,
else generate Y_{m+1} by a birth-death step as in Algorithm 7.4.

Algorithm 7.5 is reversible with respect to h, and it inherits the convergence properties of Algorithm 7.4; see Proposition 7.15 for details. The algorithm extends the method of Norman & Filinov (1969) (see also Section 4.6 in Allen & Tildesley 1987). Instead of a random scheme, we may alternate between using Algorithm 7.1 and Algorithm 7.4 in some systematic way. According to Allen & Tildesley (1987, p. 128) the values $q = 1/3$ and $p(\cdot) = 1/2$ (i.e. move and birth-death proposals are equally likely) are commonly employed in simulations of molecular systems. However, for the experiments reported in Geyer & Møller (1994), leaving out the move proposals (i.e. $q = 0$) was optimal.

If $h(\cdot)$ is multimodal, Algorithms 7.4 and 7.5 may produce poorly mixing Markov chains, and simulated tempering may be a useful alternative (see Remark 7.3).

7.1.3 Simulation of marked and multivariate point processes with a density

Following Remark 7.5 above, the algorithms in Sections 7.1.1–7.1.2 and the corresponding theory in Sections 7.3.1–7.3.2 easily extend to finite point processes defined on a general state space (Geyer & Møller 1994, Møller 1999a). In particular, we can consider a marked point process with an unnormalised density h with respect to $\text{Poisson}(T \times M, \rho)$, where $T \subset \mathbb{R}^d$, $0 < |T| < \infty$, $M \subseteq \mathbb{R}^p$, $\rho(\xi, m) = p_M(m)$, and p_M is a discrete or continuous density on M, cf. Section 6.6.1. For example, consider Algorithm 7.4 where now $q_b(x, (\xi, m))$ is the density for a birth proposal of a marked point (ξ,m) (i.e. $q_b(x,\cdot) \geq 0$ satisfies $\int_T \sum_{m \in M} q_b(x,(\xi,m))d\xi = 1$ in the discrete case and $\int_T \int_M q_b(x,(\xi,m))d\xi dm = 1$ in the continuous case). Then in the Hastings ratio (7.6) we substitute $q_b(x,(\xi,m))$ by $q_b(x,(\xi,m))/p_M(m)$, and similarly we substitute $q_b(x\setminus(\eta,m),(\eta,m))$ by $q_b(x\setminus(\eta,m),(\eta,m))/p_M(m)$ in (7.7). The algorithm is reversible with respect to h, and in the locally stable case (see Remark 7.6, letting now $\lambda^*(x,(\xi,m)) \leq \phi^*(\xi,m)$ where ϕ^* is an integrable function, i.e. $\int_T \sum_{m \in M} \phi^*(\xi,m)d\xi < \infty$ in the discrete case and $\int_T \int_M \phi^*(\xi,m)d\xi dm < \infty$ in the continous case) we obtain V-uniform ergodicity (along similar lines as in Proposition 7.14).

In the following we concentrate on the particular case of a *continuum Ising model* $X = (X_1, \ldots, X_k)$ with $k \geq 2$ different types of points, extending the density (6.36) to

$$f(x_1, \ldots, x_k) \propto \prod_{i=1}^{k} \beta_i^{n(x_i)} \prod_{\xi_1 \in x_1} \cdots \prod_{\xi_k \in x_k} \phi(\xi_1, \ldots, \xi_k) \qquad (7.12)$$

with $0 \leq \phi(\xi_1, \ldots, \xi_k) \leq 1$. Instead of a birth-death Metropolis-Hastings algorithm, it seems natural to use a k-component *Gibbs sampler* which exploits the simple structure of the conditional distributions

$$X_1|(X_2, \ldots, X_k) \sim \text{Poisson}(T, \rho_{1, X_2, \ldots, X_k})$$
$$\vdots$$
$$X_k|(X_1, \ldots, X_{k-1}) \sim \text{Poisson}(T, \rho_{k, X_1, \ldots, X_{k-1}})$$

where

$$\rho_{1, X_2, \ldots, X_k}(\xi) = \beta_1 \prod_{\eta_2 \in X_2} \cdots \prod_{\eta_k \in X_k} \phi(\xi, \eta_2, \ldots, \eta_k)$$
$$\vdots$$
$$\rho_{k, X_1, \ldots, X_{k-1}}(\xi) = \beta_k \prod_{\eta_1 \in X_1} \cdots \prod_{\eta_{k-1} \in X_{k-1}} \phi(\eta_1, \ldots, \eta_{k-1}, \xi).$$

ALGORITHM 7.6 (*Gibbs sampler for a continuum Ising model*) Using a cyclic updating scheme, generate a chain $Y_m = (Y_{m,1}, \ldots, Y_{m,k})$, $m = 0, 1, \ldots$, so that

(1) $Y_{m+1,1}|(Y_{m,2}, \ldots, Y_{m,k}) \sim \text{Poisson}(T, \rho_{1, Y_{m,2}, \ldots, Y_{m,k}})$,

(2) $Y_{m+1,2}|(Y_{m+1,1}, Y_{m,3}, \ldots, Y_{m,k}) \sim \text{Poisson}(T, \rho_{2, Y_{m+1,1}, Y_{m,3}, \ldots, Y_{m,k}})$,

$\vdots$

(k) $Y_{m+1,k}|(Y_{m+1,1}, \ldots, Y_{m+1,k-1}) \sim \text{Poisson}(T, \rho_{k, Y_{m+1,1}, \ldots, Y_{m+1,k-1}})$.

DESCRIPTION OF ALGORITHMS

REMARK 7.7 For the initial state $Y_0 = (Y_{0,1}, \ldots, Y_{0,k})$, each $Y_{0,i}$ can be any (finite or not) subset of T, because after one cycle, $f(Y_1) > 0$ (with probability one). For the generation of Poisson processes, see Section 3.2.3. Algorithm 7.6 can easily be modified to other types of updating schemes, including random updating schemes. Proposition 7.16 establishes uniform ergodicity of Algorithm 7.6.

Consider the special case of a Widom-Rowlinson penetrable spheres mixture model where

$$\phi(\xi_1, \ldots, \xi_k) = \mathbf{1}[d(x_i, x_j) > R \text{ whenever } i \neq j]$$

(extending Example 6.8 in Section 6.6.3 from 2 to k types). The Gibbs sampler mixes quickly for small and modest values of the β_i, but for large values of the β_i, the Gibbs sampler becomes stuck in configurations dominated by a single type of points (Chayes et al. 1995, Häggström et al. 1999). In the symmetric case $\beta_1 = \ldots = \beta_k = \beta$, we can improve the mixing properties by exploiting the following fact. Let $Z = \cup_{i=1}^k X_i$ and consider the connected components $C_1, \ldots, C_m$, say, of $U_{Z,R/2} = \cup_{\xi \in Z} b(\xi, R/2)$ (note that m is random). Then all points in each $Z_i = Z \cap C_i$ are of the same type I_i, say, and conditional on Z, the I_i are i.i.d., with common distribution Uniform$\{1, \ldots, k\}$. This follows from the fact that

$$f(x_1, \ldots, x_n) \propto \beta^{n(z)} \mathbf{1}[\text{all points in each } z_i \text{ are of the same type}].$$

So letting

$$\rho_{\cup_{i=2}^k X_i}(\xi) = \beta \mathbf{1}[d(\xi, \cup_{i=2}^k X_i) > R],$$

a continuum version of the famous *Swendsen-Wang algorithm* (Swendsen & Wang 1987, Chayes & Machta 1998, Häggström et al. 1999) is given in Algorithm 7.7.

ALGORITHM 7.7 (*Swendsen-Wang type algorithm in the symmetric case*) For $m = 0, 1, \ldots$, given $Y_m = (Y_{m,1}, \ldots, Y_{m,k})$

(i) generate $Y'_{m+1,1} \sim \text{Poisson}(T, \rho_{\cup_{i=2}^k Y_{m,i}})$,

(ii) generate $(Y_{m+1,1}, \ldots, Y_{m+1,k})$ by assigning i.i.d. Uniform$\{1, \ldots, k\}$ random variables to the connected components in $U_{Z_{m+1}, R/2}$ where $Z_{m+1} = Y'_{m+1,1} \cup Y_{m,2} \ldots \cup Y_{m,k}$.

The algorithm is uniformly ergodic if T is bounded (Proposition 7.16), and because of step (ii) it does not get stuck in configurations dominated by a single type of points. The superposition $Z = \cup_{i=1}^k X_i$ follows a so-called continuum random cluster model, that is (F.10) with $\gamma = 1/k$, cf. Remark F.2 in Section F.3.

7.2 Background material for Markov chains obtained by MCMC algorithms

In this section we survey a number of important convergence results for Markov chains constructed by MCMC algorithms. Since we have different uncountable state spaces for the chains in Section 7.1, we consider a framework with a general state space Ω. The theory for Markov chains on general state spaces is studied in great detail in Nummelin (1984) and Meyn & Tweedie (1993); see also Tierney (1994) and Roberts & Tweedie (2003) which are directed more to MCMC applications.

Consider a given probability distribution Π defined on Ω.[†] We can often construct a *time-homogeneous Markov chain* $Y_0, Y_1, \ldots$ with state space Ω by some MCMC algorithm so that

$$P^m(x, F) \to \Pi(F) \qquad (7.13)$$

for $F \subseteq \Omega$ and $x \in \Omega$, where

$$P^m(x, F) = P(Y_m \in F | Y_0 = x)$$

is the m-step *transition probability*. Time-homogeneity means that the *transition kernel*

$$P(x, F) = P(Y_{m+1} \in F | Y_m = x)$$

does not depend on $m \in \mathbb{N}_0$ (where $\mathbb{N}_0 = \{0, 1, 2, \ldots\}$). The distribution of Y_0 is called the *initial distribution*. By time-homogeneity and the Markov property, the distribution of $(Y_0, \ldots, Y_m)$ for any $m \in \mathbb{N}_0$ is determined by the initial distribution and the transition kernel.

Note that (7.13) implies that if $Y_0 \sim \Pi$ then $Y_m \sim \Pi$ for all $m \in \mathbb{N}_0$. Because, if e.g. Ω is discrete and $Y_0 \sim \Pi$, then for an arbitrary $y \in \Omega$,

$$P(Y_1 \in F) = \sum_x P(x, F)\Pi(\{x\}) = \sum_x P(x, F) \lim_{m \to \infty} P^m(y, \{x\})$$

$$= \lim_{m \to \infty} \sum_x P(x, F) P^m(y, \{x\}) = \lim_{m \to \infty} P^{m+1}(y, F) = \Pi(F)$$

and so by induction and time-homogeneity, $Y_m \sim \Pi$ for all $m \in \mathbb{N}$. MCMC algorithms are therefore naturally constructed so that Π is an *invariant distribution*, that is, if $Y_m \sim \Pi$ then $Y_{m+1} \sim \Pi$. Often the construction is so that *reversibility with respect to* Π holds, i.e. if $Y_m \sim \Pi$ then (Y_m, Y_{m+1}) and (Y_{m+1}, Y_m) are identically distributed, or equiva-

---

[†] For technical reasons we assume as in Nummelin (1984) and Meyn & Tweedie (1993) that Ω is equipped with a separable σ-algebra. This is satisfied for the space N_f and hence for the spaces E_n and E defined by (7.5) and (7.8), cf. Proposition B.1.

BACKGROUND MATERIAL FOR MARKOV CHAINS

lently for all $F, G \subseteq \Omega$,

$$P(Y_m \in F, Y_{m+1} \in G, Y_m \neq Y_{m+1})$$
$$= P(Y_m \in G, Y_{m+1} \in F, Y_m \neq Y_{m+1}) \quad \text{if } Y_m \sim \Pi. \tag{7.14}$$

This implies invariance, since if $Y_m \sim \Pi$,

$$P(Y_{m+1} \in F) = P(Y_m \in \Omega, Y_{m+1} \in F) = P(Y_m \in F, Y_{m+1} \in \Omega)$$
$$= \Pi(F).$$

7.2.1 Irreducibility and Harris recurrence

For $a > 0$ and $X \sim \Pi$, let $\mathcal{L}^a(\Pi)$ denote the class of functions $k : \Omega \to \mathbb{R}$ such that $\mathbb{E}|k(X)|^a < \infty$, and set $\mathcal{L}(\Pi) = \mathcal{L}^1(\Pi)$; if Π is specified by a normalised or unnormalised density h, we also write $\mathcal{L}^a(h)$ for $\mathcal{L}^a(\Pi)$, and $\mathcal{L}(h)$ for $\mathcal{L}(\Pi)$. Let $\Pi(k) = \mathbb{E}k(X)$ if $k \in \mathcal{L}(\Pi)$, and consider the *ergodic average* given by

$$\bar{k}_n = \frac{1}{n} \sum_{m=0}^{n-1} k(Y_m). \tag{7.15}$$

Below we establish two *strong laws of large numbers* (Propositions 7.3–7.4) for the estimate $\bar{k}_n$ of $\Pi(k)$ (such estimates are used several times in this book, see in particular Chapter 8). All what is needed is either irreducibility or Harris recurrence as defined in the sequel; convergence such as in (7.13) is a sufficient but not a necessary condition, cf. Section 7.2.2.

No matter whether an invariant distribution exists or not, we say that the chain is *irreducible* if there is a nonzero measure Ψ on Ω so that for any $x \in \Omega$ and $F \subseteq \Omega$ with $\Psi(F) > 0$, $P^m(x, F) > 0$ for some $m \in \mathbb{N}$; to stress the dependence on Ψ we also say that the chain is Ψ-*irreducible*. *Harris recurrence* means that the chain is Ψ-irreducible for some Ψ, and that for all $x \in \Omega$ and all $F \subseteq \Omega$ with $\Psi(F) > 0$,

$$P(Y_m \in F \text{ for some } m \,|\, Y_0 = x) = 1.$$

PROPOSITION 7.2 *If an invariant distribution Π exists, Ψ-irreducibility implies the following properties:*

(i) Π-irreducibility;

(ii) Π *is the unique invariant distribution (up to null sets);*

(iii) Π *dominates Ψ, i.e. $\Pi(F) = 0$ implies $\Psi(F) = 0$;*

(iv) *there exists some $A \subset \Omega$ with $\Pi(A) = 0$ so that the chain restricted to $\Omega \setminus A$ is Harris recurrent and $\Omega \setminus A$ is absorbing, i.e. if $Y_0 \in \Omega \setminus A$ then $Y_m \in \Omega \setminus A$ for all $m \in \mathbb{N}_0$.*

Proof. (i)–(iii) follow from Meyn & Tweedie (1993, Proposition 4.2.2 and Theorem 10.4.9). By Meyn & Tweedie (1993, Proposition 10.1.1), the chain is recurrent, that is, for any $x \in \Omega$ and $F \subseteq \Omega$ with $\Pi(F) > 0$, the mean number of times the chain visits F is infinite. Thus (iv) follows from Meyn & Tweedie (1993, Proposition 9.0.1). □

Obviously, (i) and (iii) are advantageous, since there can be much fewer sets $F \subseteq \Omega$ with $\Psi(F) > 0$ than $\Pi(F) > 0$ (this is exploited several times in Section 7.3). As requested, (ii) is telling us that if the chain has a limiting distribution, it must be Π. Finally, (iv) shows that irreducibility and Harris recurrence are essentially equivalent concepts; the disturbance is the nullset A.

PROPOSITION 7.3 If the chain is irreducible with invariant distribution Π and $k \in \mathcal{L}(\Pi)$, then there exists a set $A \subset \Omega$ so that $\Pi(A) = 0$ and with probability one, conditional on $Y_0 = x$ with $x \in \Omega \setminus A$, $\lim_{n \to \infty} \bar{k}_n = \Pi(k)$.

Proof. Combine (iv) in Proposition 7.2 with Proposition 7.4 below. □

To get rid of the nullset A in Proposition 7.3 Harris recurrence is needed:

PROPOSITION 7.4 The chain is Harris recurrent and has invariant distribution Π if and only if for all $k \in \mathcal{L}(\Pi)$, $\lim_{n \to \infty} \bar{k}_n = \Pi(k)$ with probability one and regardless of the initial distribution.

Proof. See Meyn & Tweedie (1993, Theorem 17.1.7). □

Proposition 7.4 includes the classical strong law of large numbers for i.i.d. random variables. To establish Harris recurrence, the following is useful. A set $C \subseteq \Omega$ is said to be *small* if there exist $m \in \mathbb{N}$ and a nonzero measure M on Ω such that

$$P^m(x, F) \geq M(F) \quad \text{for all } x \in C \text{ and } F \subseteq \Omega. \tag{7.16}$$

To stress the dependence on m and M we say that C is (m, M)-small. As we shall see later, small sets can be rather large; it may even occur that Ω is small.

PROPOSITION 7.5 [†] If the chain is Ψ-irreducible and there exist a small

---

[†] In Proposition 7.5 "small" may be replaced by the weaker concept of "petite", cf. Meyn and Tweedie 1993, but for simplicity we have excluded this in the formulation of the proposition. In fact, every small set is petite, and when the chain is irreducible and aperiodic, small and petite mean the same thing (Meyn & Tweedie 1993, Theorem 5.5.7).

BACKGROUND MATERIAL FOR MARKOV CHAINS

set $C \subseteq \Omega$ and a function $V : \Omega \to [1, \infty)$ such that $\{x \in \Omega : V(x) \le \alpha\}$ is small for all $\alpha > 1$ and

$$\mathbb{E}[V(Y_1)|Y_0 = x] \le V(x) \quad \text{for all } x \in \Omega \setminus C, \tag{7.17}$$

then the chain is Harris recurrent.

Proof. See Meyn & Tweedie (1993, Theorem 9.1.8). □

We call (7.17) the *drift criterion for recurrence*.

7.2.2 Aperiodicity and ergodicity

We now consider when convergence results such as (7.13) are satisfied.

For an Ψ-irreducible chain, Ω can be partitioned into sets $D_0, \ldots, D_{d-1}$ and A so that $P(x, D_j) = 1$ for $x \in D_i$ and $j = i+1 \bmod d$, and $\Psi(A) = 0$ (Meyn & Tweedie 1993, Theorem 5.4.4). If there is such a partition with $d > 1$ the chain is *periodic*, else it is *aperiodic*. We have aperiodicity if for example $P(x, \{x\}) > 0$ for some $x \in \Omega$. The following result is often useful.

PROPOSITION 7.6 An irreducible Markov chain with invariant distribution Π is aperiodic if and only if for some small C with $\Pi(C) > 0$ and some $n \in \mathbb{N}$, $P^m(x, C) > 0$ for all $x \in C$ and $m \ge n$.

Proof. See Proposition 5.3.4 in Roberts & Tweedie (2003) (since in the formulation of Proposition 7.6, "for some small" can be replaced by "for all small"). □

Clearly, periodicity implies that $P^m(x, D_0)$ cannot converge. In the sequel we therefore restrict attention to the aperiodic case.

For any initial distribution, let

$$Q^m(F) = \mathbb{E}P^m(Y_0, F) = P(Y_m \in F), \quad F \subseteq \Omega,$$

denote the marginal distribution of Y_m. In the special case where the initial distribution is concentrated at a given $x \in \Omega$, $Q^m(F) = P^m(x, F)$. To measure the distance between Q^m and Π we introduce the *total variation norm* given by

$$\|\mu - \nu\|_{\text{TV}} = \sup_{F \subseteq \Omega} |\mu(F) - \nu(F)|$$

for any two probability distributions μ and ν defined on Ω. Note that $\|\mu - \nu\|_{\text{TV}} \le 1$, $\|\mu - \nu\|_{\text{TV}} = 0$ if $\mu = \nu$, and $\|\mu - \nu\|_{\text{TV}} = 1$ if μ and ν have disjoint supports. Note also that convergence of Q^m in total variation norm to Π is a rather strong form of convergence, which implies convergence of Q^m in distribution to Π.

PROPOSITION 7.7 Suppose the chain has invariant distribution Π. Then the following properties hold:

(i) Regardless of the initial distribution, $\|Q^m - \Pi\|_{TV}$ is nonincreasing in m.

(ii) If the chain is irreducible and aperiodic, there exists $A \subset \Omega$ so that $\Pi(A) = 0$ and for all $x \in \Omega \setminus A$, $\lim_{m \to \infty} \|P^m(x, \cdot) - \Pi\|_{TV} = 0$.

(iii) The chain is Harris recurrent and aperiodic if and only if $\lim_{m \to \infty} \|Q^m - \Pi\|_{TV} = 0$ regardless of the initial distribution.

Proof. (i) follows from Meyn & Tweedie (1993, Theorem 13.3.2). Combining (iv) in Proposition 7.3 with Meyn & Tweedie (1993, Theorem 13.3.3) we obtain (ii)–(iii). □

We say that the chain is *ergodic* if it is both Harris recurrent and aperiodic. Particularly, ergodicity implies (7.13) regardless of the initial state.

7.2.3 Geometric and uniform ergodicity

We discuss now the rate of convergence, and consider finally a central limit theorem.

For $x \in \Omega$ and functions $V : \Omega \to [1, \infty)$ with $\Pi(V) < \infty$, define the V-norm of $P^m(x, \cdot) - \Pi$ by

$$\|P^m(x, \cdot) - \Pi\|_V = \frac{1}{2} \sup_{|k| \leq V} \left| \mathbb{E}[k(Y_m) | Y_0 = x] - \Pi(k) \right|$$

where the supremum is over all functions $k : \Omega \to \mathbb{R}$ with $|k(\cdot)| \leq V(\cdot)$. Note that $\|P^m(x, \cdot) - \Pi\|_V = \|P^m(x, \cdot) - \Pi\|_{TV}$ if $V = 1$, and $\|P^m(x, \cdot) - \Pi\|_{V_1} \leq \|P^m(x, \cdot) - \Pi\|_{V_2}$ if $V_1 \leq V_2$. Following Meyn & Tweedie (1993) the chain is said to be V-*geometrically ergodic* if it is Harris recurrent with invariant distribution Π and there exists a constant $r > 1$ such that for all $x \in \Omega$,

$$\sum_{m=1}^{\infty} r^m \|P^m(x, \cdot) - \Pi\|_V < \infty. \tag{7.18}$$

We say that the chain is *geometrically ergodic* if it is V-geometrically ergodic for some V. In fact (7.18) is equivalent to that

$$\|P^m(x, \cdot) - \Pi\|_V \leq A(x)\tilde{r}^m \tag{7.19}$$

where $A(x) < \infty$ and $\tilde{r} < 1$ (Meyn & Tweedie 1993, p. 355). This shows both that the rate of convergence is geometric and that the chain is ergodic.

When we have geometric ergodicity with $A(\cdot) \propto V(\cdot)$ in (7.19), the chain is said to be V-*uniformly ergodic*. This is equivalent to that

$$\lim_{m \to \infty} \sup_{x \in \Omega} \|P^m(x, \cdot) - \Pi\|_V = 0,$$

cf. Meyn & Tweedie (1993, Theorem 16.2.1). If we have geometric ergodicity for a constant function A in (7.19), the chain is said to be *uniformly ergodic*. This is then equivalent to that

$$\lim_{m \to \infty} \sup_{x \in \Omega} \|P^m(x, \cdot) - \Pi\|_{TV} = 0,$$

but even more than that is known:

PROPOSITION 7.8 Suppose the chain has invariant distribution Π. Then uniform ergodicity is equivalent to that Ω is small. If it is (n, M)-small, then

$$\|P^m(x, \cdot) - \Pi\|_{TV} \leq (1 - M(\Omega))^{m/n}, \quad m \in \mathbb{N}. \quad (7.20)$$

Proof. This follows from Meyn & Tweedie (1993, Theorem 16.2.2); see also Theorem 9.1.1 in Roberts & Tweedie (2003). □

To establish V-uniform ergodicity the following is useful.

PROPOSITION 7.9 Suppose the chain is aperiodic and irreducible with invariant distribution Π and there exist a function $V : \Omega \to [1, \infty)$, a small set $C \subseteq \Omega$, and constants $a < 1$ and $b < \infty$ such that for all $x \in \Omega$,

$$\mathbb{E}[V(Y_1)|Y_0 = x] \leq aV(x) + b\mathbf{1}[x \in C]. \quad (7.21)$$

Then $\Pi(V) < \infty$, the chain is V-uniformly ergodic, and (7.18) can be sharpened to

$$\sum_{m=1}^{\infty} r^m \|P^m(x, \cdot) - \Pi\|_V \leq \tilde{r} V(x) \quad (7.22)$$

for some constants $r > 1$ and $\tilde{r} < \infty$ and for all $x \in \Omega$.

Proof. This follows from Meyn & Tweedie (1993, Theorems 15.0.1 and 16.0.2), noticing that $\Pi(V) \leq b/(1-a)$ is finite by (7.21), and that the chain is Harris recurrent by Meyn & Tweedie (1993, Lemma 15.2.8) and Proposition 7.5. □

We call (7.21) the *geometric drift condition*. Note that the bounds in (7.19) and (7.22) provide only qualitative results, since we do not know e.g. the value of $\tilde{r}$. Quantitative results for geometrically ergodic chains can be found in e.g. Rosenthal (1995), but as discussed in Møller (1999a) the results seem of very limited use for simulation of spatial point processes. This is also the case for uniformly ergodic chains: the upper bound given by (7.20) provides only a very rough estimate of the rate of convergence, cf. Example 7.2 and Remark 7.9.

Many algorithms for spatial point processes are V-uniformly ergodic but not uniformly ergodic, cf. Remark 7.9. Fortunately, geometric ergodicity (and hence V-uniform ergodicity) implies the following *central limit theorem*.

PROPOSITION 7.10 Suppose the chain is geometrically ergodic with invariant distribution Π, and k is a real function defined on Ω so that either

(i) $k \in \mathcal{L}^{2+\epsilon}(\Pi)$ for some $\epsilon > 0$, or

(ii) the chain is reversible and $k \in \mathcal{L}^2(\Pi)$, or

(iii) the chain is V-uniformly ergodic and $k^2 \leq V$.

Define

$$\bar{\sigma}^2 = \mathrm{Var}(k(Y_0)) + 2 \sum_{m=1}^{\infty} \mathrm{Cov}(k(Y_0), k(Y_m))$$

where the variance and covariances are calculated under the assumption that $Y_0 \sim \Pi$. Then $\bar{\sigma}^2$ is well-defined and finite, and regardless of the initial distribution, $\sqrt{n}(\bar{k}_n - \Pi(k))$ converges in distribution to $N(0, \bar{\sigma}^2)$ as $n \to \infty$ (in the special case $\bar{\sigma}^2 = 0$, convergence in distribution to $N(0, 0)$ means that $\sqrt{n}(\bar{k}_n - \Pi(k))$ converges with probability one to the degenerate distribution concentrated at 0).

Proof. (i) See Chan & Geyer (1994, Theorem 1). (ii) See Roberts & Rosenthal (1997, Corollary 3). (iii) See Meyn & Tweedie (1993, Theorem 17.0.1). □

REMARK 7.8 Proposition 7.10 extends easily to the multivariate case. For example, in the bivariate case, for functions $k^{(1)}$ and $k^{(2)}$ which satisfy either (i), (ii), or (iii) in Proposition 7.10, $\sqrt{n}(\bar{k}_n^{(1)} - \Pi(k^{(1)}), \bar{k}_n^{(2)} - \Pi(k^{(2)}))$ is asymptotically normally distributed with mean $(0,0)$ and a covariance matrix with entries

$$\bar{\sigma}_{ij} = \mathrm{Cov}(k^{(i)}(Y_0), k^{(j)}(Y_0)) + \sum_{m=1}^{\infty} \mathrm{Cov}(k^{(i)}(Y_0), k^{(j)}(Y_m))$$
$$+ \sum_{m=1}^{\infty} \mathrm{Cov}(k^{(j)}(Y_0), k^{(i)}(Y_m))$$

for $i, j = 1, 2$, where the covariances are calculated under the assumption that $Y_0 \sim \Pi$.

CONVERGENCE PROPERTIES OF ALGORITHMS

7.3 Convergence properties of algorithms

In the following we investigate the convergence properties of the Markov chains generated by the Metropolis-Hastings algorithms in Section 7.1.

7.3.1 The conditional case

Consider the Markov chain generated by Algorithm 7.1, with unnormalised target density π given by (7.1) and state space E_n, assuming that $Y_0 \in E_n$, cf. Remark 7.1. For $\bar{x} \in E_n$ and $F \subseteq E_n$, the chain has transition kernel

$$P(\bar{x}, F) = \frac{1}{n}\sum_{i=1}^{n}\int_B \mathbf{1}[(x_1,\ldots,x_{i-1},\xi,x_{i+1},\ldots,x_n) \in F] q_i(\bar{x},\xi)\alpha_i(\bar{x},\xi)\mathrm{d}\xi$$
$$+ r(\bar{x})\mathbf{1}[\bar{x} \in F] \tag{7.23}$$

where the acceptance probability $\alpha_i(\bar{x},\xi)$ is defined by (7.3) and where

$$r(\bar{x}) = \frac{1}{n}\sum_{i=1}^{n}\int_B \mathbf{1}[(x_1,\ldots,x_{i-1},\xi,x_{i+1},\ldots,x_n) \in F] q_i(\bar{x},\xi)(1-\alpha_i(\bar{x},\xi))\mathrm{d}\xi$$

is the probability of remaining at x. In Proposition 7.11 below we refer to the chain as the Metropolis-Hastings chain, and in the proof we use only the part of (7.23) corresponding to accepted proposals. We also refer to the "proposal" chain which is obtained by always accepting proposals, i.e. it has transition kernel

$$Q(\bar{x}, F) = \frac{1}{n}\sum_{i=1}^{n}\int_B \mathbf{1}[(x_1,\ldots,x_{i-1},\xi,x_{i+1},\ldots,x_n) \in F] q_i(\bar{x},\xi)\mathrm{d}\xi$$

for $\bar{x} \in E_n$ and $F \subseteq E_n$.

PROPOSITION 7.11 *The following properties hold for the fixed number of points Metropolis-Hastings Algorithm 7.1.*

(i) *The Metropolis-Hastings chain is reversible with respect to π.*

(ii) *Suppose that $E_n = B^n$ and the proposal chain is Ψ-irreducible, and for all $\bar{x} = (x_1,\ldots,x_n) \in B^n$, $i \in \{1,\ldots,n\}$, and $\xi \in B$,*

$$q_i(\bar{x},\xi) > 0 \Rightarrow q_i(\bar{y},x_i) > 0 \tag{7.24}$$

where $\bar{y} = (x_1,\ldots,x_{i-1},\xi,x_{i+1},\ldots,x_n)$. Then the Metropolis-Hastings chain is Ψ-irreducible.

(iii) *Suppose that the Metropolis-Hastings chain is irreducible, and there exist $\epsilon > 0$, $x_2,\ldots,x_n \in B$, and $D \subseteq B$ such that $|D| > 0$ and for all $x_1,\xi \in D$, we have that $\pi(\bar{x}) > 0$, $\pi(\xi,x_2,\ldots,x_n) > 0$, and*

$$\min\{q_1(\bar{x},\xi), \pi(\xi,x_2,\ldots,x_n)q_1((\xi,x_2,\ldots,x_n),x_1)/\pi(\bar{x})\} \geq \epsilon \tag{7.25}$$

where $\bar{x} = (x_1, x_2, \ldots, x_n)$. Then $C = \{(x_1, x_2, \ldots, x_n) : x_1 \in D\}$ is small, and the Metropolis-Hastings chain is aperiodic.

(iv) Suppose that $E_n = B^n$ and there exists $\epsilon > 0$ such that for all $\bar{x} = (x_1, \ldots, x_n) \in B^n$, $i \in \{1, \ldots, n\}$, and $\xi \in B$,

$$\min\{q_i(\bar{x}, \xi), \pi(\bar{y})q_i(\bar{y}, x_i)/\pi(\bar{x})\} \geq \epsilon \tag{7.26}$$

where $\bar{y} = (x_1, \ldots, x_{i-1}, \xi, x_{i+1}, \ldots, x_n)$. The Metropolis-Hastings chain is then uniformly ergodic, with a total variation distance to π which is bounded by $(1 - n!(\epsilon|B|/n)^n)^{m/n}$ at times $m = 1, 2, \ldots$.

Proof. (i) For $i \in \{1, \ldots, n\}$ and $\bar{x}, \bar{y} \in E_n$ with $x_j = y_j$ for all $j \neq i$,

$$\pi(\bar{x})\frac{1}{n}q_i(\bar{x}, y_i)\alpha_i(\bar{x}, y_i) = \pi(\bar{y})\frac{1}{n}q_i(\bar{y}, x_i)\alpha_i(\bar{y}, x_i)$$

whereby (7.14) is straightforwardly verified.

(ii) Let $\Psi(F) > 0$ and $\bar{x} \in E_n$. Since the proposal chain is Ψ-irreducible, there exists an $m \in \mathbb{N}$ so that the m-step probability $Q^m(\bar{x}, F) > 0$. Consider a path of the proposal chain given by $X_0 = \bar{x} \to X_1 \to \cdots \to X_m$ with associated i.i.d. uniform visits $I_0, \ldots, I_{m-1}$ in $\{1, \ldots, n\}$, indicating which points in $X_0, \ldots, X_{m-1}$ are updated: $\xi_0|(I_0, X_0) \sim q_{I_0}(X_0, \cdot), \ldots, \xi_{m-1}|(I_{m-1}, X_{m-1}) \sim q_{I_{m-1}}(X_{m-1}, \cdot)$. Conditional on $X_0 = \bar{x}$, we have with probability one that $q_i(X_i, \xi_i) > 0$ for $i = 0, \ldots, m-1$. Then by (7.24), with probability one, $\alpha_i(X_i, \xi_i) > 0$ for $i = 0, \ldots, m-1$. Hence, using the "acceptance part" of (7.23), $P^m(\bar{x}, F) > 0$, i.e. the Metropolis-Hastings chain is Ψ-irreducible.

(iii) Define the nonzero measure

$$M(F) = (\epsilon/n)\int_D \mathbf{1}[(\xi, x_2, \ldots, x_n) \in F]d\xi$$

for $F \subseteq E_n$. Combining (7.23) and (7.25) we obtain that C is $(1, M)$-small, and for all $\bar{x} \in C$ and $m \in \mathbb{N}$,

$$P^m(\bar{x}, C) \geq (\epsilon/n)^m |D|^m > 0.$$

Hence, by Proposition 7.6, the Metropolis-Hastings chain is aperiodic.

(iv) Define the nonzero measure

$$M(F) = n!(\epsilon/n)^n \int \cdots \int \mathbf{1}[(\xi_1, \ldots, \xi_n) \in F]d\xi_1 \cdots d\xi_n$$

for $F \subseteq E_n$. Combining (7.23) and (7.26) we obtain that E_n is (n, M)-small. Hence, by Proposition 7.8, the Metropolis-Hastings chain is uniformly ergodic with a rate of convergence as stated. $\square$

EXAMPLE 7.1 (*Conditional pairwise interaction processes*) Consider a conditional pairwise interaction process with

$$\pi(\bar{x}) = \prod_{i<j} \varphi(x_i, x_j)$$

where $\varphi(\cdot,\cdot)$ is a symmetric nonnegative function, cf. (6.12). For all the examples of pairwise interaction functions in Examples 6.1–6.4 (Section 6.2.2) except those in Examples 6.1–6.2 which are of a hard core type model, we can take $E_n = B^n$. Suppose that $E_n = B^n$, and let Ψ be given by

$$\Psi(A) = \int_B \cdots \int_B \mathbf{1}[(\xi_1, \ldots, \xi_n) \in A)] \mathrm{d}\xi_1 \cdots \mathrm{d}\xi_n, \quad A \subseteq B^n$$

(i.e. Ψ is the Lebesgue measure on B^n). Consider the Metropolis algorithm (Algorithm 7.2) with uniform proposal densities $q_i = 1/|B|$, $i = 1, \ldots, n$. The Metropolis chain is then Ψ-irreducible, since the proposal chain is Ψ-irreducible and (7.24) is trivially satisfied. Suppose that B contains a ball D. Then (7.25) is satisfied if there exists $x_2 \in B$ such that $\inf_{x_1 \in D} \varphi(x_1, x_2) > 0$. This is clearly the case for all the examples of pairwise interaction functions in Examples 6.1–6.4 possibly except for the hard core cases. By (iii) in Proposition 7.11 (with $x_2 = \ldots = x_n$), the Metropolis chain is then aperiodic. As exemplified in Example 7.2 below, it can be shown that the conditions in (iv) in Proposition 7.11 are satisfied for the Strauss and multi-scale pairwise interaction functions (Examples 6.1–6.2, excluding again the hard core type models), but (7.26) is not satisfied for the models in Examples 6.3–6.4.

EXAMPLE 7.2 (*Conditional Strauss process*) Consider both the Metropolis algorithm (Algorithm 7.2) with uniform proposal densities and the Gibbs sampler (Algorithm 7.3) when the target density is a conditional Strauss process with density (6.15), i.e.

$$\pi(\bar{x}) = \gamma^{\sum_{i<j} \mathbf{1}[\|x_i - x_j\| \leq R]}$$

where $R > 0$ and $\gamma \geq 0$. Thus, for the Metropolis algorithm $q_i = 1/|B|$, while for the Gibbs sampler

$$q_i(\bar{x}, \xi) = \gamma^{\sum_{j: j \neq i} \mathbf{1}[\|\xi - x_j\| \leq R]} \Big/ \int_B \gamma^{\sum_{j: j \neq i} \mathbf{1}[\|\eta - x_j\| \leq R]} \mathrm{d}\eta.$$

We consider three cases below.

1) The case $0 < \gamma \leq 1$: We claim that both algorithms are uniformly ergodic. Note that $\gamma^{n(n-1)/2} \leq \pi(\bar{x}) \leq 1$ and for the Gibbs sampler, $q_i(\bar{x}, \xi) \geq \gamma^{n-1}/|B|$. Uniform ergodicity thus follows from (iv) in Proposition 7.11 with $\epsilon = \gamma^{n(n-1)/2}/|B|$ for the Metropolis algorithm and $\epsilon = \gamma^{(n+2)(n-1)/2}/|B|$ for the Gibbs sampler. By (7.20), the total variation distance to π at time m is bounded by $(1 - n! n^{-n} \gamma^{n^2(n-1)/2})^{m/n}$ for

the Metropolis algorithm, and by $(1 - n!n^{-n}\gamma^{n(n+2)(n-1)/2})^{m/n}$ for the Gibbs sampler. These bounds indicate that the rate of convergence can be very small for large values of n and or small values of γ, but we should of course keep in mind that the two bounds only provide conservative and rough estimates. Note that the bounds do not depend on R, though we expect that the rate of convergence is a decreasing function of R. In a comparison of the two algorithms it should also be taken into account that the Metropolis algorithm is faster to iterate when Gibbs sampling involves rejection sampling.

2) The case $\gamma > 1$: Since $\gamma^{n(n-1)/2} \geq \pi(\bar{x}) \geq 1$, we obtain immediately from (iv) in Proposition 7.11 with $\epsilon = \gamma^{-n(n-1)/2}/|B|$ that the Metropolis algorithm is uniformly ergodic. For the Gibbs sampler, uniform ergodicity is obtained with $\epsilon = \gamma^{-(n+2)(n-1)/2}/|B|$. By (7.20), the total variation distance to π at time m is bounded by $(1 - n!n^{-n}\gamma^{-n^2(n-1)/2})^{m/n}$ for the Metropolis algorithm, and by $(1 - n!n^{-n}\gamma^{-n(n+2)(n-1)/2})^{m/n}$ for the Gibbs sampler. In both cases the bound can be very large for large values of n and γ.

3) The hard core case $\gamma = 0$: If n is sufficiently large, any MCMC algorithm with only one point update at each iteration (like in Algorithms 7.1–7.3) is reducible. For example, if $B = [0,1]^2$, $n = 2$, and R is less than but very close to $\sqrt{2}$, any such algorithm can never move from states with two points close to two opposite corners of the unit square to states with two points close to the two other opposite corners. We recommend instead to use *simulated tempering*, which is better mixing (Mase et al. 2001).

EXAMPLE 7.3 (*Conditional area-interaction process*) Since the density of the conditional area-interaction process given by

$$\pi(\bar{x}) = \gamma^{-|\cup_{i=1}^n b(x_i, R)|}$$

is bounded from below by $\gamma^{-|b(0,R)|}$ if $0 < \gamma \leq 1$ and by $\gamma^{-n|b(0,R)|}$ if $\gamma \geq 1$, we find by similar arguments as in Example 7.2 that the Gibbs sampler and the Metropolis algorithm with uniform proposal densities are both uniformly ergodic.

7.3.2 The unconditional case

We start by considering the birth-death Metropolis-Hastings algorithm (Algorithm 7.4). Recall that h denotes our unnormalised target density and $\Omega = E$ given by (7.8) denotes the state space, assuming that $Y_0 \in E$, cf. Remark 7.4 in Section 7.1.2.

PROPOSITION 7.12 The Markov chain generated by Algorithm 7.4 is reversible with respect to h.

CONVERGENCE PROPERTIES OF ALGORITHMS

Proof. We have to verify (7.14). The state space E is a disjoint union of the sets $E_n = \{x \in E : n(x) = n\}$, $n = 0, 1, \ldots$. If $Y_m \in E_n$ and $Y_m \neq Y_{m+1}$, then $Y_{m+1} \in E_{n-1} \cup E_{n+1}$ for $n > 0$ and $Y_{m+1} \in E_1$ for $n = 0$. It thus suffices to verify the following identity: For $n \in \mathbb{N}_0$, $F \subseteq E_n$, and $G \subseteq E_{n+1}$,

$$\int_B \cdots \int_B \int_B 1[x \in F, x \cup \xi \in G] p(x) q_b(x, \xi) \alpha_b(x, \xi) h(x)$$
$$\exp(-|B|)/n! \, d\xi dx_1 \cdots dx_n$$

$$= \int_B \cdots \int_B \sum_{i=1}^{n+1} 1[y \setminus y_i \in F, y \in G](1 - p(y)) q_d(y, y_i) \alpha_d(y, y_i) h(y)$$
$$\exp(-|B|)/(n+1)! \, dy_1 \cdots dy_{n+1} \quad (7.27)$$

where $x = \{x_1, \ldots, x_n\}$, $y = \{y_1, \ldots, y_{n+1}\}$, and if $n = 0$ we interpret the left hand side as

$$\int_B 1[\emptyset \in F, \xi \in G] p(\emptyset) q_b(\emptyset, \xi) \alpha_b(\emptyset, \xi) h(\emptyset) \exp(-|B|) d\xi,$$

cf. (6.1). The right hand side in (7.27) is equal to

$$\int_B \cdots \int_B 1[y \setminus y_1 \in F, y \in G](1 - p(y)) q_d(y, y_1) \alpha_d(y, y_1) h(y)$$
$$\exp(-|B|)/n! \, dy_1 \cdots dy_{n+1}.$$

Setting $x = y \setminus y_1$ and $\xi = y_1$, we obtain then the left hand side in (7.27), since

$$h(x) p(x) q_b(x, \xi) \alpha_b(x, \xi) = h(x \cup \xi)(1 - p(x \cup \xi)) q_d(x \cup \xi, \xi) \alpha_d(x \cup \xi, \xi).$$

□

Define a probability measure by $\Psi(F) = 1[\emptyset \in F]$, $F \subseteq E$.

PROPOSITION 7.13 *If (7.10)–(7.11) are satisfied, the Markov chain generated by Algorithm 7.4 is Ψ-irreducible and aperiodic.*

Proof. Let $x \in E$ and set $m = n(x)$. If $m \geq 1$, then by (7.11) there exists $\eta \in x$ so that

$$P(x, \{x \setminus \eta\}) = (1 - p(x)) q_d(x, \eta) \min\left\{1, \frac{h(x \setminus \eta) p(x \setminus \eta) q_b(x \setminus \eta, \eta)}{h(x)(1 - p(x)) q_d(x, \eta)}\right\}$$

is positive, and for $m \geq 2$ we have that

$$P^m(x, \{\emptyset\}) \geq P(x, \{x \setminus \eta\}) P^{m-1}(x \setminus \eta, \{\emptyset\}).$$

Hence by induction on $m \geq 1$, we obtain that $P^m(x, \{\emptyset\}) > 0$. Consequently, the chain is Ψ-irreducible: If $\Psi(F) > 0$ then $P^m(x, F) \geq$

$P^m(x, \{\emptyset\}) > 0$ for $m \geq 1$. If $x = \emptyset$ then by (7.10), $P(\emptyset, F) \geq P(\emptyset, \{\emptyset\}) = 1 - p(\emptyset) > 0$. The latter inequality also shows that the chain is aperiodic. □

PROPOSITION 7.14 Consider the Markov chain generated by a locally stable version of Algorithm 7.4. For any constant $\beta > 1$, let $V(x) = \beta^{n(x)}$. Then

(i) every set $\{x \in E : n(x) \leq \alpha\}$ is small for $\alpha \geq 0$;

(ii) the chain is V-uniformly ergodic;

(iii) the chain is uniformly ergodic if and only if there exists $n_0 \in \mathbb{N}$ so that $n(x) \leq n_0$ whenever $h(x) > 0$, which in turn implies that the upper bound (7.20) on the total variation distance to h at time m is given by $(1 - \min\{1 - p, p/c^*\}^{n_0})^{m/n_0}$.

Proof. (i) Recall that the Hastings ratio for a birth is given by (7.9). Let $x \in E$ and set $n = n(x)$ and $\tilde{a} = \min\{1 - p, p/c^*\}$. For $n = 0$, $P(x, \{\emptyset\}) \geq 1 - p \geq \tilde{a}$. By ϕ^*-local stability, if x consists of a single point η,

$$P(x, \{\emptyset\}) = (1-p)\min\left\{1, \frac{\phi^*(\eta)p}{\lambda^*(\emptyset, \eta)(1-p)c^*}\right\} \geq \tilde{a}.$$

Similarly, for $n \geq 2$,

$$P^n(x, \{\emptyset\}) = \sum_{\eta \in x} \frac{1-p}{n} \min\left\{1, \frac{\phi^*(\eta)pn}{\lambda^*(x \setminus \eta, \eta)(1-p)c^*}\right\} P^{n-1}(x \setminus \eta, \{\emptyset\})$$

$$\geq \sum_{\eta \in x} (\tilde{a}/n) P^{n-1}(x \setminus \eta, \{\emptyset\}) \geq \tilde{a}^n,$$

using induction to obtain the last inequality. Hence, for $m \geq n$,

$$P^m(x, \{\emptyset\}) \geq P^n(x, \{\emptyset\})P(\emptyset, \{\emptyset\})^{m-n} \geq \tilde{a}^m.$$

Consequently, $\{x \in E : n(x) \leq \alpha\}$ is small for $\alpha \geq 0$: if $x \in E$ and $m \geq \alpha \geq n(x)$, then

$$P^m(x, F) \geq P^m(x, \{\emptyset\})\mathbf{1}[\emptyset \in F] \geq \tilde{a}^m \Psi(F). \qquad (7.28)$$

(ii) We verify that the conditions in Proposition 7.9 are satisfied. By Proposition 7.13, since (7.10)–(7.11) are satisfied, the chain is Ψ-irreducible and aperiodic (this follows also from the proof of (i) above). To verify the geometric drift condition, let $C = \{x \in E : n(x) \leq \alpha\}$ where $\alpha \geq 0$, and let $x \in E$ and $n = n(x)$. Then, for $n \geq 1$, considering each of the three cases birth, death, or rejection of a birth or death, we

CONVERGENCE PROPERTIES OF ALGORITHMS 131

obtain

$$\begin{aligned}
\mathbb{E}[V(Y_1)|Y_0 = x] &= \beta^{n+1} \int_B p\phi^*(\xi)/c^* \min\left\{1, \frac{\lambda^*(x,\xi)(1-p)c^*}{\phi^*(\xi)p(n+1)}\right\}d\xi \\
&+ \beta^{n-1} \sum_{\eta \in x} \frac{1-p}{n} \min\left\{1, \frac{\phi^*(\eta)pn}{\lambda^*(x \setminus \eta, \eta)(1-p)c^*}\right\} \\
&+ \beta^n \left[\int_B p\phi^*(\xi)/c^*\left(1 - \min\left\{1, \frac{\lambda^*(x,\xi)(1-p)c^*}{\phi^*(\xi)p(n+1)}\right\}\right)d\xi \right. \\
&\left. + \sum_{\eta \in x} \frac{1-p}{n}\left(1 - \min\left\{1, \frac{\phi^*(\eta)pn}{\lambda^*(x \setminus \eta, \eta)(1-p)c^*}\right\}\right)\right] \\
&= V(x)\left[\int_B p\phi^*(\xi)/c^*\left(1 + (\beta-1)\min\left\{1, \frac{\lambda^*(x,\xi)(1-p)c^*}{\phi^*(\xi)p(n+1)}\right\}\right)d\xi \right. \\
&\left. + \sum_{\eta \in x} \frac{1-p}{n}\left(1 + (1/\beta - 1)\min\left\{1, \frac{\phi^*(\eta)pn}{\lambda^*(x \setminus \eta, \eta)(1-p)c^*}\right\}\right)\right].
\end{aligned}$$

Suppose we choose α so large that $(1-p)c^*/(p(\alpha+1)) < 1$. Then by local stability, if $n > \alpha$,

$$\mathbb{E}[V(Y_1)|Y_0 = x] \leq V(x) \times$$

$$\left[\int_B p\phi^*(\xi)/c^*\left(1 + (\beta-1)\frac{(1-p)c^*}{p(\alpha+1)}\right)d\xi + \sum_{\eta \in x}\frac{1-p}{n}(1 + (1/\beta - 1))\right]$$

$$= aV(x)$$

where

$$a = p[1 + (\beta-1)(1-p)c^*/(p(\alpha+1))] + (1-p)/\beta.$$

Suppose we choose α so large that $a < 1$. Letting $b = \beta^{\alpha+1}$, if $n \leq \alpha$, then $\mathbb{E}[V(Y_1)|Y_0 = x] \leq b$ since $\beta > 1$. Thereby (7.21) is verified.

(iii) Suppose there exists $n_0 \in \mathbb{N}$ so that $n(x) \leq n_0$ whenever $h(x) > 0$. Then by (7.28), E is (n_0, M)-small with $M = \tilde{a}^{n_0}\Psi$. On the other hand, suppose that for any $n_0 \in \mathbb{N}$, there exists an x with $n(x) > n_0$ and $h(x) > 0$, and suppose that E is (m, M)-small for some $m \in \mathbb{N}$ and some finite nonzero measure M. We then obtain a contradiction, namely that $M(F) = 0$ for all $F \subseteq E$: Consider $F_n = F \cap E_n$, where $E_n = \{x \in E : n(x) = n\}$, and consider an x with $h(x) > 0$ and $n(x) - m > n$. Then, since we can at most delete one point at each transition, $P^m(x, F_n) = 0$ whereby $M(F_n) = 0$. Consequently, $M(F) = 0$. Thereby the assertions follow from Proposition 7.8. □

REMARK 7.9 Recall that nearly all Markov point process models used in practice are locally stable, cf. Chapter 6, whereby V-uniform ergodicity follows by (ii) above. V-uniform ergodicity and reversibility imply that the central limit theorem (Proposition 7.10) applies for functions k with either $k \in \mathcal{L}^2(h)$ or $|k(x)| \leq \beta^{n(x)}$ for some $\beta > 0$. By (iii), uniform ergodicity is typically only satisfied for hard core type models, and like in Example 7.2 the upper bound on the total variation distance is not useful in practice. For any birth-death type algorithm, the condition in (iii) is necessary for obtaining uniform ergodicity, cf. the proof of (iii).

We consider next the birth-death-move algorithm (Algorithm 7.5).

PROPOSITION 7.15 *The Markov chain generated by Algorithm 7.5 is*

(i) *reversible with respect to h,*

(ii) Ψ-*irreducible and aperiodic if (7.10)–(7.11) are satisfied.*

(iii) *If Algorithm 7.4 is given by a locally stable version, then (i)–(iii) in Proposition 7.14 are also true for Algorithm 7.5, except that in the uniformly ergodic case the total variation distance to h is now bounded by $(1-(1-q)\min\{1-p, p/c^*\})^{n_0})^{m/n_0}$ at times $m = 1, 2, \ldots$.*

Proof. The transition kernel of the chain generated by Algorithm 7.5 is given by
$$P(x, F) = qP_1(x, F) + (1-q)P_2(x, F)$$
where $0 \leq q < 1$, $P_1(x, F)$ is the transition kernel of Algorithm 7.1, and $P_2(x, F)$ is the transition kernel of Algorithm 7.4. Thus (i) follows immediately from Propositions 7.11 and 7.12, and (ii) follows immediately from Proposition 7.13. Finally, we obtain (iii) from the proof of Proposition 7.14, since (7.28) holds if we replace $\tilde{a}$ with $(1-q)\tilde{a}$, and since the geometric drift condition is satisfied:
$$\mathbb{E}[V(Y_1)|Y_0 = x] \leq (q + (1-q)a)V(x) + (1-q)b\mathbf{1}[x \in C]$$
where $q + (1-q)a < 1$, since $a < 1$. □

7.3.3 The case of marked and multivariate point processes

Since the theory in Sections 7.3.1–7.3.2 immediately extends to the case of marked and multivariate point processes, cf. the beginning of Section 7.1.3, we concentrate again on the particular case of a continuum Ising process.

PROPOSITION 7.16 *The Gibbs sampler for a continuum Ising process (Algorithm 7.6) and the Swendsen-Wang type algorithm for a symmetric Widom-Rowlinson penetrable spheres mixture model (Algorithm 7.7, assuming that T is bounded) are both uniformly ergodic.*

CONVERGENCE PROPERTIES OF ALGORITHMS

Proof. Consider first Algorithm 7.6. Suppose that Y_m is distributed according to the continuum Ising model. Since each of the steps (1)–(k) in Algorithm 7.6 replaces a component $Y_{m,i}$ with a draw from its conditional distribution, it follows directly that also Y_{m+1} is distributed according to the continuum Ising model, which is thus an invariant distribution for the Gibbs sampler chain. For any finite $y_1, \ldots, y_k \subset T$, since $\phi \leq 1$,

$$P(Y_{m+1} = (\emptyset, \ldots, \emptyset) | Y_m = (y_1, \ldots, y_k)) \geq \exp(-(\beta_1 + \ldots + \beta_k)|T|),$$

i.e. the state space is small, and so we have uniform ergodicity, cf. Proposition 7.8.

Consider next Algorithm 7.7. It follows as above for the Gibbs sampler that the continuum Ising model is an invariant distribution for step (i) of Algorithm 7.7. By symmetry, it is also clear that the continuum Ising model is invariant for step (ii). Assume that T is bounded. Then there is an upper bound $n < \infty$ on the number of connected components in Z_{m+1}. So conditional on Y_m, with probability at least $1/k^n$, we have that all components in Z_{m+1} are of type 1. Moreover, given that all components in Z_{m+1} are of type 1, $Y'_{m+2,1} = \emptyset$ with probability $\exp(-\beta|T|)$, since $Y'_{m+2,1} \sim \text{Poisson}(T, \beta)$ when $Y'_{m+1,2} = \ldots = Y'_{m+1,k} = \emptyset$. Thus

$$P(Y_{m+2} = (\emptyset, \ldots, \emptyset) | Y_m = (y_1, \ldots, y_k)) \geq (1/k^n) \exp(-\beta|T|),$$

i.e. the state space is small, and so uniform ergodicity is established, cf. Proposition 7.8. □

CHAPTER 8

Simulation-based inference

For many point process models, closed form expressions for expectations of statistics of interest are not available and Monte Carlo methods hence become useful for calculation of such quantities. In likelihood-based inference for point processes, a particular problem is computation of normalising constants. For a Markov point process, the normalising constant in the likelihood is given by a high dimensional integral which typically cannot be calculated explicitly. For a Cox process, the likelihood is given in terms of an expectation with respect to the unobserved random intensity function and can in fact be viewed as an unknown normalising constant. In Bayesian inference for Markov point processes, ratios of normalising constants must be evaluated in Metropolis-Hastings algorithms for posterior computation.

Section 8.1 provides an introduction to Monte Carlo methods and output analysis. The succeeding sections consider how Monte Carlo methods can be used in likelihood-based inference when normalising constants are not known. Apart from a few exceptions, the methods in this chapter are not restricted to point process applications. Section 8.1 therefore considers a general setting, where Π is a given target distribution and Y_m denotes a Markov chain with invariant distribution Π. In the remaining sections, Π is specified by a parametrised density for a point process, where in applications the normalising constant is typically unknown. Section 8.2 concerns MCMC approximations of ratios of normalising constants. This is then used in Sections 8.3–8.6 for approximate likelihood inference, including the case of missing data likelihoods. As discussed later in Section 9.3 the methods in Section 8.2 are also useful for Bayesian MCMC inference.

8.1 Introduction to Monte Carlo methods and output analysis

A *Monte Carlo method* is a particular form of numerical integration which uses samples of stochastic simulations. For recent treatments of such methods, see Robert & Casella (1999), Chen, Shao & Ibrahim (2000), and Liu (2001). Section 8.1.1 describes the basic idea of Monte Carlo estimation. When MCMC samples are used instead of samples of independent simulations at least three problems emerge. The first is assessment of convergence of Markov chains which is considered in Section 8.1.2. Section 8.1.3 addresses the second problem which is computation of the asymptotic variance of a Monte Carlo estimate. Section 8.1.4 concerns the third problem of subsampling a Markov chain.

8.1.1 Ergodic averages

Suppose that we wish to estimate $\mathbb{E}k(X)$ where X is a random variable, k is a real function, and the mean is with respect to a given distribution Π. Ordinary Monte Carlo methods use a sequence of random variables $Y_0, \ldots, Y_{n-1}$ all having distribution Π. The Monte Carlo estimate given by the ergodic average $\bar{k}_n$ in (7.15) is then unbiased. If also $Y_0, \ldots, Y_{n-1}$ are independent, then by the classical strong law of large numbers $\bar{k}_n$ is a consistent estimate, and by the classical central limit theorem $\sqrt{n}(\bar{k}_n - \mathbb{E}k(X))$ is asymptotically normally distributed provided $k(X)$ has finite variance.

However, for complex models such as most spatial point process models, it is often not feasible to make large numbers of exact and independent simulations. Suppose instead that we have generated a sample $Y_0, \ldots, Y_{n-1}$ of an irreducible Markov chain with invariant distribution Π. Then we can still use ergodic averages as MCMC estimates, since they are consistent (Propositions 7.3–7.4 in Section 7.2.1), and also often asymptotically normally distributed (Proposition 7.10 in Section 7.2.3). Note that in practice an initial part of the chain may have been discarded, cf. the discussion on burn-in in Section 8.1.2. The precision of an ergodic average can be expressed by its asymptotic variance studied in Section 8.1.3.

8.1.2 Assessment of convergence

The *burn-in* is the time $j \geq 0$ at which the marginal distribution of a Markov chain state Y_j is sufficiently close to its limit distribution for all practical purposes (provided the limit distribution exists, see Section 7.2). The states in the initial part of the chain may be far from the limiting distribution due to the choice of the value of Y_0, so to reduce the bias of Monte Carlo estimates, it is sensible to use only Y_m, $m \geq j$. Below we consider some simple graphical methods for determining the burn-in; a number of useful *convergence diagnostics* have also been proposed, see e.g. Cowles & Carlin (1996), Brooks & Gelman (1998), Brooks & Roberts (1998), and Mengersen, Robert & Guihenneuc-Jouyaux (1999).

Visual inspection of *trace plots* $k(Y_m)$, $m = 0, 1, \ldots$, for various real functions k is a commonly used method to assess if the chain has not reached equilibrium. Often we consider functions k given by some natural

MONTE CARLO METHODS AND OUTPUT ANALYSIS 137

sufficient statistics. For example, for a Strauss process X (Example 6.1, Section 6.2.2), it is natural to consider trace plots of $n(Y_m)$ and $s_R(Y_m)$ since, as explained later in Section 8.2.2, $(n(X), s_R(X))$ is the minimal canonical sufficient statistic. Trace plots of $n(Y_m)$, $s_R(Y_m)$, and other statistics are shown for example in Figures 9.3 and 10.3 (Sections 9.1.5 and 10.3.1).

Suppose trace plots are obtained from two chains of length n with different starting values. If the chains behave differently according to the trace plots for large values of $m \leq n$, the burn-in for at least one of the chains will be greater than n. For point processes, one extreme initial value is given by $\emptyset$. If the limit distribution is given by a ϕ^*-locally stable unnormalised density h, another extreme initial value is obtained by a simulation from a Poisson process D_0 with intensity function ϕ^*, because if $Y_0 \sim h$ then there is a coupling so that $P(Y_0 \subseteq D_0) = 1$, cf. Section 11.2.6. An example is given in Figure 9.3 (Section 9.1.5).

Suppose that Y_m is a reversible Markov chain obtained by one of the birth-death type algorithms in Section 7.1.2. If the chain has converged at time m, then by reversibility, $n(Y_{m+1}) - n(Y_m)$ is symmetrically distributed on $\{-1, 0, 1\}$. Hence, comparison of $\sum_{i=0}^{m-1}(n(Y_{i+1}) - n(Y_i))/m$ with zero or plotting $n(Y_{m+1}) - n(Y_m)$ versus m is useful for determining the burn-in.

REMARK 8.1 In some cases the issues of fast convergence of Markov chains and efficiency of Monte Carlo estimates somewhat conflicts, see e.g. Peskun (1973), Besag & Green (1993), and Tierney (1998). However, since there is no such conflict in the reversible case (Geyer 1992), and most of the MCMC algorithms used in this book are reversible, we do not pursue this further.

8.1.3 Estimation of correlations and asymptotic variances

Plots of estimated auto-correlations and cross-correlations for different statistics often provide good indications of the chain's *mixing behaviour*. Assume that $Y_j \sim \Pi$ for some $j \in \mathbb{N}_0$. For a given real function k with finite variance

$$\sigma^2 = \mathrm{Var}(k(Y_j)),$$

define the lag m *auto-correlation* by

$$\rho_m = \mathrm{Corr}(k(Y_j), k(Y_{j+m})), \quad m = 0, 1, \ldots.$$

Under fairly weak conditions, $\rho_m \to 0$ as $m \to \infty$. Similarly, given two functions $k^{(1)}$ and $k^{(2)}$ with finite variances, the lag m *cross-correlations* are defined by

$$\rho_m^{i_1, i_2} = \mathrm{Corr}(k^{(i_1)}(Y_j), k^{(i_2)}(Y_{j+m})), \quad m = 0, 1, \ldots, \; i_1, i_2 \in \{1, 2\}.$$

In the reversible case, $\rho_m^{1,2} = \rho_m^{2,1}$. Under fairly weak conditions, $\rho_m^{i_1,i_2} \to 0$ as $m \to \infty$. The chain is slowly respectively rapidly mixing if the correlations are slowly respectively rapidly decaying to 0.

For the estimation of ρ_m and $\rho_m^{i_1,i_2}$, let us for ease of presentation assume that $j = 0$ (in practice a burn-in $j \geq 0$ may have been used, however, the following estimates are also consistent without a burn-in). The lag m *auto-covariance* $\gamma_m = \sigma^2 \rho_m$ is estimated by the empirical auto-covariance

$$\hat{\gamma}_m = \frac{1}{n} \sum_{i=0}^{n-1-m} (k(Y_i) - \bar{k}_n)(k(Y_{i+m}) - \bar{k}_n)$$

for $m = 0, \ldots, n-1$ (an argument for using the divisor n rather than $n - m$ is given by Priestley 1981, pages 323–324). From this we obtain natural estimates $\hat{\sigma}^2 = \hat{\gamma}_0$ and $\hat{\rho}_m = \hat{\gamma}_m / \hat{\sigma}^2$. Similar methods apply for estimation of cross-correlations.

The *Monte Carlo error* of $\bar{k}_n$ can be expressed by the *Monte Carlo variance*

$$\mathrm{Var}(\bar{k}_n) = \frac{\sigma^2}{n} \left[1 + 2 \sum_{m=1}^{n-1} \left(1 - \frac{m}{n}\right) \rho_m \right].$$

Under the conditions in Proposition 7.10, the *asymptotic variance* is well defined and finite and given by

$$\bar{\sigma}^2 = \lim_{n \to \infty} n \mathrm{Var}(\bar{k}_n) = \sigma^2 \tau \qquad (8.1)$$

where

$$\tau = 1 + 2 \sum_{m=1}^{\infty} \rho_m \qquad (8.2)$$

is called the *integrated auto-correlation time*, and as $n \to \infty$, $\sqrt{n}(\bar{k}_n - \mathbb{E}k(Y_0))$ is asymptotically normally distributed with mean 0 and variance $\bar{\sigma}^2$. The asymptotic variance determines the *efficiency of a Monte Carlo estimate*. In the special i.i.d. case, $\tau = 1$. Therefore, when $\bar{k}_n$ has been used for estimating $\mathbb{E}k(X)$, the *effective sample size* is defined by $n/\tau = n\sigma^2/\bar{\sigma}^2$. Note that finiteness of τ implies by (8.2) that $\rho_m \to 0$ as $m \to \infty$.

Although $\hat{\gamma}_m$ is a consistent estimate of γ_m, it is well known that the obvious estimate $1 + 2\sum_{m=1}^{n-1} \hat{\rho}_m$ of τ is not consistent as $n \to \infty$ (Priestley 1981, page 432). One method for estimation of the asymptotic variance is the method of *batch means*: Suppose that $n = n_1 n_2$ where $n_1, n_2 \in \mathbb{N}$ and n_2 is so large that we can treat the n_1 batch mean estimates

$$\bar{k}_{n_1,n_2}^{(i)} = \frac{1}{n_2} \sum_{m=(i-1)n_2}^{in_2 - 1} k(Y_m), \quad i = 1, \ldots, n_1,$$

MONTE CARLO METHODS AND OUTPUT ANALYSIS   139

as being (approximately) uncorrelated. Note that $\bar{k}_n = \sum_{i=1}^{n_1} \bar{k}_{n_1,n_2}^{(i)}/n_1$. If also n_2 is sufficiently large, the batch mean estimates are approximately normally distributed. This suggests to estimate $\mathbb{V}\mathrm{ar}(\bar{k}_n)$ by

$$\sum_{i=1}^{n_1} (\bar{k}_{n_1,n_2}^{(i)} - \bar{k}_n)^2/(n_1(n_1-1)),$$

see e.g. Ripley (1987) for further details.

Geyer (1992) suggests another estimate, using that for an irreducible and reversible Markov chain, $\Gamma_m = \gamma_{2m} + \gamma_{2m+1}$ is a strictly positive, strictly decreasing, and strictly convex function of $m = 0, 1, \ldots$ (recall that the algorithms in e.g. Sections 7.1.1–7.1.2 are reversible). Let $l_s \leq (n-2)/2$, $s = \mathrm{pos, mon, conv}$, be the largest integers so that $\hat{\Gamma}_m = \hat{\gamma}_{2m} + \hat{\gamma}_{2m+1}$, $m = 0, \ldots, l_s$, is respectively strictly positive, strictly decreasing, or strictly convex. Then Geyer (1992) shows that the *initial sequence estimates*

$$\hat{\tau}_s = 1 + 2 \sum_{m=1}^{2L_s+1} \hat{\rho}_m,$$

where

$$L_{\mathrm{pos}} = l_{\mathrm{pos}}, \quad L_{\mathrm{mon}} = \min\{L_{\mathrm{pos}}, l_{\mathrm{mon}}\}, \quad L_{\mathrm{conv}} = \min\{L_{\mathrm{mon}}, l_{\mathrm{conv}}\},$$

provide consistent conservative estimates of τ, i.e. $\liminf_{n\to\infty}(\hat{\sigma}^2\hat{\tau}_s) \geq \bar{\sigma}^2$ for $s = \mathrm{pos, mon, conv}$, where $\hat{\tau}_{\mathrm{pos}} \geq \hat{\tau}_{\mathrm{mon}} \geq \hat{\tau}_{\mathrm{conv}}$. These estimates are examples of so-called window estimates.

8.1.4 Subsampling

Sometimes *subsampling* with a *spacing* $s \geq 2$ is used, i.e. we use only the subchain $Y_j, Y_{j+s}, Y_{j+2s}, \ldots$ for some given $j \in \mathbb{N}_0$. There may be various reasons for using a subsample: storage problems may be reduced if it is required to store a sample e.g. for plotting; trace and auto-correlation plots may be more informative; and more efficient Monte Carlo estimates may be obtained. The latter issue is discussed in Geyer (1992) who shows that for an irreducible and reversible Markov chain, if the computational cost of evaluating $k(Y_m)$ is negligible, it is optimal not to use subsampling. However, if the samples are highly auto-correlated and the evaluation of $k(Y_m)$ is expensive, then a large spacing s may be desirable. In the data examples later on in Chapters 9–10 we usually use subsampling to obtain sample sizes which are manageable for storing and plotting.

For the Markov chain $Y_j, Y_{j+s}, Y_{j+2s}, \ldots,$ we proceed as in Sections 8.1.1–8.1.3 by substituting the original chain by the subsampled chain.

For example, the asymptotic variance of the Monte Carlo average

$$\bar{k}_n^s = \sum_{m=0}^{n-1} k(Y_{j+ms})/n$$

based on the subsampled chain is given by $\sigma^2 \tau^s$, where

$$\tau^s = 1 + 2 \sum_{m=1}^{\infty} \rho_{ms}$$

is the integrated auto-correlation time for the subsampled chain.

8.2 Estimation of ratios of normalising constants

Ratios of normalising constants are often needed in statistical inference; several examples are given in the rest of this book. Sections 8.2.1–8.2.2 provide some background material, and Sections 8.2.3–8.2.5 consider various techniques for estimation of ratios of normalising constants. Particularly, techniques based on importance, bridge, and path sampling are considered.

8.2.1 Setting and assumptions

Henceforth we assume that the distribution of a point process X is specified by a density in a parametric family $\{f_\theta : \theta \in \Theta\}$, $\Theta \subseteq \mathbb{R}^p$, $p \geq 1$, of densities $f_\theta(x) = h_\theta(x)/c_\theta$ with respect to $Y \sim \text{Poisson}(S, 1)$, where $S \subset \mathbb{R}^d$ with $|S| < \infty$ and

$$c_\theta = \mathbb{E} h_\theta(Y) \tag{8.3}$$

is the normalising constant, cf. Section 6.1. For ease of presentation, unless otherwise stated, we assume that Θ is open and h_θ and c_θ are twice continuous differentiable with respect to θ viewed as a p-dimensional row vector. In some of our applications, Θ is not open but it still makes sense to consider derivatives with respect to θ (see, for instance, Example 8.1 below).

In applications, when we consider a ratio of normalising constants c_θ/c_{θ_0}, it is sometimes of interest to choose θ_0 so that c_{θ_0} is known. In such cases, θ_0 is usually chosen so that h_{θ_0} specifies the distribution of a Poisson process, see e.g. Example 9.3 in Section 9.1.4.

REMARK 8.2 For clarity and specificity, we consider point processes with densities as described above. However, the techniques in this and the following sections apply as well for general parametric families of unnormalised densities. So everything also applies for finite point processes in a general setting, including marked and multivariate point processes.

ESTIMATION OF RATIOS OF NORMALISING CONSTANTS   141

We shall often consider derivatives of $\log c_\theta$. Let $\mathbb{E}_\theta$ denote expectation with respect to $X \sim f_\theta$. By (8.3),

$$\frac{\mathrm{d}}{\mathrm{d}\theta} \log c_\theta = \frac{1}{c_\theta} \frac{\mathrm{d}}{\mathrm{d}\theta} \mathbb{E} h_\theta(Y) = \frac{1}{c_\theta} \mathbb{E} \frac{\mathrm{d}}{\mathrm{d}\theta} h_\theta(Y) = \frac{1}{c_\theta} \mathbb{E}\left[\left(\frac{\mathrm{d}}{\mathrm{d}\theta} \log h_\theta(Y)\right) h_\theta(Y)\right]$$
$$= \mathbb{E}_\theta \mathrm{d} \log h_\theta(X)/\mathrm{d}\theta \qquad (8.4)$$

noting that $h_\theta(Y) > 0$ with probability one and assuming in the second identity the legitimacy of interchange of differentiation and expectation. In fact this is legal if for all $i \in \{1, \ldots, p\}$, the partial derivative $g_i(\theta, x) = \partial h_\theta(x)/\partial \theta_i$ is *locally dominated integrable*, that is, for all $\theta \in \Theta$ exist $\epsilon > 0$ and a function $H_{\theta, i} \geq 0$ such that

$$b(\theta, \epsilon) \subseteq \Theta, \quad \mathbb{E} H_{\theta, i}(Y) < \infty, \quad H_{\theta, i}(\cdot) \geq |g_i(\tilde\theta, \cdot)| \quad \text{for all } \tilde\theta \in b(\theta, \epsilon). \qquad (8.5)$$

See also Example 8.1 and Remark 8.3 below. Letting

$$V_\theta(x) = \mathrm{d} \log h_\theta(x)/\mathrm{d}\theta,$$

we rewrite (8.4) as the following basic identity,

$$\mathrm{d} \log c_\theta/\mathrm{d}\theta = \mathbb{E}_\theta V_\theta(X). \qquad (8.6)$$

8.2.2 Exponential family models

Most of the examples of Markov point process densities considered in Chapters 6 and 9 belong to *exponential families*, i.e. the densities are of the form

$$f_\theta(x) = b(x) \exp(\theta \cdot t(x))/c_\theta, \quad x \in N_\mathrm{f}, \ \theta \in \Theta, \qquad (8.7)$$

where $b : N_\mathrm{f} \to [0, \infty)$ and $t : N_\mathrm{f} \to \mathbb{R}^p$ are functions and $\cdot$ is the usual inner product. Many such representations exist, but it is natural to require *identifiability*, that is $f_\theta \neq f_{\tilde\theta}$ whenever $\theta, \tilde\theta \in \Theta$ are different. Note that

$$V_\theta(X) = t(X) \qquad (8.8)$$

which is called the *canonical sufficient statistic* or, if the density is identifiable, the *minimal canonical sufficient statistic*. The theory of exponential families is well covered by Barndorff-Nielsen (1978), but we shall only use a few results including the following: if Θ contains an open subset of $\mathbb{R}^p$, then identifiability is equivalent to that for any $\theta_0 \in \Theta$,

$$P_{\theta_0}(\exists \theta \in \mathbb{R}^p \setminus \{0\} : \theta \cdot t(X) \text{ is constant}) = 0. \qquad (8.9)$$

EXAMPLE 8.1 (*The Strauss process as an exponential family model*) A Strauss process X (Example 6.1, Section 6.2.2) with fixed interaction range $R > 0$ is an exponential family model with $b \equiv 1$, $\theta = (\theta_1, \theta_2) =$

$(\log \beta, \log \gamma)$, $t(x) = (n(x), s_R(x))$, and $\Theta = \mathbb{R} \times (-\infty, 0]$. Identifiability follows immediately from (8.9) (provided $P_{\theta_0}(X = \emptyset) < 1$). In order to obtain an exponential family model, we have to exclude the hard core case $\gamma = 0$ and to fix R.

As the set Θ is closed, we consider derivatives with respect to θ_2 at $\theta_2 = 0$ as left derivatives, i.e. for a function $g(\theta_2)$, the derivative at $\theta_2 = 0$ is given by the limit of $(g(\theta_2) - g(0))/\theta_2$ as $\theta_2 < 0$ tends to 0 (provided that the limit exists). It can be verified that (8.6) is still true, since

$$\partial h_\theta(x)/\partial \theta_i = \begin{cases} n(x) \exp(\theta_1 n(x) + \theta_2 s_R(x)) & \text{if } i = 1 \\ s_R(x) \exp(\theta_1 n(x) + \theta_2 s_R(x)) & \text{if } i = 2 \end{cases}$$

is locally dominated integrable for $\theta \in \mathbb{R} \times (-\infty, 0)$ (this follows from Remark 8.3 below) as well as for $\theta \in \mathbb{R} \times \{0\}$ (using an appropriate modification of (8.5) because we consider left derivatives).

REMARK 8.3 Consider the exponential family density (8.7) and let

$$\Theta = \{\theta \in \mathbb{R}^p : \mathbb{E}[b(Y) \exp(\theta \cdot t(Y))] < \infty\}.$$

This is the largest set of θ's where the exponential family density is well defined (the exponential model is thus said to be full). Let $\theta \in \text{int}\Theta$, the interior of Θ, and let $t_i(x)$ denote the ith coordinate function of $t(x)$. Using well known properties for Laplace transforms, the following is obtained (for details, see e.g. Barndorff-Nielsen 1978). The set Θ is convex, $\mathbb{E}_\theta |t_i(X)|^\alpha < \infty$ for any $\alpha > 0$, and $\partial h_\theta(x)/\partial \theta_i = t_i(x) b(x) \exp(\theta \cdot t(x))$ is locally dominated integrable for $\theta \in \text{int}\Theta$. Thus (8.6) is true for $\theta \in \text{int}\Theta$.

8.2.3 Importance sampling

The simplest technique for estimation of ratios of normalising constants is importance sampling.

Let $\Omega_\theta = \{x : h_\theta(x) > 0\}$ denote the support of f_θ. Assume that $\Omega_\theta \subseteq \Omega_{\theta_0}$ for $\theta, \theta_0 \in \Theta$. For a statistic $k(X)$, using that $\Omega_\theta \subseteq \Omega_{\theta_0}$, we obtain the *importance sampling formula*,

$$\mathbb{E}_\theta k(X) = \mathbb{E}_{\theta_0}[k(X) h_\theta(X)/h_{\theta_0}(X)]/[c_\theta/c_{\theta_0}]. \qquad (8.10)$$

Letting $k \equiv 1$, this reduces to

$$c_\theta/c_{\theta_0} = \mathbb{E}_{\theta_0}[h_\theta(X)/h_{\theta_0}(X)]. \qquad (8.11)$$

Thus the ratio of normalising constants can be approximated by

$$c_\theta/c_{\theta_0} \approx \frac{1}{n} \sum_{m=0}^{n-1} h_\theta(Y_m)/h_{\theta_0}(Y_m) \qquad (8.12)$$

where $Y_0, Y_1, \ldots, Y_{n-1}$ is an MCMC sample with invariant density f_{θ_0},

ESTIMATION OF RATIOS OF NORMALISING CONSTANTS   143

see Section 8.1.1. We call θ_0 the *importance sampling parameter*, and define the *importance sampling distribution* (or weights) by

$$w_{\theta,\theta_0,n}(Y_m) = \frac{h_\theta(Y_m)/h_{\theta_0}(Y_m)}{\sum_{i=0}^{n-1} h_\theta(Y_i)/h_{\theta_0}(Y_i)}, \quad m = 0, \ldots, n-1, \qquad (8.13)$$

and the *importance sampling estimator* of $\mathbb{E}_\theta k(X)$ by

$$\mathbb{E}_{\theta,\theta_0,n} k = \sum_{m=0}^{n-1} k(Y_m) w_{\theta,\theta_0,n}(Y_m). \qquad (8.14)$$

This is a consistent estimator if the chain Y_m is irreducible or Harris recurrent, cf. Propositions 7.3–7.4 (Section 7.2.1).

REMARK 8.4 A natural requirement is that a central limit theorem (CLT) holds for $\mathbb{E}_{\theta,\theta_0,n} k$, or more precisely for

$$\sqrt{n} \mathbb{E}_{\theta,\theta_0,n} k = \frac{\frac{1}{\sqrt{n}} \sum_{m=0}^{n-1} k(Y_m) h_\theta(Y_m)/h_{\theta_0}(Y_m)}{\frac{1}{n} \sum_{m=0}^{n-1} h_\theta(Y_m)/h_{\theta_0}(Y_m)}.$$

Suppose that the chain Y_m is geometrically ergodic. Since the chain then is Harris recurrent, the denumerator above converges to $\mathbb{E}_{\theta_0}(h_\theta/h_{\theta_0}(X))$ $= c_\theta/c_{\theta_0}$ (Proposition 7.4, Section 7.2.1). A CLT holds for the numerator above if

$$\mathbb{E}_{\theta_0} |k(X) h_\theta(X)/h_{\theta_0}(X)|^{2+\epsilon} < \infty$$

for some $\epsilon > 0$ or, if the chain is reversible, with $\epsilon = 0$, cf. Proposition 7.10 in Section 7.2.3. This implies a CLT for $\sqrt{n} \mathbb{E}_{\theta,\theta_0,n} k$ with mean

$$\frac{\mathbb{E}_{\theta_0}(k(X) h_\theta h_{\theta_0}(X))}{\mathbb{E}_{\theta_0}(h_\theta/h_{\theta_0}(X))} = \frac{\mathbb{E}_\theta k(X) c_\theta/c_{\theta_0}}{c_\theta/c_{\theta_0}} = \mathbb{E}_\theta k(X)$$

and variance $\bar{\sigma}^2 (c_{\theta_0}/c_\theta)^2$, where $\bar{\sigma}^2$ is the asymptotic variance of the numerator.

For example, for the Strauss process in Example 8.1 above, if $R > 0$ is fixed and $\theta = (\log \beta, \log \gamma)$ and $\theta_0 = (\log \beta', \log \gamma')$ with $\beta, \beta' > 0$ and $\gamma, \gamma' \in (0,1]$, then

$$\mathbb{E}_{\theta_0} |k(X) h_\theta(X)/h_{\theta_0}(X)|^2 = \frac{1}{c_{\theta_0}} \mathbb{E}\left[k(Y)^2 \left(\frac{\beta^2}{\beta'}\right)^{n(Y)} \left(\frac{\gamma^2}{\gamma'}\right)^{s_R(Y)}\right].$$

This is finite if $\gamma \leq \sqrt{\gamma'}$ and if e.g. $k(x)$ is equal to 1, $n(x)$, or $s_R(x)$; in general it is infinite for $\gamma > \sqrt{\gamma'}$, cf. Example 6.1 in Section 6.2.2.

8.2.4 Bridge sampling and related methods

The approximation (8.12) is only useful for θ sufficiently close to θ_0 so that the variance $\text{Var}_{\theta_0}[h_\theta(X)/h_{\theta_0}(X)]$ is moderate. In this and the following section we describe some alternative approximations.

Suppose that θ_0 and θ are not "close" but that $f^b(x) = h^b(x)/c^b$ is a density with support Ω^b so that h_{θ_0} and h_θ are both "close" to h^b and $\Omega^b \subseteq \Omega_\theta \cap \Omega_{\theta_0}$. Applying the importance sampling formula (8.11) on $c_\theta/c_{\theta_0} = (c^b/c_{\theta_0})/(c^b/c_\theta)$ we obtain

$$c_\theta/c_{\theta_0} = \mathbb{E}_{\theta_0}\left[h^b(X)/h_{\theta_0}(X)\right]/\mathbb{E}_\theta\left[h^b(X)/h_\theta(X)\right] \qquad (8.15)$$

where the numerator and denumerator can be estimated by ergodic averages using samples from f_{θ_0} and f_θ, respectively. The density f^b serves as a "bridge" between the densities f_{θ_0} and f_θ, and (8.15) is a special case of the so-called *bridge sampling formula* (Meng & Wong 1996). In our applications we for ease of implementation let $f^b = f_\psi$ for an "intermediate parameter value" ψ between θ and θ_0. An iterative procedure for finding the optimal version of the bridge sampling formula is proposed in Meng & Wong (1996).

The idea of bridge sampling can be extended to a situation with several densities $f_{\psi_0}, \ldots, f_{\psi_{2k}}$, say, with $\psi_0 = \theta_0$, $\psi_{2k} = \theta$, and intermediate parameter values $\psi_1, \ldots, \psi_{2k-1}$. Then (8.15) extends to

$$\frac{c_\theta}{c_{\theta_0}} = \prod_{i=0}^{k-1} \frac{c_{\psi_{2i+2}}}{c_{\psi_{2i}}} = \prod_{i=0}^{k-1} \frac{\mathbb{E}_{\psi_{2i}}[h_{\psi_{2i+1}}(X)/h_{\psi_{2i}}(X)]}{\mathbb{E}_{\psi_{2i+2}}[h_{\psi_{2i+1}}(X)/h_{\psi_{2i+2}}(X)]} \qquad (8.16)$$

provided $h_{\psi_{2i}}(x) > 0$ and $h_{\psi_{2i+2}}(x) > 0$ whenever $h_{\psi_{2i+1}}(x) > 0$. Application of (8.16) requires samples from only $k+1$ densities $f_{\psi_{2i}}$, $i = 0, \ldots, k$, compared to the $2k$ samples needed if importance sampling was used to approximate the ratios $c_{\psi_{i+1}}/c_{\psi_i}$ in $c_\theta/c_{\theta_0} = \prod_{i=0}^{2k-1} c_{\psi_{i+1}}/c_{\psi_i}$.

If instead $\Omega_{\theta_0} \cup \Omega_\theta \subseteq \Omega^b$, an alternative to (8.15) is

$$c_\theta/c_{\theta_0} = \mathbb{E}^b\left[h_\theta(X)/h^b(X)\right]/\mathbb{E}^b\left[h_{\theta_0}(X)/h^b(X)\right] \qquad (8.17)$$

where $\mathbb{E}^b$ denotes expectation with respect to f^b. In this case f^b could e.g. be a mixture of f_{θ_0} and f_θ. Monte Carlo estimation using (8.17) is an example of *umbrella sampling* (Torrie & Valleau 1977).

Yet another method for estimation of normalising constants is *reverse logistic regression* (Geyer 1991, 1999). This is used in combination with umbrella sampling in Geyer & Møller (1994) for likelihood inference for the Strauss process and for estimation of normalising constants in Mase et al. (2001). The rationale behind reverse logistic regression is not intuitively obvious, and we omit the details of this method (however, see the retrospective formulation given in Kong, McCullagh, Nicolae, Tan & Meng, 2003).

8.2.5 Path sampling

Path sampling is a useful alternative to importance and bridge sampling; Gelman & Meng (1998) provide an historical account of importance, bridge, and path sampling methods. An early application of path sampling for point processes can be found in Ogata & Tanemura (1984).

Let $\theta(s)$, $0 \le s \le 1$, be a continuous differentiable path in Θ where $\theta_0 = \theta(0)$ and $\theta = \theta(1)$. From (8.6) we obtain the *path sampling identity*,

$$\log(c_\theta/c_{\theta_0}) = \int_0^1 \mathbb{E}_{\theta(s)} V_{\theta(s)}(X) \cdot \theta'(s) \mathrm{d}s \tag{8.18}$$

where $\theta'(s) = \mathrm{d}\theta(s)/\mathrm{d}s$. In practice the integral in (8.18) is approximated using a numerical quadrature rule. We use the trapezoidal rule,

$$\log(c_\theta/c_{\theta_0}) \approx \frac{1}{l}\bigg[\mathbb{E}_{\theta(s_0)} V_{\theta(s_0)}(X) \cdot \theta'(s_0)/2 + \mathbb{E}_{\theta(s_l)} V_{\theta(s_l)}(X) \cdot \theta'(s_l)/2$$
$$+ \sum_{i=1}^{l-1} \mathbb{E}_{\theta(s_i)} V_{\theta(s_i)}(X) \cdot \theta'(s_i)\bigg] \tag{8.19}$$

where $l \in \mathbb{N}$, $s_i = i/l$, and the expectations $\mathbb{E}_{\theta(s_i)} V_{\theta(s_i)}(X)$ are estimated by Monte Carlo averages. This approach is used in Berthelsen & Møller (2002a) with Monte Carlo averages $\sum_{m=0}^{n-1} V_{\theta(s_i)}(Y_m^i)/n$ obtained from independent Markov chains Y_m^i, $m = 0, 1, \ldots$, with invariant densities $f_{\theta(s_i)}$, $i = 0, \ldots, l$, using either a burn-in or a perfect simulation (Section 11.2) for the initial state Y_0^i. In our applications we for ease of implementation let $\theta(s)$ be a linear function of s; the question of choosing an optimal path $\theta(\cdot)$ is addressed in Gelman & Meng (1998), but general practically applicable schemes are not provided.

Alternatively, we may rewrite (8.18) as an expectation

$$\log(c_\theta/c_{\theta_0}) = \mathbb{E}[V_{\theta(Z)}(X) \cdot \theta'(Z)/p(Z)] \tag{8.20}$$

where Z is a random variable on $[0,1]$ with density $p(\cdot) > 0$, and where the expectation is with respect to the joint density

$$f(x,s) = f_{\theta(s)}(x)p(s).$$

The expectation can be replaced by a Monte Carlo average, however, generating simulations from $f(x,s)$ is not straightforward as we usually do not know the normalising constant for $f_{\theta(s)}$; see the discussion in Gelman & Meng (1998).

For exponential families, by (8.8), $V_{\theta(s)}(X) \cdot \theta'(s) = t(X) \cdot \theta'(s)$ whose variance is typically much smaller than the variance of $h_\psi(X)/h_\theta(X) = \exp((\psi - \theta) \cdot t(X))$ for ψ between θ_0 and θ (consider e.g. the case of the Strauss process in Example 8.1). This partly explains why path sampling in general seems more stable than importance and bridge sampling.

8.3 Approximate likelihood inference using MCMC

8.3.1 Some basic ingredients in likelihood inference

For the following sections we need to introduce the following terminology and notation.

For an observation x the *log likelihood function* is given by

$$l(\theta) = \log h_\theta(x) - \log c_\theta \tag{8.21}$$

and the *score function* by

$$u(\theta) = \mathrm{d}l(\theta)/\mathrm{d}\theta = V_\theta(x) - \mathrm{d}\log c_\theta/\mathrm{d}\theta. \tag{8.22}$$

The *observed Fisher information* is minus the second derivative of $l(\theta)$,

$$j(\theta) = -\mathrm{d}u(\theta)/\mathrm{d}\theta^\mathsf{T} = -\mathrm{d}V_\theta(x)/\mathrm{d}\theta^\mathsf{T} + \mathrm{d}^2 \log c_\theta/(\mathrm{d}\theta\mathrm{d}\theta^\mathsf{T}) \tag{8.23}$$

The first derivative of $\log c_\theta$ is given by

$$\mathrm{d}\log c_\theta/\mathrm{d}\theta = \mathbb{E}_\theta V_\theta(X) \tag{8.24}$$

provided that $\mathrm{d}\mathbb{E}h_\theta(Y)/\mathrm{d}\theta = \mathbb{E}[\mathrm{d}h_\theta(Y)/\mathrm{d}\theta],$[†] and the second derivative by

$$\mathrm{d}^2 \log c_\theta/(\mathrm{d}\theta\mathrm{d}\theta^\mathsf{T}) = \mathbb{E}_\theta\left[\mathrm{d}V_\theta(X)/\mathrm{d}\theta^\mathsf{T}\right] + \mathrm{Var}_\theta V_\theta(X) \tag{8.25}$$

provided $\mathrm{d}[\mathbb{E}\mathrm{d}h_\theta(Y)/\mathrm{d}\theta]\mathrm{d}\theta^\mathsf{T} = \mathbb{E}[\mathrm{d}^2 h_\theta(Y)/\mathrm{d}\theta\mathrm{d}\theta^\mathsf{T}].$[‡] To stress the dependence on x we shall sometimes write $l(\theta;x)$, $u(\theta;x)$, $j(\theta;x)$ for respectively $l(\theta)$, $u(\theta)$, $j(\theta)$. The (expected) *Fisher information* is given by

$$i(\theta) = \mathbb{E}_\theta j(\theta; X).$$

If $l(\theta)$ has a unique maximum $\hat{\theta}$, it is called the *maximum likelihood estimate (MLE)*. The MLE is a solution to the *likelihood equation* $u(\theta) = 0$. This is an unbiased estimating equation, since by (8.22), $\mathbb{E}_\theta u(\theta; X) = 0$.

In the exponential family case (8.7),

$$l(\theta) = \theta \cdot t(x) - \log c_\theta \tag{8.26}$$

(omitting the constant $\log b(x)$),

$$u(\theta) = t(x) - \mathbb{E}_\theta t(X), \tag{8.27}$$

and

$$j(\theta) = i(\theta) = \mathrm{Var}_\theta t(X), \tag{8.28}$$

the $p \times p$ covariance matrix of $t(X)$ when $X \sim f_\theta$. Thus $l(\theta)$ is concave.

[†] However, we do not need to worry much about this condition for the moment, since (8.24) is mainly for comparison with (8.31) in Section 8.3.2, which does not depend on the condition.

[‡] For a similar reason we do not need to worry about this condition for the moment.

APPROXIMATE LIKELIHOOD INFERENCE USING MCMC   147

REMARK 8.5 In the exponential family case, Monte Carlo computations as considered in the rest of this chapter simplify and storage problems with Markov chain samples often reduce, since (8.26) only depends on x through the canonical sufficient statistic $t(x)$.

8.3.2 Estimation and maximisation of log likelihood functions

We show now how the log likelihood function and derivatives can be approximated by importance sampling, and how the approximate log likelihood function can be maximised by Newton-Raphson. The use of bridge and path sampling is discussed briefly.

For fixed $\theta_0 \in \Theta$, the *log likelihood ratio*

$$l(\theta) - l(\theta_0) = \log(h_\theta(x)/h_{\theta_0}(x)) - \log(c_\theta/c_{\theta_0}) \qquad (8.29)$$

is approximated by

$$l_{\theta_0,n}(\theta) = \log(h_\theta(x)/h_{\theta_0}(x)) - \log \frac{1}{n} \sum_{m=0}^{n-1} h_\theta(Y_m)/h_{\theta_0}(Y_m), \qquad (8.30)$$

cf. (8.12) and (8.21). This is called the *importance sampling approximation of the log likelihood ratio*. The first derivative of (8.30) is

$$u_{\theta_0,n}(\theta) = V_\theta(x) - \mathbb{E}_{\theta,\theta_0,n} V_\theta. \qquad (8.31)$$

This coincides with the importance sampling approximation of $u(\theta)$ obtained by combining (8.14), (8.22), and (8.24). The minus second derivative of (8.30) is

$$j_{\theta_0,n}(\theta) = - \, \mathrm{d}V_\theta(x)/\mathrm{d}\theta^\mathsf{T} + \mathbb{E}_{\theta,\theta_0,n}[\mathrm{d}V_\theta/\mathrm{d}\theta^\mathsf{T} + V_\theta^\mathsf{T} V_\theta]$$
$$- \mathbb{E}_{\theta,\theta_0,n} V_\theta^\mathsf{T} \mathbb{E}_{\theta,\theta_0,n} V_\theta,$$

which in turn coincides with the importance sampling approximation of $j(\theta)$ obtained by combining (8.14), (8.23), and (8.25). These approximations reduce further in the exponential family case (8.7), where in accordance with (8.27)–(8.28),

$$u_{\theta_0,n}(\theta) = t(x) - \mathbb{E}_{\theta,\theta_0,n} t$$

and

$$j_{\theta_0,n}(\theta) = \mathrm{Var}_{\theta,\theta_0,n} t.$$

Thus the approximate log likelihood function is concave.

For a starting value $\theta_0 = \hat{\theta}^{(0)}$, the *Newton-Raphson* iterations for maximising the log likelihood are given by

$$\hat{\theta}^{(m+1)} = \hat{\theta}^{(m)} + u(\hat{\theta}^{(m)}) j(\hat{\theta}^{(m)})^{-1}, \quad m = 0, 1, \ldots.$$

Following Geyer & Thompson (1992), the similar iterations for maximising the approximate log likelihood (8.30) are given by

$$\hat{\theta}^{(m+1)} = \hat{\theta}^{(m)} + u_{\theta_0,n}(\hat{\theta}^{(m)})j_{\theta_0,n}(\hat{\theta}^{(m)})^{-1}, \quad m = 0, 1, \ldots, \quad (8.32)$$

where an MCMC sample with importance sampling parameter θ_0 is used.

The idea of combining MCMC and Newton-Raphson for finding an approximate MLE is due to Penttinen (1984). However, the Monte-Carlo Newton-Raphson scheme in Penttinen (1984) is based on (8.31) with $\theta_0 = \theta$, i.e. without importance sampling. In this case a new sample is required for each iteration, and the algorithm typically does not converge but due to Monte Carlo error ends up oscillating around the maximising value; see the discussion in Geyer (1999). Moyeed & Baddeley (1991) suggest to obtain maximum likelihood estimates using *stochastic approximation* based on the Robbins-Monro method. Geyer (1999) concludes that stochastic approximation may be useful to obtain starting values for Newton-Raphson maximisation but not to obtain a precise approximation of the maximum likelihood estimate.

The importance sampling approximations are only useful for $\hat{\theta}^{(m)}$ sufficiently close to θ_0. We may therefore use a *"trust region"* procedure (Geyer & Thompson 1992) where the importance sampling parameter θ_0 is replaced by the output $\hat{\theta}^{(m+1)}$ of the Newton-Raphson iteration whenever this output falls outside a trust region around θ_0; hence a new MCMC sample with importance sampling parameter $\hat{\theta}^{(m+1)}$ is needed. Another approach is to replace θ_0 with $\hat{\theta}^{(m+1)}$ and to generate a new MCMC sample if the importance sampling distribution (8.13) is far away from a uniform distribution. Here we may require that the importance sampling weights do not fall below $(1/n)/T$ for some *"trust factor"* $T \geq 1$, since if $\theta = \theta_0$, the importance sampling weights are equal to $1/n$.

Newton-Raphson is particularly feasible for exponential family models since both l and $l_{\theta_0,n}$ are concave. In the nonexponential family case, the (approximate) likelihood may be multimodal, and so it may be needed to check if a global maximum is obtained. For this approximations of the log likelihood function based on bridge or path sampling become useful.

EXAMPLE 8.2 Recall that the Strauss process with varying interaction range R is not an exponential family model, cf. Example 8.1. For (β, R) fixed and $\theta = \log \gamma \in (-\infty, 0]$, since (8.6) holds by Example 8.1, (8.18) yields that

$$\log(c_\theta/c_0) = \int_0^\theta \mathbb{E}_s \sum_{\{\xi,\eta\} \subseteq X} 1[\|\xi - \eta\| \leq R] \, ds \quad (8.33)$$

where c_0 is the known normalising constant of Poisson(S, β). Thereby an estimate of $\log c_\theta$ can be obtained. Repeating this for different values

MONTE CARLO ERROR   149

of (β, R), the entire likelihood surface, the likelihood ratio statistic for a specified hypothesis (e.g. that $\gamma = 1$), etc., can be approximated. For details, see Berthelsen & Møller (2003) who also determine the distribution of the approximate likelihood ratio statistic by making independent perfect simulations. In Example 9.3 (Section 9.1.4) we use a similar approach to compute a profile likelihood function for an inhomogeneous multitype Strauss model for varying values of the interaction range.

8.4 Monte Carlo error for path sampling and Monte Carlo maximum likelihood estimates

The Monte Carlo error of the importance sampling approximation of c_θ/c_{θ_0} in (8.29)–(8.30) and of the Monte Carlo estimate of the numerator in the bridge sampling formula (8.15) can be assessed using the methods in Section 8.1.3. The denominator in the bridge sampling estimate converges to c^b/c_θ so the Monte Carlo variance of the bridge sampling estimate is given by the Monte Carlo variance of the numerator divided by $(c^b/c_\theta)^2$ (see also Remark 8.4).

When path sampling is implemented using a one-dimensional quadrature rule such as e.g. (8.19), it is not straightforward to assess the error, as error bounds for the trapezoidal rule involve the supremum of the second derivative of the integrand in (8.18) (Burden & Faires 2001) which is not explicitly known. At least it may be helpful to consider the Monte Carlo errors of the estimates of the integrand $\mathbb{E}_{\theta(s)} V_{\theta(s)}(X) \cdot \theta'(s)$ at the quadrature points $s_i = i/l$, $i = 0, \ldots, l$. In practice we experiment with varying numbers of l (the number of quadrature points) and n (the length of the Monte Carlo samples) until stable results are obtained.

It is our experience that path sampling provides much more accurate estimates than bridge sampling with the same sampling effort. In particular, bridge sampling is much more sensitive to the choice of burn-in since a few initial values in the burn-in period may completely corrupt the Monte Carlo estimate, see Example 10.4 in Section 10.3. For path sampling where Monte Carlo estimates are computed on the log scale, often a small burn-in period seems sufficient (Berthelsen & Møller 2003), and in the path sampling applications in this book we in fact do not bother to use a burn-in.

When a Monte Carlo estimate of the likelihood is maximised, we obtain a *Monte Carlo MLE* $\hat{\theta}_n$ which differ from the exact MLE $\hat{\theta}$. Since $\hat{\theta}_n$ is not given by an explicitly known function of a Monte Carlo average, it is not straightforward to compute its Monte Carlo error. Asymptotic normality of the Monte Carlo error $\sqrt{n}(\hat{\theta}_n - \hat{\theta})$ as $n \to \infty$ is established in Geyer (1994) in the case of the importance sampling approximation (8.30). Geyer (1994) also gives a method for estimating the asymptotic variance of $\sqrt{n}(\hat{\theta}_n - \hat{\theta})$, which is used in a practical example for point processes in Geyer (1999).

8.5 Distribution of estimates and hypothesis tests

In classical branches of statistics, $(\hat{\theta} - \theta)j(\hat{\theta})^{1/2}$ is typically asymptotically $N_p(0, I)$-distributed, where θ denotes the true unknown parameter value. Moreover, if a hypothesis H_1 is specified by $\theta \in \Theta_1 \subset \Theta$ and $\hat{\theta}_1$ is the MLE under H_1, then often the *log likelihood ratio statistic* $-2 \log Q = -2(l(\hat{\theta}_1) - l(\hat{\theta}))$ is asymptotically $\chi^2(p - p_1)$-distributed, where p_1 is the number of free parameters under H_1. Consider $H_1 : \theta L = 0$ where L is a $p \times d_1$ matrix of full rank $d_1 \leq p$. Then a computationally simpler alternative to the likelihood ratio statistic is the *Wald statistic*,

$$(\hat{\theta}L)(L^\mathsf{T} j(\hat{\theta})^{-1} L)^{-1}(\hat{\theta}L)^\mathsf{T} \qquad (8.34)$$

which does not involve $\hat{\theta}_1$ and which is typically asymptotically $\chi^2(d_1)$-distributed.

If the MLEs and $-2 \log Q$ are approximated by MCMC methods, we need first to find approximate Monte Carlo MLEs $\hat{\theta}_n$ and $\hat{\theta}_{n,1}$, using e.g. Newton-Raphson as in Section 8.3.2, and next to estimate $c(\hat{\theta}_n)/c(\hat{\theta}_{n,1})$. For a hypothesis $H_1 : \theta L = 0$ as above, it is computationally much simpler to use (8.34) for which an MCMC estimate is obtained by replacing $\hat{\theta}$ and $j(\hat{\theta})$ with $\hat{\theta}_n$ and $j_{\theta_0,n}(\hat{\theta}_n)$. The approximate observed information $j_{\theta_0,n}(\hat{\theta}_n)$ is available for free as a by-product when using Newton-Raphson, cf. (8.32), and there is no need for estimating a ratio of normalising constants.

For spatial point processes, asymptotic results would be based on increasing size of the observation window (or, alternatively, increasing intensity). For the estimate $\hat{\rho} = N(W)/|W|$ of the intensity of a homogeneous Poisson process observed on a window W, we e.g. have that $(|W|/\hat{\rho})^{1/2}(\hat{\rho} - \rho)$ converges in distribution to $N(0, 1)$ as $|W| \to \infty$ (this follows since $P(\hat{\rho} \to \rho) = 1$ and $N(W)$ is approximately $N(\rho|W|, \rho|W|)$-distributed). Asymptotic results for inhomogeneous spatial Poisson processes can be found in Rathbun & Cressie (1994a) who extend results in Kutoyants (1984). Asymptotic normality of maximum likelihood estimates for Markov point processes is considered in Jensen (1993), but under rather restrictive assumptions, see also Section 9.1.3. We also do not know of rigorous asymptotic results for spatial Cox processes. So in practice, possibly apart from Poisson processes, we may have to rely on simulation studies. For example, for a Strauss process with $\theta = (\beta, \gamma, R)$, Berthelsen and Møller (2003) simulate the distribution for $-2 \log Q$ under the Poisson hypothesis $H_1 : \gamma = 1$, and compare this with a χ^2-distribution; see also Chapters 9 and 10.

APPROXIMATE MISSING DATA LIKELIHOODS   151

When reliable asymptotic results are not available, we may prefer to use a *parametric bootstrap* (Efron & Tibshirani 1993). The idea is to compute approximate p-values, confidence intervals, etc., using a sample $Y_0, \ldots, Y_{n-1}$ from the density $f_{\hat\theta_1}$ or $f_{\hat\theta_{n,1}}$. If e.g. large values of a statistic $T(x)$ are critical for H_1, then $\sum_{i=0}^{n-1} \mathbf{1}[T(Y_i) \geq T(x)]/n$ is a bootstrap estimate of the p-value. In the case of a simple hypothesis $H_1 : \Theta_1 = \{\theta_1\}$, we may alternatively use an *exact Monte Carlo p-value*. Under H_1, $X, Y_0, \ldots, Y_{n-1}$ are identically distributed. If $X, Y_0, \ldots, Y_{n-1}$ are also independent, then (apart from the possibility of ties) the rank R of $T(X)$ among $T(X), T(Y_0), \ldots, T(Y_{n-1})$ is uniform on $\{1, \ldots, n+1\}$. So if the observed value of R is equal to k we can declare $k/(n+1)$ to be an exact p-value. Besag & Clifford (1989) generalise Monte Carlo p-values to the case where the sample $Y_0, \ldots, Y_{n-1}$ is obtained using MCMC. If n is large, the parametric bootstrap p-value and the exact Monte Carlo p-value are effectively equal.

8.6 Approximate missing data likelihoods and maximum likelihood estimates

Consider point processes X and Y on spaces S_1 and S_2 with joint distribution specified by a density f_θ with respect to (X_1, X_2) where X_1 and X_2 are independent Poisson processes. That is,

$$P(X \in F, Y \in G) = \mathbb{E}\big(\mathbf{1}[X_1 \in F, X_2 \in G] f_\theta(X_1, X_2)\big)$$

for events F and G of locally finite point configurations in S_1 and S_2, respectively. The definition of S_1 and S_2 depends on the context. In Sections 10.3.1–10.3.2, X is the observed part of a Cox process within an observation window $S_1 = W$ and Y is an unobserved process of "mother points" on a certain extended region S_2. In Section 9.1.1, X is the observed part on $S_1 = W$ and Y is the unobserved part on $S_2 = S \backslash W$ of a Markov point process defined on a space S, where $W \subseteq S$ is again the observation window. As noticed in Remark 8.6 below, the choice of (X_1, X_2) is not important for maximum likelihood inference, and for MCMC simulations and computations we can chose (X_1, X_2) as we like (except of course that for any $\theta \in \Theta$, the distribution of (X, Y) has to be absolutely continuous with respect to the distribution of (X_1, X_2)).

In this section we suppose that only $X = x$ is observed. The likelihood is then the marginal density

$$f_\theta(x) = \mathbb{E} f_\theta(x, X_2) \tag{8.35}$$

of X with respect to X_1. If the likelihood cannot be calculated analytically, we may use the Monte Carlo methods in Section 8.2, since $f_\theta(x)$ may be regarded as the normalising constant of the conditional density

$f_\theta(y|x) \propto f_\theta(x,y)$. This follows from (8.3) when $f_\theta(x,\cdot)$ is considered as the unnormalised density for Y given $X = x$. Section 8.6.1 describes importance, bridge, and path sampling in this case. Section 8.6.2 considers estimation of derivatives for missing data likelihoods, and how approximate maximum likelihood estimates can be obtained using Newton-Raphson. Section 8.6.3 discusses the Monte Carlo EM algorithm.

More background on Monte Carlo maximum likelihood for missing data situations can be found in Gelfand & Carlin (1993) and Geyer (1994, 1999).

8.6.1 Importance, bridge, and path sampling for missing data likelihoods

Because of the interpretation of $f_\theta(x)$ as the normalising constant of the unnormalised density $f_\theta(x,\cdot)$, we obtain immediately the formulae below which are used for estimating ratios of "normalising constants" $f_\theta(x)/f_{\theta_0}(x)$. Sometimes $f_\theta(x,y) \propto h_\theta(x,y)$ depends on a normalising constant c_θ which we may need to estimate as well, see for example Section 9.1.1. This point is also discussed below.

By the importance sampling formula (8.11),

$$f_\theta(x)/f_{\theta_0}(x) = \mathbb{E}_{\theta_0}\left[f_\theta(x,Y)/f_{\theta_0}(x,Y)\big|X=x\right],$$

so that the likelihood ratio $f_\theta(x)/f_{\theta_0}(x)$ can be approximated by

$$\frac{1}{n}\sum_{m=0}^{n-1}\frac{f_\theta(x,Y_m)}{f_{\theta_0}(x,Y_m)} \qquad (8.36)$$

where $Y_0, Y_1, \ldots$ is an MCMC sample with invariant density $f_{\theta_0}(\cdot|x)$. The ratio $f_\theta(x,y)/f_{\theta_0}(x,y)$ may depend on the ratio c_θ/c_{θ_0} of unknown normalising constants which need to be estimated using a Monte Carlo method.

For missing data likelihoods, the bridge sampling formula (8.15) becomes

$$\frac{f_\theta(x)}{f_{\theta_0}(x)} = \mathbb{E}_{\theta_0}\left[\frac{f^b(x,Y)}{f_{\theta_0}(x,Y)}\bigg|X=x\right]\bigg/\mathbb{E}_\theta\left[\frac{f^b(x,Y)}{f_\theta(x,Y)}\bigg|X=x\right] \qquad (8.37)$$

assuming that $f^b(x,y) > 0$ implies both $f_\theta(x,y) > 0$ and $f_{\theta_0}(x,y) > 0$. Here the bridge density f^b is typically given by f_{θ^b} for an intermediate parameter value θ^b between θ and θ_0, and as above there may be a need for estimating c_θ/c_{θ_0}.

Let

$$V_{\theta,x}(y) = \mathrm{d}\log f_\theta(x,y)/\mathrm{d}\theta.$$

APPROXIMATE MISSING DATA LIKELIHOODS

The path sampling identity (8.18) becomes

$$\log(f_\theta(x)/f_{\theta_0}(x)) = \int_0^1 \mathbb{E}_{\theta(s)}[V_{\theta(s),x}(Y)|X=x] \cdot \theta'(s) \mathrm{d}s \qquad (8.38)$$

assuming like in (8.4) the legitimacy of interchange of differentiation and expectation, that is, $\mathrm{d}\mathbb{E}[f_\theta(x,X_2)]/\mathrm{d}\theta = \mathbb{E}[\mathrm{d}f_\theta(x,X_2)/\mathrm{d}\theta]$. Sometimes we may need to replace the derivative $\mathrm{d}\log c_\theta(s)/\mathrm{d}\theta(s)$ entering in $V_{\theta(s),x}(Y)$ with a Monte Carlo estimate using unconditional samples from $f_{\theta(s)}$.

REMARK 8.6 The Poisson processes X_1 and X_2 are only used for specifying the joint densities $f_\theta(x,y)$, $\theta \in \Theta$, while derivatives of the log likelihood such as the score function and the observed information do not depend on the choice of X_1 and X_2. Also the ratio $f_\theta(x)/f_{\theta_0}(x)$ and hence (8.36)–(8.38) do not depend on the choice of X_1 and X_2. So from a computational point of view, we can chose X_1 and X_2 as we like. For instance, when an MCMC sample from $f_{\theta_0}(\cdot|x)$ is needed, we may let the intensity functions for X_1 and X_2 depend on θ_0.

EXAMPLE 8.3 Baddeley et al. (2000) consider a situation where Y is a Strauss process and X is obtained by a thinning of Y. The retention probabilities are up to a constant α estimated using a nonparametric kernel estimate, and α and the parameters of the Strauss process are estimated by maximising the likelihood approximated using (8.36).

8.6.2 Derivatives of missing data likelihoods and approximate maximum likelihood estimates

Assuming again that $\mathrm{d}\mathbb{E}[f_\theta(x,X_2)]/\mathrm{d}\theta = \mathbb{E}[\mathrm{d}f_\theta(x,X_2)/\mathrm{d}\theta]$, we obtain from (8.35) that

$$\mathrm{d}f_\theta(x)/\mathrm{d}\theta = \mathbb{E}[\mathrm{d}f_\theta(x,X_2)/\mathrm{d}\theta] = \mathbb{E}[V_{\theta,x}(X_2)f_\theta(x,X_2)].$$

So in analogy with (8.24), the score function $u(\theta) = \mathrm{d}\log f_\theta(x)/\mathrm{d}\theta$ is given by

$$u(\theta) = \mathbb{E}_\theta\big[V_{\theta,x}(Y)\big|X=x\big]. \qquad (8.39)$$

Similarly, under appropriate assumptions concerning interchange of differentiation and expectation, we obtain in analogy with (8.25) the observed information,

$$j(\theta) = -\mathbb{E}_\theta\big[\mathrm{d}V_{\theta,x}(Y)/\mathrm{d}\theta^\mathsf{T}\big|X=x\big] - \mathbb{V}\mathrm{ar}_\theta\big[V_{\theta,x}(Y)\big|X=x\big]. \qquad (8.40)$$

If $f_\theta(x,y)$ belongs to an exponential family model with canonical sufficient statistic $t(x,y)$, (8.39)–(8.40) reduce to

$$u(\theta) = \mathbb{E}_\theta[t(x,Y)|X=x] - \mathbb{E}_\theta[t(X,Y)] \qquad (8.41)$$

and
$$j(\theta) = \text{Var}_\theta[t(X,Y)] - \text{Var}_\theta[t(x,Y)|X=x]. \tag{8.42}$$

The score function and observed information can be estimated along the same lines as in Section 8.3.2, but now using samples from the conditional density $f_{\theta_0}(\cdot|x)$. Specifically, let

$$w_{\theta,\theta_0,n,x}(Y_m) = \frac{f_\theta(x,Y_m)/f_{\theta_0}(x,Y_m)}{\sum_{i=0}^{n-1} f_\theta(x,Y_i)/f_{\theta_0}(x,Y_i)}, \quad m = 0,\ldots,n-1,$$

be the importance sampling distribution, where $Y_0,\ldots,Y_{n-1}$ is an MCMC sample from $f_{\theta_0}(\cdot|x)$. Then for a statistic $k(Y)$,

$$\mathbb{E}_{\theta,\theta_0,n,x} k = \sum_{m=0}^{n-1} k(Y_m) w_{\theta,\theta_0,n,x}(Y_m) \tag{8.43}$$

is the importance sampling estimate of $\mathbb{E}_\theta[k(Y)|X=x]$. So Monte Carlo approximations of $u(\theta)$ and $j(\theta)$ are immediately obtained by replacing the expectations in (8.39)–(8.40) with expectations with respect to the importance sampling distribution. Possibly unknown ratios c_θ/c_{θ_0} (cf. Section 8.6.1) cancel out in the computation of the importance sampling distribution, but we may still need both conditional and unconditional samples of Y as exemplified by (8.41) and (8.42).

The importance sampling approximations of the score function and the observed information can be used for Newton-Raphson maximisation of the missing data likelihood. As demonstrated in Section 10.3 we may combine this with bridge or path sampling techniques.

8.6.3 Monte Carlo EM algorithm

Instead of Newton-Raphson we may use the *EM algorithm* for maximisation of a missing data likelihood. If $\tilde{\theta}^{(m)}$ is the output of the mth EM iteration, the next value $\tilde{\theta}^{(m+1)}$ is obtained by maximising the conditional expectation

$$\mathbb{E}_{\tilde{\theta}^{(m)}}[\log f_\theta(x,Y)|X=x]$$

of the log "full data" likelihood. This corresponds to solving

$$\mathbb{E}_{\tilde{\theta}^{(m)}}[V_{\theta,x}(Y)|X=x] = 0 \tag{8.44}$$

with respect to θ. In the *Monte Carlo EM algorithm* this equation is approximated by

$$\mathbb{E}_{\tilde{\theta}^{(m)},\tilde{\theta}^{(m)},n,x}[V_{\theta,x}(Y)] = 0. \tag{8.45}$$

An introduction to the EM algorithm and its Monte Carlo variant can e.g. be found in Robert & Casella (1999).

Geyer (1999) discusses the advantages of the importance sampling and Newton-Raphson approach compared with the Monte Carlo ver-

APPROXIMATE MISSING DATA LIKELIHOODS

sion of the EM algorithm. One advantage is that the samples generated from $f_{\theta_0}(\cdot|x)$ may be reused in the Newton-Raphson updates as long as $\theta^{(m)}$ and θ_0 are not too distant. For the Monte Carlo EM algorithm a new sample is needed in each iteration. Further, as demonstrated in Section 10.3.2, it is sometimes hard to solve the EM equations (8.44)–(8.45). It is moreover known that the EM algorithm often converges slowly. For example, Newton-Raphson is considerably faster than Monte Carlo EM in the numerical comparison in Example 10.4. Finally, as a by-product of the importance sampling/Newton-Raphson approach we obtain an estimate of the observed information.

CHAPTER 9

Inference for Markov point processes

The basic problem with likelihood-based inference for a Markov point process is the presence of a normalising constant which cannot be evaluated explicitly, cf. Section 6.1. Parameter estimation is therefore based on approximations of the normalising constant or on methods where knowledge of it is not needed. Pseudo likelihood and Takacs-Fiksel estimation, for example, are based on the Papangelou conditional intensity (6.6) where the normalising constant cancels out. With modern computers, accurate Monte Carlo estimation of the normalising constant is quite feasible, since the local Markov property reduces the computational complexity of generating MCMC samples, and since exponential family properties often simplify the Monte Carlo computations and reduce storage problems with Markov chain samples, cf. Remarks 6.2, 7.4, and 8.5 in Sections 6.3.1, 7.1.2, and 8.3.1. For Cox processes, simulation-based inference is often considerably more involved (see Chapter 10) and this is why we consider first the case of Markov point processes. The minimum contrast method considered for Cox processes in Section 10.1 is not very helpful in the case of Markov point processes where closed form expressions for summary statistics like the K, F, and G-functions are typically not available, see Section 6.4.2.

Section 9.1 covers maximum likelihood inference which is the main topic of this chapter. Section 9.2 is concerned with pseudo likelihood estimation which has been the common method for estimation and will probably continue to be of importance as a computationally easy alternative. Takacs-Fiksel estimation is also briefly considered. The rather sparse literature on Bayesian inference for Markov point processes is reviewed in Section 9.3.

The starting point in parametric inference for a Markov point process is the choice of a parametrised interaction function ϕ_θ, $\theta \in \Theta \subseteq \mathbb{R}^p$. In most of this chapter (unless otherwise stated), X denotes a Markov point process with interaction function ϕ_θ defined on some space $S \subseteq \mathbb{R}^d$ which is either bounded or equal to $\mathbb{R}^d$; in the latter case we consider a Gibbs point process as in Section 6.4.1. Further, $x \neq \emptyset$ denotes a realisation of X_W where $W \subseteq S$ is a bounded observation window. Everything can easily be extended to a general setting, including marked and mul-

9.1 Maximum likelihood inference

tivariate point processes as demonstrated in some of the examples of applications.

Depending on the relation between S and the observation window W, the likelihood function may take several forms which are discussed in Section 9.1.1. Conditional inference based on conditioning on $N(W) = n(x)$ is discussed in Section 9.1.2. Section 9.1.3 reviews asymptotic results, and Section 9.1.4 considers how Monte Carlo methods can be used to implement likelihood inference. Section 9.1.5 contains examples of applications.

9.1.1 Likelihood functions for Markov point processes

For a Markov point process with interaction function ϕ_θ, the log likelihood function $l(\theta)$ may take several forms depending on whether S coincides with the observation window W or not, and on whether S is bounded or unbounded.

Suppose that S is bounded. The density of X with respect to Poisson$(S, 1)$ is then

$$f_\theta(x) \propto h_\theta(x) = \prod_{y \subseteq x:\, y \neq \emptyset} \phi_\theta(y) \qquad (9.1)$$

with $h_\theta(\emptyset) = 1$, where the normalising constant

$$c_\theta = 1/\phi_\theta(\emptyset) = \sum_{n=0}^{\infty} \exp(-|S|)/n! \int_S \cdots \int_S h_\theta(\{x_1, \ldots, x_n\}) \mathrm{d}x_1 \cdots \mathrm{d}x_n$$

typically does not have a closed form expression. If $W = S$, then

$$l(\theta) = \log h_\theta(x) - \log c_\theta. \qquad (9.2)$$

If $W \subset S$, the log likelihood function is given by the more complicated expression

$$l_{\mathrm{mis}}(\theta) = \log \mathbb{E} f_\theta(x \cup Y_{S \setminus W}) = \log \mathbb{E} h_\theta(x \cup Y_{S \setminus W}) - \log c_\theta \qquad (9.3)$$

where $Y \sim \mathrm{Poisson}(S, 1)$. We then face a "missing data" problem, since $\mathbb{E} h_\theta(x \cup Y_{S \setminus W})$ in general cannot be calculated explicitly. It can, however, be estimated using the Monte Carlo methods in Section 8.6 (see also Section 9.1.4).

If $S = \mathbb{R}^d$, the distribution of X is typically specified in terms of conditional densities $f_{\theta,B}$ for bounded $B \subset \mathbb{R}^d$, where ∂B is bounded and

$$X_B | X_{\mathbb{R}^d \setminus B} = z_{\mathbb{R}^d \setminus B} \sim X_B | X_{\partial B} = z_{\partial B} \sim f_{\theta,B}(\cdot | z_{\partial B})$$

MAXIMUM LIKELIHOOD INFERENCE   159

for locally finite point configurations $z \subset \mathbb{R}^d$, see Section 6.4. The log likelihood is then

$$\log \mathbb{E}_\theta f_{\theta,W}(x|X_{\partial W}) \qquad (9.4)$$

which in general is intractable because the marginal distribution of $X_{\partial W}$ is not available. The log likelihood may be approximated by considering a typically large but bounded region $\tilde{S} \subset \mathbb{R}^d$ with $W \subseteq \tilde{S}$ and assuming that $X_{\partial \tilde{S}} = \emptyset$; this case is known in statistical physics as the *free boundary condition*. Here $X_{\tilde{S}}|X_{\partial \tilde{S}} = \emptyset$ has density

$$f_{\theta,\tilde{S}}(\tilde{x}|\emptyset) \propto h_{\theta,\tilde{S}}(\tilde{x}|\emptyset) = \prod_{y \subseteq \tilde{x}:\, y \neq \emptyset} \phi_\theta(y) \qquad (9.5)$$

for finite point configurations $\tilde{x}$ in $\tilde{S}$, and we approximate (9.4) by

$$\log \mathbb{E}_\theta[f_{\theta,W}(x|X_{\partial W})|X_{\partial \tilde{S}} = \emptyset] = \log \mathbb{E}[h_{\theta,\tilde{S}}(x \cup Y_{\tilde{S} \setminus W}|\emptyset)] - \log c_{\theta,\tilde{S}}(\emptyset) \qquad (9.6)$$

where $Y \sim \text{Poisson}(\tilde{S}, 1)$ and $c_{\theta,\tilde{S}}(\emptyset)$ is the normalising constant for $h_{\theta,\tilde{S}}(\cdot|\emptyset)$. MLEs based on (9.6) for different increasing choices of $\tilde{S}$ may be compared to assess whether a suitable approximation of (9.4) is obtained. Since (9.6) is effectively given by (9.2) if $\tilde{S} = W$ and by (9.3) if $\tilde{S} \supset W$ (letting $S = \tilde{S}$ and using (9.5) in (9.2) or (9.3)), we concentrate on (9.2) and (9.3)) in the sequel.

We refer to $l(\theta)$ in (9.2) as the *free boundary log likelihood* and to $l_{\text{mis}}(\theta)$ in (9.3) as the *missing data log likelihood*. When the marginal density of the observed data is unknown as in (9.3)–(9.4), it has been the custom (in analogy with estimation of nonparametric summary statistics) to speak of problems with *edge effects*. The use of the likelihoods given by (9.3) and (9.6) with $\tilde{S} \supset W$ are then known as *missing data approaches* to handling edge effects for Markov point processes.

If $S = \mathbb{R}^2$ and X is a stationary Gibbs point process in the plane with translation invariant interaction function, a *toroidal approximation* may be used to mimic stationarity. Then $\tilde{S} \supseteq W$ is chosen to be a rectangle wrapped on a torus so that points at opposite edges may be considered to be neighbours. More specifically, consider for simplicity the case of a pairwise interaction process with first order interaction function $\phi_\theta(\xi) = \phi_{1,\theta}(\xi) = \beta$ and second order interaction function $\phi_\theta(\{\xi,\eta\}) = \phi_{2,\theta}(\|\xi - \eta\|)$, and let $\tilde{S}$ be a rectangle of side lengths D_1 and D_2. We then let

$$\tilde{f}_\theta(\tilde{x}) \propto \tilde{h}_\theta(\tilde{x}) = \beta^{n(\tilde{x})} \prod_{\{\xi,\eta\} \subseteq \tilde{x}} \phi_{2,\theta}(\|\xi - \eta\|_{\text{torus}}) \qquad (9.7)$$

for finite point configurations $\tilde{x}$ in $\tilde{S}$, where

$$\|(\xi_1 - \eta_1, \xi_2 - \eta_2)\|_{\text{torus}}$$
$$= \left(\min\{|\xi_1 - \eta_1|, D_1 - |\xi_1 - \eta_1|\}^2 + \min\{|\xi_2 - \eta_2|, D_2 - |\xi_2 - \eta_2|\}^2 \right)^{1/2}$$

is distance on the torus. The toroidal approximation of the log likelihood is

$$\log \mathbb{E}\tilde{f}_\theta(x \cup Y_{\tilde{S}\setminus W}) \tag{9.8}$$

where $Y \sim \text{Poisson}(\tilde{S}, 1)$. Using (9.8) with $\tilde{S} = W$ is the *toroidal approach*,[†] and we then call (9.8) the *toroidal log likelihood* and denote this by $\tilde{l}(\theta)$. For convenience (9.8) with $W \subset \tilde{S}$ is called the *missing data toroidal log likelihood*, and we denote this by $\tilde{l}_{\text{mis}}(\theta)$. Though it typically does not make sense to allow points at opposite edges to interact, in practice MLEs obtained from $\tilde{l}(\theta)$, $\tilde{l}_{\text{mis}}(\theta)$, and $l_{\text{mis}}(\theta)$ may be quite similar, see Examples 9.1 and 9.2 in Section 9.1.5.

When ϕ_θ is of range $R_\theta \leq \tilde{R} < \infty$ for all $\theta \in \Theta$, an alternative to (9.3), (9.4), (9.6), and (9.8) is the *conditional approach*, where we use the conditional density

$$f_{\theta, W_{\ominus \tilde{R}}}(x_{W_{\ominus \tilde{R}}} | x_{\partial W_{\ominus \tilde{R}}}) \propto h_{\theta, W_{\ominus \tilde{R}}}(x_{W_{\ominus \tilde{R}}} | x_{\partial W_{\ominus \tilde{R}}}) = \prod_{\substack{y \subseteq x_{W_{\ominus \tilde{R}}} \cup x_{\partial W_{\ominus \tilde{R}}}: \\ y_{W_{\ominus \tilde{R}}} \neq \emptyset}} \phi_\theta(y). \tag{9.9}$$

Here $W_{\ominus \tilde{R}} = \{\xi \in W : b(\xi, \tilde{R}) \subseteq W\}$ so that $\partial W_{\ominus \tilde{R}} \subseteq W \setminus W_{\ominus \tilde{R}}$, and we assume that $W_{\ominus \tilde{R}}$ has positive volume (see Remark 6.6 in Section 6.3.3). Since the conditional density (9.9) is known up to a normalising constant $c_\theta(x_{\partial W_{\ominus \tilde{R}}})$, we avoid the missing data problem. We refer to

$$l_{\ominus \tilde{R}}(\theta) = \log f_{\theta, W_{\ominus \tilde{R}}}(x_{W_{\ominus \tilde{R}}} | x_{W \setminus W_{\ominus \tilde{R}}}) \tag{9.10}$$

as the *conditional log likelihood*. In many applications $R_\theta = R$ does not depend on θ, and it is natural to chose $\tilde{R} = R$; an exception is given in Example 9.3 (Section 9.1.5).

Similar considerations as above apply for the likelihood function of a marked point process defined on a space $T \times M$ where we observe marked points within $A \subseteq T \times M$, where typically $A = W_T \times W_M$ with $W_T \subseteq T$ bounded and where $W_M \subseteq M$ often coincides with M, see Examples 9.2–9.3.

In Section 9.1.4 we discuss in more detail how the various likelihood functions defined above can be estimated using the Monte Carlo methods from Sections 8.3 and 8.6.

[†] This approach is common in statistical physics where large interacting particle systems are considered.

MAXIMUM LIKELIHOOD INFERENCE   161

9.1.2 Conditioning on the number of points

Consider a Markov point process X on $S = W$ with unnormalised density

$$\beta^{n(x)} \prod_{y \subseteq x: n(y) > 1} \phi_\gamma(y)$$

where $\beta > 0$ and $\gamma \in \mathbb{R}^{p-1}$. When interest is focused on the "interaction" parameter γ, a common approach is to condition on $n(X) = n(x)$. The resulting unnormalised conditional density does not depend on β and is given by

$$\prod_{y \subseteq x: n(y) > 1} \phi_\gamma(y).$$

Conditioning on the number of points simplifies matters when cluster expansions (see Section 9.1.4) are used to estimate the normalising constant and enables the use of the classical "fixed number of points Metropolis-Hastings algorithms" in Section 7.1.1. For inference on the interaction parameter γ in a Strauss process, a simulation study in Geyer & Møller (1994) suggests that $n(X)$ is an approximate ancillary statistic, since the MLEs of γ obtained from the unconditional and the conditional likelihood (given $n(X) = n(x)$) are very similar. Under somewhat restrictive conditions, Mase (1991) shows asymptotic equivalence of the unconditional and conditional MLEs of the interaction parameter for a pairwise interaction process. However, with modern MCMC methods the practical advantages of conditioning on $n(x)$ are minor. Moreover, the estimate of β may be of intrinsic interest so that conditioning is not an option. Further discussion on the matter of conditioning on $n(x)$ can be found in Gates & Westcott (1986), Ripley (1988), Geyer & Møller (1994), and Møller (1999a).

9.1.3 Asymptotic properties of maximum likelihood estimates

Asymptotic normality of likelihood estimates is considered for the free boundary likelihood (9.2) in Mase (1991) and for the conditional likelihood (9.10) in Jensen (1993). In both papers the results on asymptotic normality are obtained for increasing size of the observation window and rely on rather restrictive assumptions. For a Strauss process, for example, Geyer & Møller (1994) note that in order to apply the results in Jensen (1993), the interaction parameter γ has to be very close to one so that the process is almost Poisson. Mase (1991) considers point processes with sparse realisations, i.e. where the intensity of points is low. Mase (2002) establishes strong consistency (i.e. convergence with probability one) of the conditional MLE from Section 9.1.2, using the so-called Gibbs variational formula (Georgii 1994, 1995) which holds for so-called regular and nonintegrably divergent log pairwise interaction functions (this e.g. includes the Strauss interaction function).

9.1.4 Monte Carlo maximum likelihood

Ogata & Tanemura (1984) and Penttinen (1984) are pioneering works on Monte Carlo maximum likelihood estimation for Markov point processes. Ogata & Tanemura (1981, 1984, 1985, 1986) also use so-called cluster or virial expansions for estimation of normalising constants for a variety of pairwise interaction point processes. The expansions are based on conditioning on the number of observed points and assuming that X is concentrated on sparse point configurations, see e.g. Ripley (1988). A simulation experiment in Diggle et al. (1994) shows that the expansions are not reliable in cases with strong interaction.

With modern powerful computers, accurate estimates of the likelihood function and its derivatives can be obtained using the Monte Carlo methods in Sections 8.3 and 8.6. The Markov point process densities considered in the data examples in this chapter all belong to exponential families so that particularly simple forms of the Monte Carlo estimates are obtained. When θ is of the form (ψ, R) so that an exponential family density is obtained for fixed R, it may be helpful to use the profile likelihood approach described in Remark 9.1 below.

In the case of the free boundary, toroidal, and conditional log likelihood functions $l(\theta)$, $\tilde{l}(\theta)$, and $l_{\ominus \tilde{R}}(\theta)$, we obtain maximum likelihood estimates using Newton-Raphson as described in Section 8.3.2 (with h_θ given by either (9.1), (9.7), or (9.9)). We can often benefit from concavity of these log likelihood functions. Subsequently we use path sampling (Section 8.2.5) to compute ratios of normalising constants in likelihood ratio test statistics for various hypotheses of interest.

For the missing data log likelihood function $l_{\mathrm{mis}}(\theta)$, we use the methods in Section 8.6 (with $X = X_W$ and $Y = X_{S \setminus W}$ in Section 8.6). Working with the exponential family case, we obtain rather simple expressions (8.41) and (8.42) for the score function and observed Fisher information used in the Newton-Raphson iterations. Further, Monte Carlo samples are needed both from the unconditional distribution of X and from the conditional distribution of $X_{S \setminus W}$ given $X_W = x$. For estimation of likelihood ratios, we use path sampling as described in Section 8.6.1, where in the exponential family case, (8.38) becomes

$$\log(f_\theta(x)/f_{\theta_0}(x)) = \int_0^1 u(\theta(s); x) \cdot \theta'(s) \mathrm{d}s$$

where

$$u(\theta; x) = \mathbb{E}_\theta[t(x \cup X_{S \setminus W}) | X_W = x] - \mathbb{E}_\theta t(X)$$

MAXIMUM LIKELIHOOD INFERENCE

is the score function and $t(X)$ is the canonical sufficient statistic, cf. (8.41). Monte Carlo estimation of the missing data toroidal log likelihood function $\tilde{l}_{\mathrm{mis}}(\theta)$ is handled analogously.

REMARK 9.1 (*Estimation of interaction radius*) Many examples of Markov point process models belong to exponential families at least when parameters controlling the interaction radius are fixed. One example is provided by the Strauss process which is an exponential family model when the interaction radius R is fixed (Example 8.1, Section 8.2.2). Maximisation of the likelihood with respect to R for a Strauss process is not straightforward, since the likelihood is not differentiable and log concave as a function of R (the likelihood is a piecewise constant function of R).

It may then be helpful to use a profile likelihood approach: Suppose that the parameter vector is $\theta = (\psi, R)$ and that an exponential family is obtained when R is fixed. First compute *maximum profile likelihood estimates*

$$\hat{\psi}_k = \arg\max_{\psi} l(\psi, R_k)$$

for a range $R_1, \ldots, R_K$ of R values (for specificity we consider here the free boundary likelihood, but any kind of likelihood function may of course be considered). Second use e.g. path sampling to compute log likelihood ratios $l(\hat{\psi}_k, R_k) - l(\theta_0)$, using a common reference parameter θ_0 (often this corresponds to a Poisson process), and return $\hat{\theta}_i = (\hat{\psi}_i, R_i)$ as the (approximate) MLE if $l(\hat{\psi}_i, R_i) - l(\theta_0)$ is maximal among $l(\hat{\psi}_k, R_k) - l(\theta_0)$, $k = 1, \ldots, K$.

9.1.5 Examples

In the following examples when generating MCMC samples used for estimation of $l(\theta)$, $\tilde{l}(\theta)$, or $l_{\ominus \tilde{R}}(\theta)$, we use the locally stable version of Algorithm 7.4 (see Remark 7.6, Section 7.1.2) with $p = 1/2$, and with an appropriate modification for the multitype process in Example 9.3, cf. Section 7.1.3. By Proposition 7.14 (Section 7.3.2) the algorithm is V-uniformly ergodic. Further, when generating conditional samples used for estimation of $l_{\mathrm{mis}}(\theta)$ or $\tilde{l}_{\mathrm{mis}}(\theta)$, we use the empty point configuration on $S \setminus W$ or $\tilde{S} \setminus W$ as the initial state, and we use a slight modification of the locally stable version of Algorithm 7.4 with a uniform proposal density on $S \supset W$ or $\tilde{S} \supset W$ for birth proposals (so proposed points which fall within W are always rejected). Proposition 7.14 easily extends to this case, so we have again V-uniformly ergodicity. The programming work is moderate, since all the Markov point processes considered are pairwise interaction processes. We can thus use the same basic code for all the examples and just change the definitions of the first and second order

	$\log\beta$	$\log\gamma_1$	$\log\gamma_2$	$\log\gamma_3$	$\log\gamma_4$
$l(\theta)$	−1.78	−3.24	−1.03	−0.27	0.00
$l_{\text{mis}}(\theta)$	−0.84	−3.58	−1.38	−0.55	−0.12
$\tilde{l}_{\text{mis}}(\theta)$	−0.86	−3.63	−1.35	−0.55	−0.13
$\tilde{l}(\theta)$	−0.95	−3.53	−1.34	−0.55	−0.11
$l_{\ominus\tilde{R}}(\theta)$	−0.64	−3.26	−1.46	−0.64	−0.14

Table 9.1 *Maximum likelihood estimates for the multiscale model using different likelihood functions.*

interaction functions. Moreover, given the MCMC samples, we can use the same code for the Newton-Raphson and path sampling computations due to the common exponential family structure.

EXAMPLE 9.1 (*Maximum likelihood estimation for Norwegian spruces: I*) For the spruce data (Example 1.4), $W = [0, 56] \times [0, 38]$ and we consider two different models: First a multiscale process (6.17) where we ignore the stem diameters, and second in Example 9.2 below a biologically more realistic model where overlap of the influence zones of the trees is penalised (see Figure 1.4).

For the multiscale process, let in (6.17) $k = 4$, $R_i = 1.1 \times i$, $0 < \gamma_i \leq 1$, $i = 1, \ldots, 4$. The range of interaction is thus less or equal to 4.4 — a value suggested by the estimated summary statistics in Figure 4.3 (Section 4.3.4). Background knowledge on how S should be defined is not available, so it is not clear which likelihood function from Section 9.1.1 is most appropriate from a biological point of view.

Table 9.1 shows the MLE of $\theta = (\log\beta, \log\gamma_1, \log\gamma_2, \log\gamma_3, \log\gamma_4)$ obtained using either $l(\theta)$, $l_{\text{mis}}(\theta)$ with $S = [-10, 66] \times [-10, 48]$, $\tilde{l}_{\text{mis}}(\theta)$ with $\tilde{S} = [-10, 66] \times [-10, 48]$, $\tilde{l}(\theta)$ with $\tilde{S} = W$, or $l_{\ominus\tilde{R}}(\theta)$ with $\tilde{R} = 4.4$. For the Monte Carlo estimation, we use unconditional and conditional samples of length 8000 obtained by subsampling each 200th of 1,600,000 updates of Algorithm 7.4. Note that the MLE based on $l(\theta)$ differs markedly from the four other MLEs which are quite similar.

In the following we restrict attention to results obtained by the missing data toroidal log likelihood. The left plot in Figure 9.1 shows the estimated L-function together with envelopes calculated from 39 approximately independent simulations of the estimated multiscale model (the 39 simulations are obtained by subsampling from an MCMC sample of length 1,600,000). The L-function does not reveal a lack of fit for the multiscale process.

For the null hypothesis $\gamma_1 = \gamma_2 = \gamma_3 = \gamma_4 = \gamma$ corresponding to a Strauss process with $R = 4.4$, we obtain the MLE $(-1.61, -0.33)$ for

MAXIMUM LIKELIHOOD INFERENCE

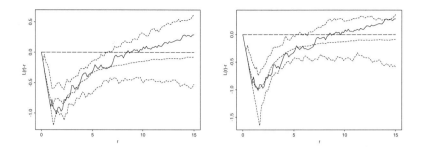

Figure 9.1 *Left: estimated $(L(r) - r)$-function for spruces (solid line) and average and envelopes calculated from 39 simulations of fitted multiscale process (dashed line). Long dashed line shows the theoretical value of $(L(r)-r)$. Right: as left but for fitted overlap model (Example 9.2).*

$(\log \beta, \log \gamma)$. The left plot in Figure 9.2 shows the interaction function corresponding to the estimated multiscale and Strauss process, respectively. The log likelihood ratio statistic $-2 \log Q$ for the null hypothesis

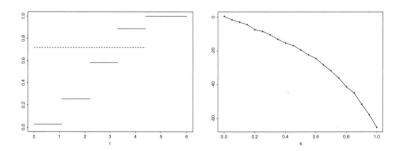

Figure 9.2 *Left: estimated multiscale interaction function $\phi_2(r)$ (see (6.16)) as a function of distance r (solid line). The dashed line shows the estimate under the null hypothesis (the Strauss model). Right: Monte Carlo estimates of $u(\theta(k/20)) \cdot \theta(k/20)$, $k = 0, \ldots, 20$, and curve corresponding to trapezoidal approximation.*

is computed using path sampling. Specifically, in (8.38) θ_0 and θ are equal to the MLE for the multiscale and Strauss model, respectively, 21 quadrature points are used, and the integrand at each quadrature point

$$u(\theta(k/20)) \cdot \theta(k/20)$$
$$= (\mathbb{E}_{\theta(k/20)}[t(x \cup \tilde{X}_{\tilde{S}\setminus W})|\tilde{X}_W = x] - \mathbb{E}_{\theta(k/20)}t(\tilde{X})) \cdot \theta(k/20),$$

$k = 0, \ldots, 20$, is estimated using unconditional and conditional samples of length 2000 obtained with a spacing of 400 (here $\tilde{X}$ is the toroidal version of the multiscale model on $\tilde{S}$, cf. (9.7)). The right plot in Figure 9.2 shows the Monte Carlo estimates of the integrand at the quadrature points together with the trapezoidal approximation. The estimated value of $-2\log Q$ is 47. The Wald statistic (8.34) for the null hypothesis $\theta L = 0$ where

$$L^\mathsf{T} = \begin{bmatrix} 0 & 1 & 0 & 0 & -1 \\ 0 & 0 & 1 & 0 & -1 \\ 0 & 0 & 0 & 1 & -1 \end{bmatrix},$$

is equal to 29. Both $-2\log Q$ and the Wald statistic are highly significant according to standard asymptotic results, cf. Section 8.5. Alternatively, as described in Section 8.5, we may consider a parametric bootstrap. We here restrict attention to the Wald statistic for which path sampling is not required. Bootstrap values of the Wald statistic calculated from 100 approximately independent simulations under the estimated Strauss process (further details are given in the paragraph below) fall between 0.11 and 10.5, so also the bootstrap provides strong evidence against the Strauss model.

Figure 9.3 shows trace plots and estimated auto-correlations for time series of the number of points $n(Y_m)$ and $R = 4.4$ close pairs $s_{4.4}(Y_m)$ for a Markov chain Y_m with invariant distribution given by the toroidal version of the estimated Strauss process on $\tilde{S}$. The time series are obtained by subsampling the updates of Algorithm 7.4, using a spacing equal to 200 and using different initial states. In the left plots, the initial state is given by the empty point configuration and in the middle plots by a realisation of the "dominating" Poisson$(\exp(-1.61), \tilde{S})$-process (see Sections 8.1.2 and 11.2.6). The effect of the initial state disappears quickly, and it appears that the auto-correlation dies out at lag 6 or 7 corresponding to around 1300 updates of Algorithm 7.4. The 100 approximately independent simulations used for the bootstrap are obtained by subsampling, using a spacing equal to 16,000, where for each subsampled realisation Y_m, we use $Y_m \cap W$ for the bootstrap.

EXAMPLE 9.2 (*Maximum likelihood estimation for Norwegian spruces: II*) The multiscale model considered in Example 9.1 is not so satisfactory from a biological point of view, since it does not take into account the influence zones. In the following we specify a more realistic marked pairwise interaction point process model (see Section 6.6).

A spruce location ξ and the associated radius m_ξ is viewed as a marked point $(\xi, m_\xi) \in T \times M$, where $T \subseteq \mathbb{R}^2$ is bounded and specified in different ways below, and $M \subseteq (0, \infty)$ is the mark space. For convenience we restrict the mark space to a bounded interval $M = [m_{\text{lo}}, m_{\text{up}}]$, where we have estimated $m_{\text{lo}} = 0.8$ and $m_{\text{up}} = 1.85$ by the minimal and maximal

MAXIMUM LIKELIHOOD INFERENCE

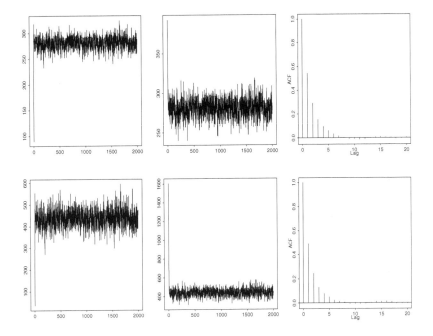

Figure 9.3 *Left and middle plots: trace plots for the number of points (upper plots) and the number of $R = 4.4$ close pairs (lower plots) for the toroidal version of the fitted Strauss process. The initial state is empty in the left plots and given by a realisation of a Poisson process in the middle plots. Right plots: corresponding auto-correlation plots.*

observed radii for the influence zones.[†] For marked points (ξ, m_ξ) and (η, m_η) in $T \times [m_{\text{lo}}, m_{\text{up}}]$, we define the first order interaction function by $\phi_{\theta,1}((\xi, m_\xi)) = \beta_0 > 0$ if $m_\xi - m_{\text{lo}} \leq (m_{\text{up}} - m_{\text{lo}})/6$ and

$$\phi_{\theta,1}((\xi, m_\xi)) = \beta_k > 0 \quad \text{if } k/6 < \frac{m_\xi - m_{\text{lo}}}{m_{\text{up}} - m_{\text{lo}}} \leq (k+1)/6, \quad k = 1, \ldots, 5,$$

and the second order log interaction function by

$$\log \phi_{\theta,2}(\{(\xi, m_\xi), (\eta, m_\eta)\}) = |b(\xi, m_\xi) \cap b(\eta, m_\eta)| \log \gamma \qquad (9.11)$$

where $0 < \gamma \leq 1$. The parameter vector is $\theta = (\log \beta_0, \ldots, \log \beta_5, \log \gamma)$. The function $\phi_{\theta,1}$ allows modelling of different intensities of points with different values of the marks (cf. Figure 4.13) and $\phi_{\theta,2}$ models repulsion

---

[†] The estimates for m_{lo} and m_{up} are in fact MLEs. We have chosen these estimates since we have no biological knowledge about how to choose a smaller value of m_{lo} and a larger value of m_{up}.

	$\log \beta_0$	$\log \beta_1$	$\log \beta_2$	$\log \beta_3$	$\log \beta_4$	$\log \beta_5$	$\log \gamma$
$l(\theta)$	-1.24	-0.36	0.51	-0.43	-0.84	-0.55	-1.08
$l_{\mathrm{mis}}(\theta)$	-1.17	-0.29	0.63	-0.29	-0.63	-0.30	-1.18
$\tilde{l}_{\mathrm{mis}}(\theta)$	-1.15	-0.28	0.62	-0.31	-0.65	-0.36	-1.16
$\tilde{l}(\theta)$	-1.20	-0.31	0.59	-0.334	-0.71	-0.40	-1.15
$l_{\ominus 2m_{\mathrm{up}}}(\theta)$	-1.02	-0.41	0.60	-0.67	-0.58	-0.22	-1.13

Table 9.2 *Maximum likelihood estimates for the overlap model using different likelihood functions.*

due to overlapping influence zones. We refer to this marked pairwise interaction point process on $T \times [m_{\mathrm{lo}}, m_{\mathrm{up}}]$ as the overlap model.

For convenience we may view the overlap model as an ordinary (unmarked) pairwise interaction point process with density

$$f_\theta(\{(\xi_1, m_1), \ldots, (\xi_n, m_n)\}) \propto$$

$$\prod_{i=1}^{n} \phi_{\theta,1}((\xi_i, m_i)) \prod_{1 \leq i < j \leq n} \phi_{\theta,2}(\{(\xi_i, m_i), (\xi_j, m_j)\})$$

with respect to Poisson($T \times M$, 1). Recall that this is Markov with respect to the overlapping disc relation (Example 6.7, Section 6.6.3). For simulation we can immediately use the locally stable version of Algorithm 7.4 with q_b given by a uniform density on $T \times [m_{\mathrm{lo}}, m_{\mathrm{up}}]$.

Proceeding as for the multiscale process, we obtain different types of MLEs corresponding to different choices of T: for the free boundary likelihood, and the toroidal likelihood, $T = W = [0, 56] \times [0, 38]$ (for a toroidal version of $\phi_{2,\theta}$ we use that (9.11) only depends on (ξ, η) through $\|\xi - \eta\|$); for the conditional likelihood, $T \supseteq W$ and we base inference on the conditional density for the overlap process on $W_{\ominus 2m_{\mathrm{up}}} \times M$ given the process on $W \setminus W_{\ominus 2m_{\mathrm{up}}} \times M$ (trees with centres outside W do not interact with trees with centres in $W_{\ominus 2m_{\mathrm{up}}}$); for the missing data likelihood and the missing data toroidal likelihood, $T = [-10, 66] \times [-10, 48]$. The various MLEs given in Table 9.2 are rather similar.

We finally concentrate on the estimated overlap model using the missing data toroidal approach. A model check using the L-function does not reveal problems with the fitted overlap model, see the right plot in Figure 9.1. If we want to compare the fit of the multiscale model and the overlap model, a likelihood ratio cannot directly be applied, since the models are defined on different spaces. However, we can turn the multiscale process into a marked point process by introducing "imaginary" independent uniform marks in $[m_{\mathrm{lo}}, m_{\mathrm{up}}]$. We then compute likelihood ratios between a common Poisson($W \times [m_{\mathrm{lo}}, m_{\mathrm{up}}], \rho$) reference process

MAXIMUM LIKELIHOOD INFERENCE   169

and the marked multiscale model and overlap model, respectively. Letting $\rho = 134/|W \times [m_\text{up}, m_\text{lo}]|$, we obtain the values -30 and -60 for the log likelihood ratios involving the fitted multiscale model and the fitted overlap model, respectively. The log likelihood ratio between the fitted overlap model and the fitted (marked) multiscale model thus becomes 30. So the fit appears to be substantially better with the overlap model even when taking into account the larger number of parameters.

EXAMPLE 9.3 (*Maximum likelihood estimation for mucous cells*) For the mucous membrane cells (Example 1.3) where $W = [0,1] \times [0,0.7]$, we consider an inhomogeneous multitype Strauss process on $W \times \{1,2\}$ (see Section 6.6.1), where the mark value $m = 1$ corresponds to ECL cells and $m = 2$ to "other cells". The first order log interaction function is given by

$$\log \phi_{\theta,1}((\xi, m)) = \sum_{i=0}^{4} \psi_i^{(m)} \xi_2^i$$

(with $\xi = (\xi_1, \xi_2) \in W$), and the second order interaction function is

$$\phi_{\theta,2}(\{(\xi, m_1), (\eta, m_2)\}) = \gamma^{\mathbf{1}[\|\xi - \eta\| \leq R]}.$$

The parameter vector is $\theta = (\psi, R)$ with $\psi = (\psi_0^{(1)}, \ldots, \psi_4^{(1)}, \psi_0^{(2)}, \ldots, \psi_4^{(2)}, \log \gamma)$. The model is an extension of the parametric inhomogeneous Poisson model from Example 4.4 (Section 4.3.4), since we use different fourth order polynomials to model the inhomogeneity for the two different types of points and include the possibility of repulsive interaction between the points (see Figures 4.5 and 4.6).

In this case the missing data likelihood approach is not so appealing, since it is not clear how to define $\phi_{\theta,1}((\xi, m))$ when ξ does not belong to W. Further, the toroidal likelihood does not seem natural due to the inhomogeneity. So we restrict attention to the conditional approach and, following Remark 9.1, compute profile conditional MLEs $\hat{\psi}_k$ for varying values $R_k = 0.004 + k0.001$, $k = 1, \ldots, 6$, of R. To enable a meaningful comparison of the various profile conditional likelihoods, we use a common value 0.01 for $\tilde{R}$ in the multitype version of (9.10).

MCMC samples of length 8000 are obtained by subsampling each 1000th state of Markov chains of length 8,000,000 generated by the locally stable version of Algorithm 7.4 extended to the marked case (see Section 7.1.3). Specifically, a birth proposal of a marked point (ξ, m) is generated from a uniform density on $W_{\ominus 0.01} \times \{1,2\}$, the Hastings ratio for the addition of a marked point (ξ, m) to a marked point pattern y is

$$r_b(y, (\xi, m)) = \phi_{\theta,1}((\xi, m)) \gamma^{\sum_{(\eta, l) \in y} \mathbf{1}[\|\xi - \eta\| < R]} |W_{\ominus 0.01}| / (n(y) + 1),$$

and the Hastings ratio for the deletion of a uniformly chosen marked point (ξ, m) from y is $r_d(y, (\xi, m)) = 1/r_b(y \setminus (\xi, m), (\xi, m))$. Extending the proof of Proposition 7.14 (Section 7.3.2) to the multitype case, the algorithm can be shown to be geometrically ergodic.

The left plot in Figure 9.4 shows log conditional likelihood ratios $l(\hat{\psi}_k, R_k) - l(\psi_0, 0.01)$, $k = 1, \ldots, 6$, where we use as a reference parameter ψ_0 the MLE for the multitype Poisson process obtained by fixing $\gamma = 1$. The log likelihood ratios are obtained using path sampling with 30 quadrature points and samples of length 2000 obtained with a spacing equal to 1000. The maximum log likelihood ratio is obtained for $R = 0.008$ (i.e. $k = 4$) for which the corresponding MLEs are given by

$$(\hat{\psi}_0^{(1)}, \ldots, \hat{\psi}_4^{(1)}) = (6.51, -2.86, -40.05, 123.83, -105.75),$$
$$(\hat{\psi}_0^{(2)}, \ldots, \hat{\psi}_4^{(2)}) = (7.52, 8.40, -70.82, 165.80, -121.53),$$

and $\hat{\gamma} = 0.07$ (which yields rather strong repulsive interaction). The right plot in Figure 9.4 shows the estimated polynomials $\sum_{i=1}^{4} \psi_i^{(1)} \xi_2^i$ and $\sum_{i=1}^{4} \psi_i^{(2)} \xi_2^i$ (note that we here exclude the levels $\psi_0^{(1)}$ and $\psi_0^{(2)}$).

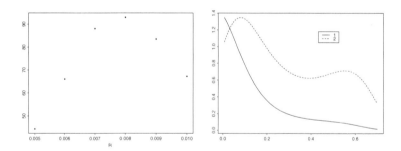

Figure 9.4 Left: profile conditional log likelihood ratios for varying values of R. Right: estimated polynomials $\sum_{i=1}^{4} \psi_i^{(m)} \xi_2^i$ for ECL cells (type $m = 1$; full line) and other cells (type $m = 2$; dashed line).

The feature of primary biological interest is the inhomogeneity. We therefore proceed to test the null hypothesis $(\psi_1^{(1)}, \ldots, \psi_4^{(1)}) = (\psi_1^{(2)}, \ldots, \psi_4^{(2)}) = (\psi_1, \ldots, \psi_4)$, i.e. that the first order interaction functions for the two types of points are proportional. We continue to work with $R = 0.008$ and $\tilde{R} = 0.01$, and obtain the MLEs $\hat{\psi}_0^{(1)} = 5.41$, $\hat{\psi}_0^{(2)} = 7.72$, $(\hat{\psi}_1, \ldots, \hat{\psi}_4) = (6.37, -63.39, 153.30, -114.02)$, and $\hat{\gamma} = 0.07$. Furthermore, $-2 \log Q$ computed using path sampling and the Wald statistic

(8.34) for the null hypothesis $\psi L = 0$ where

$$L^\mathsf{T} = \begin{bmatrix} 0 & 1 & 0 & 0 & 0 & 0 & -1 & 0 & 0 & 0 & 0 \\ 0 & 0 & 1 & 0 & 0 & 0 & 0 & -1 & 0 & 0 & 0 \\ 0 & 0 & 0 & 1 & 0 & 0 & 0 & 0 & -1 & 0 & 0 \\ 0 & 0 & 0 & 0 & 1 & 0 & 0 & 0 & 0 & -1 & 0 \end{bmatrix},$$

are 36 and 31, respectively. These values provides strong evidence against the null hypothesis according to standard asymptotic results. Using a parametric bootstrap with 100 approximately independent simulations generated under the null hypothesis, we obtain values of the Wald statistic between 0.2 and 15. The observed Wald statistic is thus also highly significant according to the bootstrap.

9.2 Pseudo likelihood

The derivation of the pseudo likelihood function is discussed in Section 9.2.1, while computational aspects are considered in Section 9.2.2. Asymptotic properties of pseudo likelihood estimates are reviewed in Section 9.2.3, and the relation to Takacs-Fiksel estimation is considered in Section 9.2.4.

9.2.1 Pseudo likelihood functions

The idea of maximum pseudo likelihood is due to Besag (1975) who first introduced the concept for Markov random fields (MRFs) in order to avoid the unknown normalising constant. The first application for point processes is in Besag (1977a), where for a Strauss process, the distribution of the counts of points within small cells is approximated by an auto-Poisson MRF (Besag 1974), and the pseudo likelihood function for this MRF is used as an approximate pseudo likelihood function for the Strauss process. Assuming that the points in each cell are uniformly dispersed given the auto-Poisson counts, Besag, Milne & Zachary (1982) show convergence to the Strauss process as the cell sizes decrease to zero. Based on these ideas, a general expression of the pseudo likelihood for point processes is stated in Ripley (1988). Below the pseudo likelihood for point processes is derived using a direct argument from Jensen & Møller (1991). Baddeley & Turner (2000) provide an extensive treatment of both practical and theoretical aspects of pseudo likelihood estimation.

Consider for the moment the case where $S = W$ and X has a Ruelle stable density f_θ given by (9.1). For any $A \subseteq S$, define the *pseudo likelihood* on A by

$$PL_A(\theta; x) = \exp(-|A|) \lim_{i \to \infty} \prod_{j=1}^{m_i} f_{\theta, A}(x_{A_{ij}} | x_{S \setminus A_{ij}}). \tag{9.12}$$

where $\{A_{ij} : j = 1, \ldots, m_i\}$, $i = 1, 2, \ldots$, are nested subdivisions of A (i.e. $A = \cup_{j=1}^{m_i} A_{ij}$, where $A_{ij} \cap A_{ij'} = \emptyset$ for $j \neq j'$ and $A_{ij} \subseteq A_{i-1,j'}$ for some j') such that

$$m_i \to \infty \quad \text{and} \quad m_i [\max_{1 \leq j \leq m_i} |A_{ij}|]^2 \to 0 \quad \text{as } i \to \infty.$$

For example, if S is rectangular, A_{ij} may be a rectangular cell of size $|S|/m_i$. Furthermore, $f_{\theta,A}(x_{A_{ij}}|x_{S \setminus A_{ij}})$ denotes the conditional density for $X_{A_{ij}}$ given that $X_{S \setminus A_{ij}} = x_{S \setminus A_{ij}}$, cf. Proposition 6.1 in Section 6.1, setting $f_{\theta,A}(x_{A_{ij}}|x_{S \setminus A_{ij}}) = 0$ if $f_\theta(x_{S \setminus A_{ij}}) = 0$. The constant $\exp(-|A|)$ in (9.12) is included for convenience, and the product is analogous to the pseudo likelihood function for an MRF. By Theorem 2.2 in Jensen & Møller (1991), the limit in (9.12) exists and

$$PL_A(\theta; x) = \exp\left(-\int_A \lambda_\theta^*(x, \xi) d\xi\right) \prod_{\xi \in x_A} \lambda_\theta^*(x \setminus \xi, \xi) \qquad (9.13)$$

where λ_θ^* is the Papangelou conditional intensity associated to f_θ. Note that $PL_A(\theta; x)$ does not depend on the normalising constant c_θ, since λ_θ^* does not depend on c_θ.

REMARK 9.2 When $S = W$ we usually take $A = W$ and refer then to the *free boundary pseudo likelihood* (in analogy with the likelihood (9.2)). In the case where $W \subset S$ (where S is possibly unbounded) and all the ϕ_θ have range of interaction less than some $\tilde{R} < \infty$, we may let $A = W_{\ominus \tilde{R}}$ as for the conditional likelihood (9.10); we refer then to the *conditional pseudo likelihood*. We can then replace S in (9.12) by W and still obtain (9.13), cf. Remark 6.6 in Section 6.3.3. For $A = W$ rectangular, a *toroidal pseudo likelihood* is obtained using the toroidal version of f_θ defined on W, cf. (9.7) with $\tilde{S} = W$. These and other approaches of handling edge effects are discussed in more detail in Baddeley & Turner (2000).

The *maximum pseudo likelihood estimate (MPLE)* is found by maximising (9.13). Intuitively, as the pseudo likelihood only depends on the local dependence structure, global information may better be taken into account when using the likelihood function. Optimality of the MLE and inefficiency (i.e. larger asymptotic variance) of the MPLE is established in Mase (1992) for a certain class of pairwise interaction processes. Note that (9.13) agrees with the likelihood function for a Poisson process when $\lambda_\theta^*(x, \xi)$ depends only on ξ, so for point processes with weak interaction, the MPLE and the MLE may be expected to be close.

If $\Theta \subseteq \mathbb{R}^p$ is open, the MPLE is a solution to the *pseudo likelihood equation* $(\mathrm{d}/\mathrm{d}\theta)\log PL_A(\theta;x) = 0$ which is equivalent to

$$\int_A \frac{\mathrm{d}}{\mathrm{d}\theta}\log\lambda_\theta^*(x,\xi)\mathrm{d}\xi = \sum_{\xi\in x_A}\frac{\mathrm{d}}{\mathrm{d}\theta}\log\lambda_\theta^*(x\setminus\xi,\xi) \qquad (9.14)$$

provided we can interchange the order of integration and differentiation (see Section 8.2.1). Assume that the density f_θ belongs to an exponential family, i.e. f_θ is of the form (8.7). Then

$$\lambda_\theta^*(x,\xi) = b(x,\xi)\exp(\theta\cdot t(x,\xi)) \qquad (9.15)$$

where $b(x,\xi) = b(x\cup\xi)/b(x)$ and $t(x,\xi) = t(x\cup\xi) - t(x)$. So the pseudo likelihood equation (9.14) becomes

$$\int_A t(x,\xi)\exp(\theta\cdot t(x,\xi))\mathrm{d}\xi = \sum_{\xi\in x_A} t(x\setminus\xi,\xi). \qquad (9.16)$$

Furthermore, Proposition 2.3 in Jensen & Møller (1991) states that $PL_A(\theta;x)$ is log concave, and gives a condition for strict concavity.

REMARK 9.3 (*Estimation of interaction radius*) The log linear form (9.15) of the conditional intensity for exponential family models is advantageous for computational reasons, see Section 9.2.2. Suppose that the parameter vector is $\theta = (\psi, R)$ and that an exponential family is obtained when R is kept fixed. Then for similar reasons as in Remark 9.1, a profile pseudo likelihood approach may be useful (see Baddeley & Turner 2000 and Example 9.5) where maximum profile pseudo likelihood estimates

$$\hat{\psi}_k = \arg\max_\psi PL_A(\psi, R_k; x)$$

are obtained for a range $R_1, \ldots, R_K$ of R values.

REMARK 9.4 Theorem 2.2 in Jensen & Møller (1991) concerns a general setting for point processes defined on a general space S. In particular, the pseudo likelihood for a marked Markov point process (see Section 6.6) is constructed analogously to the unmarked case. Let $A \subseteq T \times M$ and let p be a discrete or continuous density on M, using the same notation as in Section 6.6.1. In the discrete case, (9.13) is modified to

$$PL_A(\theta;x) = \exp\left(-\int\sum_{m\in M}\lambda_\theta^*(x,(\xi,m))\mathbf{1}[(\xi,m)\in A]p(m)\mathrm{d}\xi\right)$$
$$\prod_{(\xi,m)\in x_A}\lambda_\theta^*(x\setminus(\xi,m),(\xi,m)). \qquad (9.17)$$

In the continuous case, the sum is replaced by an integral over M.

9.2.2 Practical implementation of pseudo likelihood estimation

Maximum pseudo likelihood estimation based on (9.13) is computationally equivalent to maximum likelihood estimation in an inhomogeneous Poisson process. In practice the integrals in (9.13) and (9.14) are approximated by numerical methods. As noticed in Berman & Turner (1992) and Baddeley & Turner (2000), standard software such as Splus and R for fitting generalised linear models can be used to provide an approximate MPLE as follows.

Suppose we partition A into a finite number of cells C_i, and let $c_i \in C_i$ denote a given "centre point". Let u_j, $j = 1, \ldots, m$, denote a list of these centre points and the points in x_A. Then the integral in (9.13) is approximated by

$$\int_A \lambda_\theta^*(x, \xi) \mathrm{d}\xi \approx \sum_{j=1}^m \lambda_\theta^*(x \setminus u_j, u_j) w_j, \qquad (9.18)$$

where $w_j = |C_i|/(1 + n(x_{C_i}))$ if $u_j \in C_i$, and so

$$\log PL_A(\theta; x) \approx \sum_{j=1}^m (y_j \log \lambda_j^* - \lambda_j^*) w_j, \qquad (9.19)$$

where $y_j = 1[u_j \in x]/w_j$ and $\lambda_j^* = \lambda_\theta^*(x \setminus u_j, u_j)$. Note that the approximation (9.18) involves a "discontinuity error", since for $u_j \in x_A$, $\lambda_\theta^*(x \setminus u_j, u_j)$ is in general not equal to the limit of $\lambda_\theta^*(x, \xi)$ as $\xi \to u_j$, cf. the discussion in Baddeley & Turner (2000). The motivation for including x_A in $\{u_1, \ldots, u_m\}$ and thus making the "discontinuity error" is that the right side of (9.19) is formally equivalent to a weighted log likelihood of independent Poisson variables y_j with means λ_j^* and weights w_j. If λ_j^* is of the log linear form (9.15) with $b(x \setminus u_j, u_j) > 0$, $j = 1, \ldots, m$, i.e.

$$\log \lambda_j^* = \log b(x \setminus u_j, u_j) + \theta \cdot t(x \setminus u_j, u_j),$$

then (9.19) can easily be maximised using standard software for generalised linear models, taking $\log b(x \setminus u_j, u_j)$ as an offset term (using the terminology of Splus).

A user-friendly implementation of pseudo likelihood estimation is given in the R package spatstat by A.J. Baddeley and R. Turner (see Section A.3). Among other features this package provides a convenient notation for separate modelling of large scale inhomogeneity (possibly depending on a number of covariates) and translation invariant interaction between points, see Section 6.5.1.

EXAMPLE 9.4 (*Pseudo likelihood for Norwegian spruces*) When considering pseudo likelihood estimation for the Norwegian spruces we restrict attention to the multiscale model from Example 9.1. We first use the

PSEUDO LIKELIHOOD

	$\log \beta$	$\log \gamma_1$	$\log \gamma_2$	$\log \gamma_3$	$\log \gamma_4$
Free boundary	-1.34	-3.24	-1.24	-0.45	-0.03
Free boundary (default)	-1.51	-2.88	-1.16	-0.40	-0.03
Toroidal	-0.84	-3.35	-1.38	-0.62	-0.15
Conditional	-0.91	-2.68	-1.43	-0.62	-0.12

Table 9.3 *MPLEs for varying choice of A (see Remark 9.2) and numerical approximation (see the text).*

approximation (9.18) with $A = W = [0, 56] \times [0, 38]$ partitioned into 56×38 quadratic cells C_i, $i = (k, l) \in \{0, \ldots, 55\} \times \{0, \ldots, 37\}$, each of unit area and with centre points $c_{(k,l)} = (k + 0.5, l + 0.5)$. We have a log linear model with $\theta = (\log \beta, \log \gamma_1, \ldots, \log \gamma_4) \in \mathbb{R} \times (-\infty, 0]^4$, $b(x \setminus u_j, u_j) = 1$, and $t(x \setminus u_j, u_j) = (1, s[j, 1], \ldots, s[j, 4])$, where $s[j, i]$ denotes the number of points $\xi \in x \setminus u_j$ with $R_{i-1} < \|\xi - u_j\| \leq R_i$, $i = 1, \ldots, 4$ (where $R_0 = 0$). An approximate MPLE is obtained using the Splus routine glm() with the call

```
glm(y~s[,1]+...+s[,4],family=poisson(link=log),weights=w),
```

and with y, w, and s constructed as above. For comparison we also use spatstat with the default quadrature scheme (see the second row in Table 9.3). Comparing the two first rows in Table 9.3, the difference between the two sets of approximate MPLEs is due to different numerical errors in the approximation (9.18). In order to correct for edge effects, we also use the toroidal pseudo likelihood with $A = W$, and the conditional pseudo likelihood with $A = W_{\ominus 4.4}$ (see Remark 9.2). Using spatstat we obtain the estimates in the third and fourth rows in Table 9.3. There is not a striking pattern of differences between the set of MLEs in Table 9.1 and the MPLEs in Table 9.3. As for the MLEs, the free boundary MPLEs differ from the edge corrected estimates.

EXAMPLE 9.5 (*Pseudo likelihood for mucous membrane cells*) For the mucous cell data, we consider the inhomogeneous multitype Strauss process from Example 9.3. Following Remark 9.3, the left plot in Figure 9.5 shows maximum profile pseudo likelihoods for varying values of R (where the computations are carried out using spatstat). All the MPLEs are obtained using the conditional approach with $A = W_{\ominus 0.01}$. The maximum profile pseudo likelihood is obtained for $R = 0.007$ for which the corresponding profile pseudo likelihood estimates of $(\psi_0^{(1)}, \ldots, \psi_4^{(1)})$, $(\psi_0^{(2)}, \ldots, \psi_4^{(2)})$, and γ are given by $(6.26, 12.47, -132.51, 321.67, -244.63)$, $(7.05, 13.13, -92.57, 205.64, -146.70)$, and 0.024. The right plot in Figure 9.5 is as the right plot in Figure 9.4 but using the (approximate)

MPLE. The MLE and the MPLE of the inhomogencity parameters $\psi_i^{(m)}$, $i = 0, \ldots, 4$, $m = 1, 2$, differ markedly, but the resulting first order interaction functions shown in Figure 9.4 and Fig 9.5 are quite similar.

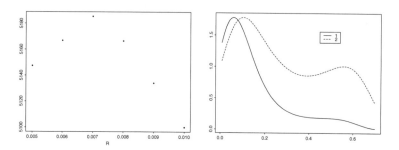

Figure 9.5 *Left: profile log pseudo likelihoods for varying values of R. Right: polynomials $\sum_{i=1}^{4} \psi_i^{(m)} \xi_2^i$ for types $m = 1$ (full line) and $m = 2$ (dashed line) with coefficients given by the MPLE.*

9.2.3 Consistency and asymptotic normality of pseudo likelihood estimates

Asymptotic results for MPLEs are more exhaustive than for MLEs (see Section 9.1.3). They have mainly been established within the framework of exponential family Markov point processes defined by a translation invariant relation $\sim$ on $\mathbb{R}^d$ and with translation invariant interaction functions of finite range of interaction. That is, for some $R < \infty$, $\|\xi - \eta\| \leq R$ whenever $\xi \sim \eta$, and the density f_θ is of the exponential form (8.7) with

$$b(x) = \prod_{y \subseteq x} \phi(y) \quad \text{and} \quad t(x) = \sum_{y \subseteq x} \psi(y),$$

where $\phi(y)$ and $\psi(y)$ are translation invariant functions defined for all finite $y \subset \mathbb{R}^d$, where $\phi(y) \in [0, \infty)$, $\psi(y) \in \mathbb{R}^p$, and $\phi(y) = 1$ and $\psi(y) = 0$ if any two points in y are not neighbours. The asymptotic results are obtained for pseudo likelihoods $PL_{A_n}(\theta; x)$ where $A_n \subset \mathbb{R}^d$, $n = 1, 2, \ldots$, is an increasing sequence of bounded subsets tending to $\mathbb{R}^d$. Jensen & Møller (1991) verify consistency (i.e. convergence in probability of the MPLE towards the true parameter value) under a certain condition of identifiability and assuming that either

$$\phi(y) \leq 1 \quad \text{and} \quad \theta \cdot \psi(y) \leq 0$$

PSEUDO LIKELIHOOD   177

whenever $n(y) \geq 2$ (i.e. the point process is purely repulsive), or that X has a hard core (see e.g. Example 6.1 in Section 6.2.2) and

$$\phi(y) \leq K \quad \text{and} \quad \|\psi(y)\| \leq K$$

for some $K > 0$ and whenever $n(y) \geq 2$. Jensen & Møller (1991) further discuss how their results can be applied to marked Markov point processes.

Specialising to homogeneous pairwise interaction Markov point processes of the exponential family type above with $\phi(\cdot) = 1$, $\psi(\xi) = (1, 0)$, $\psi(\{\xi, \eta\}) = (0, \psi_0(\xi - \eta))$, and $\theta \in (0, \infty)^2$, Jensen & Künsch (1994) establish asymptotic normality of the MPLE essentially when either $-\infty < \psi_0(\cdot) \leq 0$ (the repulsive case) or when X has a hard core and $\psi_0(\cdot) = \tilde{\psi}_0(\|\cdot\|) < K$ for some real function $\tilde{\psi}_0$ and a constant K. A practical problem is that the asymptotic covariance matrix cannot be explicitly evaluated. Jensen & Künsch (1994) obtain an estimate based directly on the observed data, while Baddeley & Turner (2000) estimate the asymptotic covariance matrix using a parametric bootstrap.

Strong consistency of pseudo likelihood estimates for pairwise interaction Gibbs point processes is verified in Mase (1995) under certain conditions ensuring the existence of an infinite stationary Gibbs process (see Section 6.4.1). Mase (1999) extends results on consistency and asymptotic normality in Mase (1995) and Jensen & Künsch (1994) to the case of marked Gibbs point processes.

9.2.4 Relation to Takacs-Fiksel estimation

Considering a stationary Gibbs point process with translation invariant interaction function of finite interaction range R (Section 6.4.1), we obtain from (6.30),

$$\mathbb{E}_\theta \Big[\sum_{\xi \in X \cap W} h(\xi, X \setminus \xi) \Big] / |W| = \mathbb{E}_\theta \big[h(\eta, X) \lambda_\theta^*(X, \eta) \big] \quad (9.20)$$

where η is an arbitrary fixed point in W and $h : \mathbb{R}^d \times N_{\text{lf}} \to \mathbb{R}$ (assuming the expectations exist). A system of Takacs-Fiksel estimating equations (Takacs 1986, Fiksel 1984a) is obtained by replacing the right and left hand side in (9.20) by unbiased estimates $R(h)$ and $L(h)$ for a number $k \geq p$ of functions $h = h_i$, $i = 1, \ldots, k$. Thereby k unbiased estimating equations are obtained. An estimate of θ is obtained by minimising $\sum_{i=1}^k (L(h_i) - R(h_i))^2$. If we e.g. take $h_i(\xi, x) = n(x_{b(\xi, r_i)})$ for some $r_i > 0$, then the left hand side is simply $\rho^2 K(r_i)$ for which an unbiased estimator can be obtained as described in Section 4.3.2. Because of the translation invariance, the right hand side can be estimated unbiasedly using minus sampling in a similar manner as for the F or G-

function in Section 4.3.6. That is, using the notation from Section 4.3.6, $\mathbb{E}\left[n(X_{b(\eta,r_i)})\lambda_\theta^*(X,\eta)\right]$ is estimated by

$$\sum_{\eta \in I_{\max(R,r_i)}} n(x_{b(\eta,r_i)})\lambda_\theta^*(x,\eta)/\#I_{\max(R,r_i)}$$

More generally, from the integral characterisation (6.30) we immediately obtain for any Gibbs point process that

$$\sum_{\xi \in X} h(\xi, X \setminus \xi) = \int_S h(\xi, X)\lambda_\theta^*(X,\xi)\mathrm{d}\xi \quad (9.21)$$

is an unbiased estimating equation for any real function $h : \mathbb{R}^d \times N_{\mathrm{lf}} \to \mathbb{R}$ (assuming again that the expectations exist). The ith component $i = 1,\ldots,p$ of the pseudo likelihood estimating equation (9.14) emerges as the special case with h in (9.21) replaced by

$$h_i(\xi, X) = \mathbf{1}[\xi \in A](\mathrm{d}/\mathrm{d}\theta_i)\log\lambda_\theta^*(X_W, \xi),$$

provided $\lambda_\theta^*(X,\xi)$ only depends on X through X_W for $\xi \in A$.

For a marked point process, Takacs-Fiksel estimating equations can be obtained similarly. In the case of a marked Gibbs point process X on $\mathbb{R}^d \times M$ with continuous marks, for example, (6.30) takes the form

$$\mathbb{E}\sum_{(\xi,m) \in X} h((\xi,m), X \setminus (\xi,m))$$

$$= \int_{\mathbb{R}^d}\int_M \mathbb{E}\left[h((\xi,m), X)\lambda_\theta^*(X, (\xi,m))\right]p(m)\mathrm{d}\xi\mathrm{d}m$$

where p is a density on M. A similar equation is valid in the case of discrete marks (replacing $\int_M$ by a sum).

9.2.5 Time-space processes

Suppose that $d \geq 2$ and $S = W = B \times [0,T]$ where $B \subset \mathbb{R}^{d-1}$ is bounded and $0 < T < \infty$, and view X as a time-space process X_t, $0 \leq t \leq T$, where $X_t \subset B$ is the point process which consists of all points appearing before or at time t,

$$X_t = \{(\psi_1,\ldots,\psi_{d-1}) : (\psi_1,\ldots,\psi_d) \in X, \ \psi_d \leq t\}.$$

For $\psi = (\psi_1,\ldots,\psi_d) \in S$ with $t = \psi_d > 0$ and for finite $x \subset S \setminus \psi$ such that all $\eta = (\eta_1,\ldots,\eta_d) \in x$ satisfies $\eta_d < t$, let $\lambda(x,\psi)$ denote the *conditional intensity* of the event that $X_t = X_{t-} \cup (\psi_1,\ldots,\psi_{d-1})$ given the "history" $\mathcal{H}_{t-}$ up to times but not including t, i.e. given that the path X_s, $0 \leq s < t$, agrees with x. Here X_{t-} is the limit of X_s as $s < t$ increases to t, and the intuitive content of $\lambda(x,\psi)$ is expressed by

$$\lambda(x,\psi)|B| \approx \mathbb{E}(N(B)|\mathcal{H}_{t-})$$

BAYESIAN INFERENCE   179

where $B \subset S$ is an infinitesimally small ball containing ψ; for a formal definition of $\lambda(x,\psi)$, see Daley & Vere-Jones (2003). Notice that this conditional intensity should be carefully distinguished from the Papangelou conditional intensity $\lambda^*(x,\psi)$. Letting $A = B \times (0, T]$ and replacing λ^* with λ in (9.13), we obtain the likelihood function for the conditional process X_t, $0 < t \leq T$, given X_0. This indicates that maximum likelihood inference for time-space processes is easier than for spatial point processes if the time-space process is specified in terms of the conditional intensity λ.

9.3 Bayesian inference

The literature on Bayesian analysis for Markov point processes is rather scarce. Below we comment on a few contributions.

As for maximum likelihood estimation, the main obstacle is again the unknown normalising constant in the likelihood. Suppose that $S = W$, let $f_\theta = h_\theta/c_\theta$ denote the density of X, and assume that a prior density π is imposed on the unknown parameter θ. The posterior density of θ is given by

$$f(\theta|x) \propto h_\theta(x)\pi(\theta)/c_\theta$$

which is in general analytically intractable at least due to the presence of the unknown c_θ. The methods from Section 8.2 therefore become useful e.g. for computing ratios of normalising constants c_θ/c'_θ in posterior odds ratios $f(\theta'|x)/f(\theta|x)$ (such ratios e.g. appear when the posterior distribution is sampled using a Metropolis-Hastings algorithm, see e.g. Heikkinen & Penttinen 1999 and Berthelsen & Møller 2001). In the missing data case where $W \subset S$, it is typically relevant to use a data augmentation approach and consider the joint posterior of $(\theta, X_{S \setminus W})$.

Heikkinen & Penttinen (1999) suggest a Bayesian smoothing technique for estimation in pairwise interaction processes, where the likelihood function is approximated by the multiscale point process (6.16) having a large number of fixed change points $R_1, \ldots, R_k$. For convenience, they condition on the observed number $n(x)$ of points. A Gaussian Markov chain prior for $\theta = (\log \gamma_1, \ldots, \log \gamma_k)$ is chosen so that large differences $|\log \gamma_i - \log \gamma_{i-1}|$ are penalised. As the full posterior analysis is considered to be too demanding, they concentrate on finding the posterior mode, using ideas from Monte Carlo likelihood estimation as given in Penttinen (1984) and Geyer & Thompson (1992).

Berthelsen & Møller (2003) consider a similar situation, but without conditioning on $n(x)$ and without fixing the number k of change points and the values of $R_1, \ldots, R_k$ for a given k; instead a Poisson process prior is imposed on $R_1, \ldots, R_k$. Ratios of normalising constants of the likelihood term are computed using path sampling so that a full Bayesian

MCMC analysis is possible. They also compare perfect simulation (see Section 11.2) and ordinary MCMC methods for various applications.

Lund, Penttinen & Rudemo (1999) consider a situation where an unobserved point process X is degraded by independent thinning, random displacement, a simple censoring mechanism, and independent superpositioning with a Poisson process of "ghost points"; this is related to aerial photographs of trees disturbed by an image analysis process, cf. Lund & Rudemo (2000). A known pairwise interaction prior on X is imposed in Lund et al. (1999), so its normalising constant is unimportant when dealing with the posterior distribution for X and certain other model parameters given the observed degraded point pattern. Perfect simulation for this posterior is discussed in Lund & Thönnes (2004) and Møller (2001).

CHAPTER 10

Inference for Cox processes

In this chapter we consider parametric inference for Cox processes. As noted in Section 5.2, the likelihood for a Cox process is in general not available in closed form. It has therefore been common to use so-called minimum contrast methods (see e.g. Diggle 1983) where parameter estimates are obtained by matching nonparametric estimates of summary statistics with their theoretical expressions involving the unknown parameters. The likelihood can be estimated using Monte Carlo methods, but the computational formulas are not so neat as for Markov point processes where exponential family structure often simplifies computations, see Chapter 9. Another difficulty is that the unobserved random intensity function often depends on an infinite number of random variables, and therefore needs to be approximated by truncation or discretisation. Moreover, MCMC simulation is in general more time consuming than for the case of Markov point processes. We, however, expect that maximum likelihood estimation and Bayesian inference using MCMC will be of increasing importance for Cox processes as more computing power becomes available.

Minimum contrast estimation is considered in Section 10.1. MCMC algorithms for conditional simulation and prediction of the unobserved intensity function are discussed in Section 10.2. These algorithms also provide samples for Monte Carlo estimation of likelihood functions in Section 10.3 and form the basis of MCMC algorithms for Bayesian inference, see Section 10.4.

Throughout this chapter $x \neq \emptyset$ denotes a realisation of a Cox process X observed within a bounded planar window $W \subset \mathbb{R}^2$. Due to the computational complexity of maximum likelihood estimation and Bayesian inference for Cox processes, we use the modestly sized and simply structured hickory data (Example 1.1) throughout the chapter to illustrate methodology. No attempt is made to relate the obvious clustering of the hickory trees to e.g. environmental conditions or dynamics of hickory tree reproduction. The presented analyses of the hickory data are therefore probably not so interesting from a biological point of view.

10.1 Minimum contrast estimation

A computationally easy approach to parameter estimation is *minimum contrast estimation* (Diggle 1983, Møller et al. 1998) where the theoretical expression for a summary statistic depending on unknown parameters are compared with a nonparametric estimate of the summary statistic. For computational efficiency it is required that the theoretical expression of the summary statistic is easy to evaluate. In the following we consider g and K which have closed form theoretical expressions for many examples of Cox processes.

Consider for example the planar Thomas process with unknown parameters $\kappa, \alpha, \omega^2 > 0$ (Example 5.3, Section 5.3, with $d = 2$). If g given by (5.9) is used, a minimum contrast estimate $(\hat\kappa, \hat\omega)$ is chosen to minimise

$$\int_{a_1}^{a_2} \{\hat g(r) - 1 - \exp(-r^2/(4\omega^2))/(4\pi\omega^2\kappa)\}^2 \mathrm{d}r \qquad (10.1)$$

where $0 \leq a_1 < a_2$ are user-specified parameters (see Example 10.1 below). With

$$A(\omega^2) = \int_{a_1}^{a_2} \left(\exp(-r^2/(4\omega^2))/(4\pi\omega^2)\right)^2 \mathrm{d}r$$

and

$$B(\omega^2) = \int_{a_1}^{a_2} (\hat g(r) - 1) \exp(-r^2/(4\omega^2))/(4\pi\omega^2)\mathrm{d}r,$$

we obtain

$$\hat\omega^2 = \arg\max\left[B(\omega^2)^2/A(\omega^2)\right], \quad \hat\kappa = \left[A(\hat\omega^2)/B(\hat\omega^2)\right].$$

Inserting $\hat\kappa$ and the natural intensity estimate $\hat\rho = n(x)/|W|$ into the equation $\rho = \alpha\kappa$, we finally obtain

$$\hat\alpha = \hat\rho/\hat\kappa.$$

Another contrast is obtained by replacing $\hat g(r) - 1$ and $\exp(-r^2/(4\omega^2))/(4\pi\omega^2\kappa)$ in (10.1) by $\hat K(r) - \pi r^2$ and $[1 - \exp(-r^2/(4\omega^2))]/\kappa$, cf. (5.10). Diggle (1983) suggests a third contrast

$$\int_{a_1}^{a_2} \{(\hat K(r))^b - (\pi r^2 + [1 - \exp(-r^2/(4\omega^2))]/\kappa)^b\}^2 \mathrm{d}r$$

where $b > 0$ is a user-specified parameter recommended to be between 0.25 and 0.5. This approach requires numerical minimisation with respect to κ and ω jointly.

Minimum contrast estimation for SNCPs (including Neyman-Scott processes) follows the same lines as for the Thomas process. Brix (1999) reports that minimum contrast estimation for SNGCPs is numerically more stable if g is used instead of K.

MINIMUM CONTRAST ESTIMATION

Minimum contrast estimation for LGCPs and space-time LGCPs is considered in Møller et al. (1998), Brix & Møller (2001), and Brix & Diggle (2001). Consider a stationary LGCP with constant mean function $m(\cdot) = \beta \in \mathbb{R}$, variance $\sigma^2 > 0$, and correlation function of the power exponential type (5.22), with unknown $\alpha > 0$ and known $\delta \in (0,2]$. Because of the simple relationship (5.24), the minimum contrast estimate $(\hat{\alpha}, \hat{\sigma}^2)$ is chosen to minimise

$$\int_{a_1}^{a_2} \left\{ \left(\log \hat{g}(r)\right)^b - \left(\sigma^2 \exp(-(r/\alpha)^\delta)\right)^b \right\}^2 dr \qquad (10.2)$$

where $b > 0$ is a user-specified parameter. As for the Thomas process we can easily find $(\hat{\alpha}, \hat{\sigma}^2)$ and combine this with (5.24) and the intensity estimate $\hat{\rho} = n(x)/|W|$ to obtain $\hat{\beta} = \log(\hat{\rho}) - \hat{\sigma}^2/2$.

We are not aware of theoretical results concerning the distribution of minimum contrast estimates for Cox processes. Møller et al. (1998) contains a small simulation study for an LGCP.

EXAMPLE 10.1 (*Minimum contrast estimation for hickory trees*) For the hickory tree data (Example 1.1) modelled by either a Thomas process, an SNGCP, or an LGCP, we base minimum contrast estimation on the g-function with $\hat{g}$ given as in the right plot in Figure 4.8. We choose $a_1 = 0.6$m due to the large variability of $\hat{g}(r)$ for small distances r, and we take $a_2 = 20$m.

For the Thomas process we obtain $\hat{\kappa}=0.0013$, $\hat{\alpha} = 4.39$, and $\hat{\omega}^2 = 16.6$. The estimated expected number of mother points in the observation window is then $\hat{\kappa} 120^2 = 18.72$.

For the Poisson-gamma case of an SNGCP from Example 5.6 (Section 5.4.3) with k given by a normal density as for the Thomas process, we obtain $\hat{\kappa} = 0.0013$, $\hat{\tau} = 0.23$, and $\hat{\omega}^2 = 16.6$. Note that $120^2 \hat{\kappa}/\hat{\tau}$ is equal to the observed number of points. The g-functions for the Thomas process and the Poisson-gamma process are identical, cf. (5.8) and (5.18), and we therefore obtain the same estimates of κ for these two processes.

Consider finally a stationary LGCP with mean and covariance function as above. The parameter δ determines the behaviour of g near zero, but $\hat{g}$ does not carry much information on δ due to the large variability near zero. We let $b = 1$ and $\delta = 1$ in (10.2) corresponding to the exponential correlation function, and obtain $\hat{\beta} = -6.1$, $\hat{\sigma}^2 = 1.9$, and $\hat{\alpha} = 8.1$.

Figure 10.1 shows $\hat{g}$ and an nonparametric estimate of the empty space function F (Section 4.2.3) with envelopes calculated under the various fitted models. No inconsistencies between the data and the fitted models are revealed apart for the LGCP where $\hat{F}(r)$ falls a bit below the envelopes for a range of distances r.

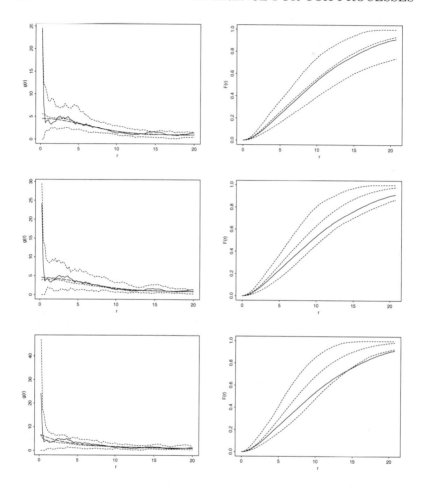

Figure 10.1 *Upper left:* $\hat{g}(r)$ *for the hickory data (solid line), envelopes and average calculated from 39 simulations under the fitted Thomas process (dashed line), and theoretical value of $g(r)$ for fitted Thomas process (long-dashed line). Upper right:* $\hat{F}(r)$ *(solid line), and envelopes and average calculated from 39 simulations under the fitted Thomas process (dashed line). Middle plots: as upper plots but for SNGCP. Lower plots: as upper plots but for LGCP.*

10.2 Conditional simulation and prediction

Given a bounded region $B \supseteq W$, we may be interested in predicting $Z_B = \{Z(\xi)\}_{\xi \in B}$ or $X_{B \setminus W}$ or, more generally, in computing the conditional distributions of Z_B and $X_{B \setminus W}$ given $X_W = x$. In the sequel we mainly focus on the conditional distribution of Z_B, since $X_{B \setminus W}$ is conditionally independent of X_W given Z_B, and $X_{B\setminus W}$ is an inhomogeneous Poisson process given Z_B which may be simulated by the methods in Section 3.2.3.

The conditional distribution of Z_B is not analytically tractable for the model classes considered in Chapter 5, so in the following we discuss MCMC algorithms for conditional simulation. The intensity function Z_B depends on some stochastic process like the point processes C for a Neyman-Scott process or Φ for an SNCP or the random field Y_B for an LGCP. In general such stochastic processes do not have finite representations, so certain approximations of the random intensity functions are required, cf. Sections 5.5 and 5.7. In the following, with an abuse of notation, we also denote by X the Cox processes obtained with an approximate random intensity function.

For an intensity function ρ on B, let

$$p(x|\rho) = \exp\left(|W| - \int_W \rho(\xi)\mathrm{d}\xi\right) \prod_{\xi \in x} \rho(\xi) \qquad (10.3)$$

be the density of a Poisson process on W with intensity function ρ (restricted to W) with respect to Poisson$(W, 1)$, cf. (3.11). The meaning of a density for a bivariate point process, which is considered a couple of times in the following, is explained in Section 8.6.

10.2.1 Conditional simulation for Neyman-Scott processes

Before considering the general and more technical case of SNCPs, it is illuminating to consider the case of a Neyman-Scott process (Section 5.3). Then we approximate Z_B by

$$Z_{\mathrm{NS}}(\xi|C^{\mathrm{ext}}) = \alpha \sum_{c \in C^{\mathrm{ext}}} k(\xi - c), \quad \xi \in B,$$

where $B_{\mathrm{ext}} \supseteq B$ is a bounded region chosen as in Section 5.5 and $C^{\mathrm{ext}} = C \cap B_{\mathrm{ext}}$. We stress the dependence of C^{ext} in the notation for the approximate random intensity function, since we consider conditional simulation of the finite point process C^{ext}. Let X_1 and X_2 be independent standard Poisson processes on W and B_{ext}, respectively. The joint density of (X_W, C^{ext}) with respect to (X_1, X_2) is

$$p(x, \psi) = p(x|Z_{\mathrm{NS}}(\cdot|\psi))\kappa^{n(\psi)} \exp(|B_{\mathrm{ext}}|(1 - \kappa)) \qquad (10.4)$$

for finite point configurations $x \subset W$ and $\psi \subset B_{\mathrm{ext}}$, and the conditional density of C^{ext} given $X_W = x$ is

$$p(\psi|x) \propto p(x|Z_{\mathrm{NS}}(\cdot|\psi))\kappa^{n(\psi)}$$

which can be simulated using Algorithm 7.4 or Algorithm 7.5 in Section 7.1.2. If we let $p(\psi) = 1/2$, $q_b(\psi, \cdot) = 1/|B_{\text{ext}}|$, and $q_d(\psi, \cdot) = 1/n(\psi)$ in Algorithm 7.4, the Hastings ratio for a birth $\psi \to \psi \cup \xi$ is

$$r_b(\psi, \xi) = \frac{p(\psi \cup \xi | x) |B_{\text{ext}}|}{p(\psi | x)(n(\psi) + 1)}$$

where

$$\frac{p(\psi \cup \xi | x)}{p(\psi | x)} = \kappa \exp\left(-\alpha \int_W k(\xi - \eta) \mathrm{d}\eta\right) \prod_{\eta \in x}\left(1 + \frac{k(\xi - \eta)}{\sum_{c \in \psi} k(c - \eta)}\right)$$

if $\psi \neq \emptyset$ (as the data $x \neq \emptyset$, $p(\emptyset | x) = 0$, and so by convention, $r_b(\emptyset, \xi) = 1$). Convergence properties of this algorithm follow along similar lines as in Section 10.2.2 for the more general case of SNCPs.

10.2.2 Conditional simulation for SNCPs

Consider now an SNCP model for X. As in Section 5.5 we replace Φ by the truncated finite Poisson process $\Phi^{\text{trunc}} = \Phi \cap (B_{\text{ext}} \times (\epsilon, \infty))$. The approximate random intensity function is then

$$Z_{\text{SN}}(\xi | \Phi^{\text{trunc}}) = \sum_{(c,\gamma) \in \Phi^{\text{trunc}}} \gamma k(\xi - c), \quad \xi \in B. \tag{10.5}$$

Suppose that $X_1 \sim \text{Poisson}(W, 1)$ is independent of $X_2 \sim \text{Poisson}(B_{\text{ext}} \times (\epsilon, \infty), \tilde{\zeta})$, where $\tilde{\zeta}$ is a given intensity function so that both $\tilde{\zeta}(c, \gamma) > 0$ whenever $\zeta(c, \gamma) > 0$ and $\int_{B_{\text{ext}}} \int_\epsilon^\infty \tilde{\zeta}(c, \gamma) \mathrm{d}c \mathrm{d}\gamma < \infty$. Except for these two restrictions the choice of $\tilde{\zeta}$ may be arbitrary, e.g. $\tilde{\zeta} = \zeta$ or if ζ is given in terms of a parametric model ζ_θ we may let $\tilde{\zeta} = \zeta_{\theta_0}$ for a fixed parameter θ_0, cf. Remark 8.6. The joint density of $(X_W, \Phi^{\text{trunc}})$ with respect to (X_1, X_2) is

$$p(x, \phi) = p(x | Z_{\text{SN}}(\cdot | \phi)) \times$$
$$\exp\left(-\int_{B_{\text{ext}}} \int_\epsilon^\infty (\zeta(c, \gamma) - \tilde{\zeta}(c, \gamma)) \mathrm{d}c \mathrm{d}\gamma\right) \prod_{(c,\gamma) \in \phi} \frac{\zeta(c, \gamma)}{\tilde{\zeta}(c, \gamma)}$$

for finite point configurations $x \subset W$ and $\phi \subset B_{\text{ext}} \times (\epsilon, \infty)$, and the conditional density of Φ^{trunc} given $X_W = x$ is

$$p(\phi | x) \propto p(x | Z_{\text{SN}}(\cdot | \phi)) \prod_{(c,\gamma) \in \phi} \zeta(c, \gamma) / \tilde{\zeta}(c, \gamma). \tag{10.6}$$

Below we give details of Algorithm 7.4 (Section 7.1.2) applied to simulation of (10.6). Since ζ is in general not a constant function, we appeal to Remark 7.5 (Section 7.1.2). Suppose ϕ is the current state of the

Metropolis-Hastings chain, and let as in Remark 5.2 (Section 5.5),

$$q(t) = \int_{B_{\text{ext}}} \int_t^\infty \zeta(c,\gamma) \mathrm{d}c \mathrm{d}\gamma. \quad (10.7)$$

With $p(\phi) = 1/2$, $q_b(\phi, \cdot) = \zeta(\cdot)/q(\epsilon)$, and $q_d(\phi, \cdot) = 1/n(\phi)$, the Hastings ratio for a birth $\phi \to \phi \cup (c, \gamma)$ becomes

$$r_b(\phi, (c,\gamma)) = \frac{p(\phi \cup (c,\gamma)|x) q(\epsilon)}{p(\phi|x)(n(\phi)+1)} \quad (10.8)$$

where

$$\frac{p(\phi \cup (c,\gamma)|x)}{p(\phi|x)} = \zeta(c,\gamma)/\tilde{\zeta}(c,\gamma) \times$$

$$\exp\left(-\gamma \int_W k(c,\xi)\mathrm{d}\xi\right) \prod_{\xi \in x} \left(1 + \frac{\gamma k(c,\xi)}{\sum_{(\tilde{c},\tilde{\gamma}) \in \phi} \tilde{\gamma} k(\tilde{c},\xi)}\right) \quad (10.9)$$

if $\phi \neq \emptyset$ (and $r_b(\emptyset, (c,\gamma)) = 1$). The integral in (10.9) can easily be calculated for the uniform kernel (I) and the Gaussian kernel (II) (Example 5.3, Section 5.3) if e.g. W is rectangular.

The algorithm is reversible by Proposition 7.12 in Section 7.3.2. The condition (7.11) does not hold for the conditional density (10.6) since $p(\emptyset|x) = 0$ when $x \neq \emptyset$, so Proposition 7.13 does not establish irreducibility and aperiodicity. We here argue informally why these properties nevertheless hold. Consider any pair of states ϕ and ϕ' with $p(\phi|x) > 0$ and $p(\phi'|x) > 0$. Since $p(\phi|x) > 0$ implies $p(\phi \cup (c,\gamma)|x) > 0$ for $(c,\gamma) \in B_{\text{ext}} \times (\epsilon, \infty)$, the Metropolis-Hastings chain can move first from ϕ to $\phi \cup \phi'$ and next by reversibility from $\phi \cup \phi'$ to ϕ'. So the chain is irreducible. It is also aperiodic, since there is positive probability of staying at ϕ. Moreover, Proposition 5 in Møller (2003b) states that the chain is geometrically ergodic under weak conditions on the kernel k. The conditions are for example satisfied for uniform and Gaussian kernels.

Alternative and more complicated algorithms using auxiliary techniques or based on spatial birth-death processes have been proposed in Wolpert & Ickstadt (1998) and van Lieshout & Baddeley (2002). The Metropolis-Hastings algorithm described above is proposed because of its simplicity and since its theoretical properties are well understood (Møller 2003b).

REMARK 10.1 (*Computational complexity*) If the integral $\int_W k(c,\xi)\mathrm{d}\xi$ is easy to compute, the main problem in the calculation of the Hastings ratio (10.8) is to compute

$$\prod_{\xi \in x} \left(1 + \gamma k(c,\xi)/\sum_{(\tilde{c},\tilde{\gamma}) \in \phi} \tilde{\gamma} k(\tilde{c},\xi)\right)$$

which in general is $O(n(x))$ if one keeps track of $\sum_{(\tilde{c},\tilde{\gamma})\in\phi}\tilde{\gamma}k(\tilde{c},\xi)$ during the iterations (that a real function $f(n)$ is $O(n)$ means that $f(n)/n$ is bounded for $n \to \infty$). This computational problem may be reduced if $k(c,\cdot)$ for $c \in B_{\text{ext}}$ has a bounded support, since some of the terms $\gamma k(c,\xi)/\sum_{(\tilde{c},\tilde{\gamma})\in\phi}\tilde{\gamma}k(\tilde{c},\xi)$, $\xi \in x$, may vanish. In general the computational burden for SNCPs is greater than for simulation of Markov point processes where the local Markov property simplifies the calculation of the Hastings ratio.

EXAMPLE 10.2 (*Conditional simulation of Poisson-gamma processes*) For a Poisson-gamma process, $\epsilon > 0$ and

$$\zeta(c,\gamma) = \kappa \exp(-\tau\gamma)/\gamma$$

and so (10.7) is given by

$$q(t) = \kappa |B_{\text{ext}}| \int_t^\infty \exp(-\tau\gamma)/\gamma \, \mathrm{d}\gamma.$$

If it is proposed to add a point, then a simulation from the birth density $q_b(\cdot) = \zeta(\cdot)/q(\epsilon)$ can be obtained by $(c,\gamma) = (c, q^{-1}(u))$ where $c \sim \text{Uniform}(B_{\text{ext}})$ and $u \sim \text{Uniform}((0, q(\epsilon)))$ are independent. Alternatively, we can reparametrise and consider conditional simulation of $\Psi = \{(c, q(\gamma)) : (c,\gamma) \in \Phi \cap (B_{\text{ext}} \times (\epsilon, \infty))\}$ which unconditionally is Poisson$(B_{\text{ext}} \times (0, q(\epsilon)), 1/|B_{\text{ext}}|)$. The conditional density of Ψ with respect to Poisson $(B_{\text{ext}} \times (0, q(\epsilon)), 1)$ is

$$p_\Psi(\psi|x) \propto p(x|Z_{\text{SN}}(\cdot|\{(c, q^{-1}(u)) : (c,u) \in \psi\})|B_{\text{ext}}|^{-n(\psi)}$$

and in this case we can use Algorithm 7.4 without appeal to Remark 7.5. The Hastings ratio for a birth proposal of a new point (c,w) generated from the uniform distribution on $B_{\text{ext}} \times (0, q(\epsilon))$ becomes

$$\frac{p\big(x|Z_{\text{SN}}(\cdot|\{(c, q^{-1}(u)) : (c,u) \in \psi \cup (c,w)\}\big)q(\epsilon)}{p\big(x|Z_{\text{SN}}(\cdot|\{(c, q^{-1}(u)) : (c,u) \in \psi\}\big)(n(\psi) + 1)},$$

which is in fact equivalent to the ratio (10.8) in the Poisson-gamma case with $\tilde{\zeta} = \zeta$. Finally, conditional samples of Ψ can be transformed into conditional samples of Φ^{trunc} using q^{-1}.

EXAMPLE 10.3 (*Conditional simulation for hickory trees*) For the hickory trees, let $B = W = [10, 130]^2$ and choose $B_{\text{ext}} = [0, 140]^2$. For each location ξ in the regular 100×100 lattice

$$I = \{(10.6 + 1.2m, 10.6 + 1.2n) : m = 0, \ldots, 99, n = 0, \ldots, 99\} \subset B$$

we consider Monte Carlo estimation of the conditional expectations $\mathbb{E}[Z_{\text{NS}}(\xi|C^{\text{ext}})|x]$ and $\mathbb{E}[Z_{\text{SN}}(\xi|\Phi^{\text{trunc}})|x]$ for the fitted Thomas and Poisson-gamma processes from Example 10.1 (where we let $\epsilon = 0.001$ in

CONDITIONAL SIMULATION AND PREDICTION

the Poisson-gamma case). For both processes we generate 1000 conditional simulations of C^{ext} and Φ^{trunc} by subsampling each 1000th state of Markov chains generated using 1,000,000 updates of Algorithm 7.4. This rather large spacing is used, as the evaluation of $(Z_{\text{NS}}(\xi|\psi))_{\xi \in I}$ and $(Z_{\text{SN}}(\xi|\phi))_{\xi \in I}$ for subsampled Markov chain states ψ and ϕ is time consuming. The acceptance rates for adding/deleting mother points are 0.26 for the Thomas process and 0.87 for the Poisson-gamma process. The acceptance rate is higher for the Poisson-gamma process, since it is often proposed to add a point (c, γ) with a small γ, and the addition of such a point does not change much the conditional density of x and is hence likely to get accepted.

The estimated conditional expectations

$$(\mathbb{E}[Z_{\text{NS}}(\xi|C^{\text{ext}})|x])_{\xi \in I} \quad \text{and} \quad (\mathbb{E}[Z_{\text{SN}}(\xi|\Phi^{\text{trunc}})|x])_{\xi \in I}$$

are shown as grey scale plots in Figure 10.2 where the minimal/maximal values are 0.00/0.06 and 0.00/0.10, respectively. Estimated conditional

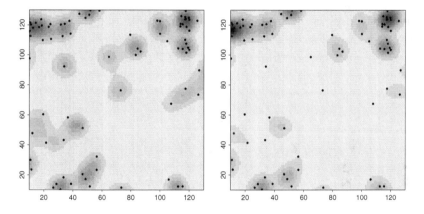

Figure 10.2 *Left: conditional expectation of $Z_{\text{NS}}(\cdot|C^{\text{ext}})$ for a Thomas process. Right: conditional expectation of $Z_{\text{SN}}(\cdot|\Phi^{\text{trunc}})$ for a Poisson-gamma process. The same grey scales are used in the two plots and the dots show the observed points.*

expectations for four specific locations in I are given in Table 10.1 together with Monte Carlo standard errors obtained as described in Section 8.1.3 using the initial monotone sequence estimate of the asymptotic variance. Table 10.1 also contains estimates of the conditional expectations of the integrated random intensity functions

$$IZ_{\text{NS}} = \int_W Z_{\text{NS}}(\eta|C^{\text{ext}}) d\eta \quad \text{and} \quad IZ_{\text{SN}} = \int_W Z_{\text{SN}}(\eta|\Phi^{\text{trunc}}) d\eta.$$

The estimated intensity functions differ considerably: $\mathbb{E}[Z_{\text{NS}}(\cdot|C^{\text{ext}})|x]$

	(33.6,92.4)	(81.6,42)	(117.6,123.6)	(42.0,51.6)	IZ
Th	0.02	0.00014	0.06	0.03	97.0
	(0.001)	(0.000058)	(0.00059)	(0.00070)	(0.19)
Pg	0.005	0.001	0.08	0.01	83.3
	(0.00024)	(0.000070)	(0.003)	(0.001)	(0.42)

Table 10.1 *Estimated conditional expectations and Monte Carlo standard errors (in parentheses) for the random intensity functions of a Thomas process (Th) and a Poisson-gamma process (Pg) at four different locations and for the integrated random intensity function (IZ).*

e.g. has a clear peak at the spatial location $(33.6, 92.4)$ caused by just one observed point. This peak is not present in $\mathbb{E}[Z_{\mathrm{SN}}(\cdot)|\Phi^{\mathrm{trunc}})|x]$. Moreover, the estimated conditional expectations of IZ_{NS} and IZ_{SN} are 97.0 for the Thomas model and 83.4 (close to the observed number of points) for the Poisson-gamma model. The Monte Carlo standard errors in Table 10.1 show that the differences are not just due to Monte Carlo error.

The random intensity function for the Poisson-gamma model is composed of Gaussian kernels of varying heights given by the γ values, cf. (10.5). This offers more flexibility than for the Thomas model where e.g. the presence of isolated observed points provokes pronounced peaks in the conditional expectation, see the left plot in Figure 10.2.

10.2.3 Conditional simulation for LGCPs

Consider now an LGCP with log Gaussian intensity function $Z_B = \exp(Y_B)$ where we use a notation as in Section 5.6. The conditional distribution of Y_B is determined by the consistent set of finite-dimensional conditional densities of $(Y(s_1), \ldots, Y(s_n))$ given $X_W = x$ for $n \in \mathbb{N}$ and pairwise distinct $s_1, \ldots, s_n \in B$, that is,

$$f_{(s_1,\ldots,s_n)}(y_1, \ldots, y_n | x) \propto \mathbb{E}[p(x| \exp(Y_W))|Y(s_1) = y_1, \ldots, Y(s_n) = y_n] \\ \times f_{(s_1,\ldots,s_n)}(y_1, \ldots, y_n) \quad (10.10)$$

where $f_{(s_1,\ldots,s_n)}(\cdot)$ denotes the marginal density of $(Y(s_1), \ldots, Y(s_n))$. The conditional density (10.10) is not analytically tractable and application of MCMC methods is not straightforward due to the presence of the random field Y_W which involves an infinite number of random variables. In practice, we therefore approximate Y_B by a step function with step heights $\tilde{Y}$ as in Section 5.7, and consider simulation of $\Gamma | X_W = x$ where Γ and $\tilde{Y}$ are related by the transformation (5.25) (see also Remark 5.3

CONDITIONAL SIMULATION AND PREDICTION 191

in Section 5.7). The conditional density of Γ given x is

$$\pi(\gamma|x) \propto \exp(-\|\gamma\|^2/2) p(x|\exp(y^{\text{step}}))$$
$$= \exp\big(-\|\gamma\|^2/2 + |W| + \sum_{\eta \in I}(\tilde{y}_\eta n_\eta - A_\eta \exp(\tilde{y}_\eta))\big) \quad (10.11)$$

where, in accordance with (5.25), $\tilde{y} = (\tilde{y}_\eta)_{\eta \in I_{\text{ext}}} = \gamma Q + \mu_{\text{ext}}$, y^{step} is the step function with steps given by $\tilde{y}$, $n_\eta = \text{card}(x \cap C_\eta)$, and $A_\eta = |W \cap C_\eta|$. The gradient of the log conditional density is

$$\nabla(\gamma) = \nabla \log \pi(\gamma|x) = -\gamma + \big(n_\eta - A_\eta \exp(\tilde{y}_\eta)\big)_{\eta \in I_{\text{ext}}} Q^{\mathsf{T}},$$

and differentiating once more, the conditional density of Γ given x is seen to be strictly log-concave.

For simulation from $\Gamma|X_W = x$, Møller et al. (1998) use a *Langevin-Hastings algorithm* or *Metropolis-adjusted Langevin algorithm* as introduced in the statistical community by Besag (1994) (see also Roberts & Tweedie 1996) and earlier in the physics literature by Rossky, Doll & Friedman (1978). This is a Metropolis-Hastings algorithm with proposal distribution $N_d(\gamma + (h/2)\nabla(\gamma), hI_d)$ where d is the cardinality of I_{ext} and $h > 0$ is a user-specified proposal variance.

ALGORITHM 10.1 (*Langevin-Hastings algorithm*) Let $R_m \sim \text{Uniform}([0, 1])$ and $U_m \sim N_d(0, I_d)$, $m = 0, 1, \ldots$, be mutually independent. Given the mth state $\Gamma_m = \gamma$ of the Langevin-Hastings chain, the next state Γ_{m+1} is generated as follows:

1. Set $\Gamma_{\text{prop}} = \gamma + (h/2)\nabla(\gamma) + \sqrt{h} U_m$.
2. Let

$$\Gamma_{m+1} = \begin{cases} \Gamma_{\text{prop}} & \text{if } R_m \leq r(\gamma, \Gamma_{\text{prop}}) \\ \gamma & \text{otherwise} \end{cases}, \quad (10.12)$$

where

$$r(\gamma, \Gamma_{\text{prop}}) = \frac{\pi(\Gamma_{\text{prop}}|x) q(\Gamma_{\text{prop}}, \gamma)}{\pi(\gamma|x) q(\gamma, \Gamma_{\text{prop}})} \quad (10.13)$$

is the Hastings ratio and $q(\gamma, \cdot)$ is the density of the proposal distribution $N_d(\gamma + (h/2)\nabla(\gamma), hI_d)$.

Theoretical results in Roberts & Rosenthal (1998) and Breyer & Roberts (2000) suggest that we should tune h to obtain acceptance rates around 0.57. The use of the gradient in the proposal distribution may lead to much better mixing properties when compared to the standard alternative of a random walk Metropolis algorithm where the proposal distribution is $N_d(\gamma, hI_d)$, see Christensen, Møller & Waagepetersen (2001) and Christensen & Waagepetersen (2002). On the other hand, regarding speed of convergence to the stationary distribution, the Langevin-

Hastings algorithm is more sensitive to the choice of initial state Γ_0, see Christensen, Roberts & Rosenthal (2003) and Example 10.6.

A truncated version of the Langevin-Hastings algorithm is obtained by replacing the gradient $\nabla(\gamma)$ in the proposal distribution by

$$\nabla_{\text{trunc}}(\gamma) = -\gamma + \left(n_\eta - \min\{H, A_\eta \exp(\tilde{y}_\eta)\}\right)_{\eta \in I_{\text{ext}}} Q^{\mathsf{T}} \quad (10.14)$$

where $H > 0$ is a user-specified parameter. During the simulations, $A_\eta \exp(\tilde{y}_\eta)$ will roughly be of the same magnitude as n_η, so we may e.g. take H to be twice the maximal n_η, $\eta \in I$. As shown in Møller et al. (1998) the *truncated Langevin-Hastings algorithm* is V-uniformly ergodic with $V(\gamma) = \exp(s\|\gamma\|)$ for any $s > 0$ provided $0 < h < 2$. The truncated Langevin-Hastings algorithm is used as a component in a hybrid MCMC algorithm for posterior simulation in Example 10.6.

10.3 Maximum likelihood inference

For a Cox process the unobserved intensity function Z (or equivalently C, Φ, or Y for a Neyman-Scott process, SNCP, or LGCP) may be considered as "missing data". Since the marginal density of the data x can typically not be evaluated analytically, cf. (5.5), Monte Carlo methods for computation of missing data likelihoods are needed. Sections 10.3.1-10.3.2 consider in detail how the Monte Carlo methods from Section 8.6 can be applied to likelihood inference for an example of a Neyman-Scott process and an example of an SNCP. The case of an LGCP is briefly considered in Section 10.3.3.

In the sequel, in order to obtain Monte Carlo samples, we consider approximations of Z_W similar to those used in the simulation algorithms from Section 10.2. As in Section 10.2 we denote by X also the Cox processes obtained with the approximate random intensity functions.

10.3.1 Likelihood inference for a Thomas process

Consider as in Section 10.1 a Thomas process X with unknown parameter $\theta = (\kappa, \alpha, \omega)$, and let $B_{\text{ext}} \supseteq W$ and C^{ext} be specified as in Section 10.2.1 with $B = W$. The approximate random intensity function is

$$Z_{\text{Th}}(\xi | C^{\text{ext}}; \alpha, \omega) = (\alpha/(2\pi\omega^2)) \sum_{c \in C^{\text{ext}}} \exp\left(-\|\xi - c\|^2/(2\omega^2)\right) \quad (10.15)$$

and the joint density of X_W and C^{ext} is

$$p_\theta(x, \psi) = p(x | Z_{\text{Th}}(\cdot | \psi, \alpha, \omega)) \kappa^{n(\psi)} \exp(|B_{\text{ext}}|(1 - \kappa)), \quad (10.16)$$

MAXIMUM LIKELIHOOD INFERENCE

cf. (10.4). The likelihood is $L(\theta) = \mathbb{E}p_\theta(x, X_2)$ where $X_2 \sim \text{Poisson}(B_{\text{ext}}, 1)$, and the likelihood can be estimated using the methods from Section 8.6 with $X = X_W$ and $Y = C^{\text{ext}}$.

Below, in connection with path sampling and Newton-Raphson maximisation, derivatives of the log joint density (10.16) are needed. The derivatives with respect to κ and α are

$$\mathrm{d}\log p_\theta(x, \psi)/\mathrm{d}\kappa = n(\psi)/\kappa - |B_{\text{ext}}| \qquad (10.17)$$

and

$$\mathrm{d}\log p_\theta(x, \psi)/\mathrm{d}\alpha = \frac{1}{\alpha}\left(n(x) - \int_W Z_{\text{Th}}(\xi|\psi; \alpha, \omega)\mathrm{d}\xi\right), \qquad (10.18)$$

cf. (10.3) and (10.15)–(10.16). The derivative $\mathrm{d}\log p_\theta(x, c)/\mathrm{d}\omega$ is more complicated and must be evaluated using numerical integration.

To obtain an approximate MLE of θ we compute maximum profile likelihood estimates $(\hat{\kappa}_l, \hat{\alpha}_l)$, $l = 1, \ldots, K$, obtained by maximising profile likelihoods

$$L_l(\kappa, \alpha) = L(\kappa, \alpha, \omega_l), \quad l = 1, \ldots, K,$$

for a range of ω-values $\omega_1, \ldots, \omega_K$. The profile likelihood approach is advantageous because it allows a graphical inspection of the two-dimensional profile likelihood functions $L_l(\kappa, \alpha)$ and since we avoid the use of derivatives with respect to ω. For each l, we maximise L_l using Newton-Raphson (see Section 8.6.2), and we next use path sampling to compute the profile likelihood function $L_l(\kappa, \alpha)$ in order to assess whether a global maximum $(\hat{\kappa}_l, \hat{\alpha}_l)$ has been obtained. This is needed since the likelihood function for a Thomas process may not be unimodal. We finally compare the likelihoods of $\hat{\theta}_l = (\hat{\kappa}_l, \hat{\alpha}_l, \omega_l)$, $l = 1, \ldots, K$, using bridge sampling, and thereby determine an approximate MLE of θ.

Let $\mathbb{E}_{\theta,x}$ denote expectation with respect to the conditional distribution of C^{ext} given $X_W = x$, with conditional density $p_\theta(\cdot|x) \propto p_\theta(x, \cdot)$. Define $\mathbb{V}\text{ar}_{\theta,x}$ and $\mathbb{C}\text{ov}_{\theta,x}$ analogously and let

$$IZ_{\text{Th}} = IZ_{\text{Th}}(C^{\text{ext}}; \alpha, \omega) = \int_W Z_{\text{Th}}(\xi|C^{\text{ext}}; \alpha, \omega)\mathrm{d}\xi.$$

By (8.39)–(8.40) and (10.17)–(10.18), the score function and the observed information for the profile likelihood $L_l(\kappa, \alpha)$ are given by

$$u_l(\kappa, \alpha) = \left(\mathbb{E}_{\theta_l,x}n(C^{\text{ext}})/\kappa - |B_{\text{ext}}|, (n(x) - \mathbb{E}_{\theta_l,x}IZ_{\text{Th}})/\alpha\right)$$

and

$$j_l(\alpha, \kappa) = \begin{pmatrix} (\mathbb{E}_{\theta_l,x}n(C^{\text{ext}}) - \mathbb{V}\text{ar}_{\theta_l,x}n(C^{\text{ext}}))/\kappa^2 & \mathbb{C}\text{ov}_{\theta_l,x}(n(C^{\text{ext}}), IZ_{\text{Th}})/(\alpha\kappa) \\ \mathbb{C}\text{ov}_{\theta_l,x}(n(C^{\text{ext}}), IZ_{\text{Th}})/(\alpha\kappa) & (n(x) - \mathbb{V}\text{ar}_{\theta_l,x}IZ_{\text{Th}})/\alpha^2 \end{pmatrix}$$

where $\theta_l = (\kappa, \alpha, \omega_l)$. The conditional moments in the score function and observed information are estimated using importance sampling (8.43).

Let $\Delta(\kappa, \alpha) = (\kappa - \kappa_0, \alpha - \alpha_0)$ and $(\kappa(t), \alpha(t)) = (\kappa_0, \alpha_0) + t\Delta(\kappa, \alpha)$ for $t \in [0, 1]$. The path sampling identity (8.38) becomes

$$\log(L_l(\kappa, \alpha)/L_l(\kappa_0, \alpha_0)) = \int_0^1 u_l(\kappa(t), \alpha(t)) \cdot \Delta(\kappa, \alpha) dt \quad (10.19)$$

where, using the trapezoidal rule, the integral is approximated by

$$\sum_{i=0}^{m-1} \left[u_l(\kappa(i/m), \alpha(i/m)) + u_l(\kappa((i+1)/m), \alpha((i+1)/m))\right] \cdot \Delta(\kappa, \alpha)/(2m)$$

for an $m \in \mathbb{N}$. In order to compute the likelihood over a two-dimensional grid $\{(\kappa_0 + \delta_\kappa i, \alpha_0 + \delta_\alpha j) : i = 0, \ldots, n_\kappa, j = 0, \ldots, n_\alpha\}$, we combine path sampling computations for the paths $(\kappa_0 + tn_\kappa \delta_\kappa, \alpha_0)$ and $(\kappa_0 + i\delta_\kappa, \alpha_0 + tn_\alpha \delta_\alpha)$, $t \in [0, 1]$, $i = 0, \ldots, n_\kappa$.

By the bridge sampling formula (8.37),

$L(\theta)/L(\theta_0) =$

$$\mathbb{E}_{\theta_0, x}[p^b(x, C^{\text{ext}})/p_{\theta_0}(x, C^{\text{ext}})]/\mathbb{E}_{\theta, x}[p^b(x, C^{\text{ext}})/p_\theta(x, C^{\text{ext}})] \quad (10.20)$$

where we use the bridge sampling density $p^b(x, \psi) = p_{\theta^b}(x, \psi)$ with $\theta^b = (\theta_0 + \theta)/2$. Estimates of

$$\frac{L(\hat{\theta}_l)}{L(\hat{\theta}_1)} = \prod_{i=2}^{l} \frac{L(\hat{\theta}_i)}{L(\hat{\theta}_{i-1})} \quad l = 2, \ldots, m, \quad (10.21)$$

are obtained by replacing the factors on the right hand side by bridge sampling estimates. If the bridge sampling estimate of a factor $L(\hat{\theta}_i)/L(\hat{\theta}_{i-1})$ has a very large Monte Carlo variance, we consider additional intermediate parameter values $\tilde{\theta}_{i,j}$, $j = 1, \ldots, J_i$, between $\hat{\theta}_{i-1}$ and $\hat{\theta}_i$, and use bridge sampling for the factors in the product on the right hand side of

$$\frac{L(\hat{\theta}_i)}{L(\hat{\theta}_{i-1})} = \frac{L(\tilde{\theta}_{i,1})}{L(\hat{\theta}_{i-1})} \left(\prod_{j=2}^{J_i} \frac{L(\tilde{\theta}_{i,j})}{L(\tilde{\theta}_{i,j-1})}\right) \frac{L(\hat{\theta}_i)}{L(\tilde{\theta}_{i,J_i})}.$$

In the following example we compare the Newton-Raphson procedure with Monte Carlo EM. Using (10.17) (10.18) and the fact that $Z_{\text{Th}}(\xi|C^{\text{ext}}; \alpha, \omega)/\alpha$ does not depend on α, the EM update (8.44) reduces to

$(\tilde{\kappa}_l^{(m+1)}, \tilde{\alpha}_l^{(m+1)}) =$

$$\left(\frac{\mathbb{E}_{(\tilde{\kappa}_l^{(m)}, \tilde{\alpha}_l^{(m)}, \omega_l), x} n(C^{\text{ext}})}{|B_{\text{ext}}|}, \frac{n(x)\tilde{\alpha}_l^{(m)}}{\mathbb{E}_{(\tilde{\kappa}_l^{(m)}, \tilde{\alpha}_l^{(m)}, \omega_l), x} \int_W Z_{\text{Th}}(\xi|C^{\text{ext}}; \tilde{\alpha}_l^{(m)}, \omega_l) d\xi}\right)$$

MAXIMUM LIKELIHOOD INFERENCE

where the conditional expectations are replaced by Monte Carlo averages, cf. (8.45).

EXAMPLE 10.4 (*Likelihood estimation for hickory trees*) For the hickory tree data we consider five ω values $\omega_l = 1 + l$, $l = 1,\ldots,5$. All the Markov chains for conditional samples are initialised using unconditional realisations of C^{ext}. Trace plots for the statistics $n(C^{\text{ext}})$, IZ_{Th}, $\log[\prod_{\xi \in x} Z_{\text{Th}}(\xi | C^{\text{ext}}, \hat{\alpha}_1, \omega_1)]$, and the averaged cumulated differences $\sum_{i=0}^{m}(n(C_{i+1}^{\text{ext}}) - n(C_i^{\text{ext}}))/m$ for a Markov chain $C_0^{\text{ext}}, C_1^{\text{ext}}, \ldots$ with invariant density $p_{\hat{\theta}_1}(\cdot|x)$ are shown in Figure 10.3. The lower left and lower right trace plots illustrate that at least 3000 iterations are needed to reach equilibrium.

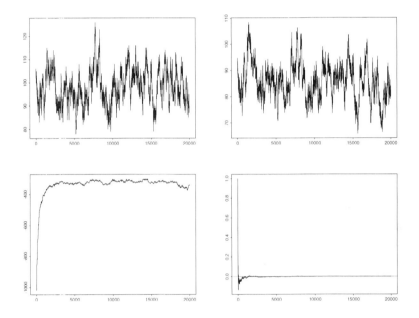

Figure 10.3 *Trace plots for* $n(C_m^{\text{ext}})$ *(upper left),* IZ_{Th} *(upper right),* $\log[\prod_{\xi \in x} Z_{\text{Th}}(\xi | C_m^{\text{ext}}; \hat{\alpha}_1, \omega_1)]$ *(lower left),* $m = 0, \ldots, 19999$, *and averaged cumulated differences* $\sum_{i=0}^{m}(n(C_{i+1}^{\text{ext}}) - n(C_i^{\text{ext}}))/m$ *(lower right),* $m = 0, \ldots, 19998$.

Maximum profile likelihood estimates $(\hat{\kappa}_l, \hat{\alpha}_l)$ are found using Newton-Raphson with initial values given by the minimum contrast estimates of (κ, α) from Example 10.1 (except for $l = 1$ and $l = 2$ where negative values for α are produced in the first Newton-Raphson iteration, and the initial value $(0.005, 1)$ is used instead). In the Newton-Raphson procedure we use a trust factor (see Section 8.3.2) equal to 100 and conditional samples of length 10,000 obtained by subsampling each 10th state of Markov chains of length 100,000. No burn-in is used, since the importance weights only depend on $n(C^{\text{ext}})$ and $\exp(-IZ_{\text{Th}})$, cf. (10.3) and (10.16), and the corresponding trace plots in Figure 10.3 show that the initial values for these statistics are close to the equilibrium distribution. For the same reason no burn-in is used for the path sampling estimation, where Monte Carlo samples are obtained by subsampling each 10th state of Markov chains of length 200,000.

Bridge sampling estimates of likelihood ratios are computed using samples of length 49,950 obtained from Markov chains of length 5,000,000 by omitting a burn-in of the first 5000 states and subsampling each 100th of the remaining states. Here we need conditional samples of the statistics $n(C^{\text{ext}})$, $\prod_{\xi \in x} Z_{\text{Th}}(\xi | C^{\text{ext}}; \alpha, \omega)$, and $\exp(-IZ_{\text{Th}})$, cf. (10.3), (10.16), (10.20), and (10.21). For the sample used in the lower left plot of Figure 10.3, the difference between the initial value $\log[\prod_{\xi \in x} Z_{\text{Th}}(\xi | C_0^{\text{ext}}; \hat{\alpha}_1, \omega_1)]$ and the average value of $\log[\prod_{\xi \in x} Z_{\text{Th}}(\xi | C_m^{\text{ext}}; \hat{\alpha}_1, \omega_1)]$ after the burn-in is around 700. The difference becomes astronomical when exponentiated, and this indicates that using a burn-in is essential for the bridge sampling estimation.

Monte Carlo maximum profile likelihood estimates and log likelihood ratios are given in Table 10.2. The maximal likelihood is obtained for $\hat{\theta}_3$ which is rather different from the minimum contrast estimate from Example 10.1. Figure 10.4 shows the path sampling estimate of the rather flat profile log likelihood when $l = 3$.

l	1	2	3	4	5
ω_l	2	3	4	5	6
$\hat{\kappa}_l$	0.0050	0.0034	0.0026	0.0021	0.0017
$\hat{\alpha}_l$	1.2	1.8	2.5	3.2	4.0
$\log(L(\hat{\theta}_l)/L(\hat{\theta}_1))$	0	4.9	6.3	6.2	5.8

Table 10.2 *Monte Carlo maximum profile likelihood estimates $\hat{\kappa}_l$ and $\hat{\alpha}_l$ for five ω values and Monte Carlo estimates of log likelihood ratios.*

We finally compare the Newton-Raphson procedure with Monte Carlo EM. For the Newton-Raphson procedure with $l = 3$ altogether four Monte Carlo samples are needed. For the Monte Carlo EM algorithm we use the same initial values and Monte Carlo samples of the same size as for the Newton-Raphson procedure. The Monte Carlo EM algorithm comes close to the Monte Carlo maximum profile likelihood estimates $(\hat{\kappa}_3, \hat{\alpha}_3)$ in Table 10.2 after around 15 iterations, each requiring generation of a Monte Carlo sample, and then it starts to oscillate around

MAXIMUM LIKELIHOOD INFERENCE

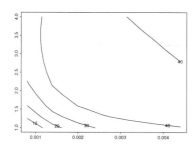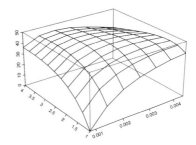

Figure 10.4 *Contour and perspective plot of profile log likelihood* $\log L_3(\kappa, \alpha)$.

the Monte Carlo maximum likelihood estimate. So Newton-Raphson is in this case about four times quicker than Monte Carlo EM.

10.3.2 Likelihood inference for a Poisson-gamma process

For the Poisson-gamma process (Example 5.6, Section 5.4.3) we let the kernel k be given by the density of $N_2(0, \omega^2 I_2)$ as for the Thomas process. The parameter vector is thus $\theta = (\kappa, \tau, \omega)$. We further consider the approximation (10.5) with Φ replaced by the truncated process Φ^{trunc} and $B = W$, and specify X_1 and X_2 as in Section 10.2.2 with $\tilde{\zeta}$ of the form $\tilde{\zeta}(c, \gamma) = \tilde{\chi}(\gamma)$ where $\tilde{\chi}$ is positive and with finite integral on the interval (ϵ, ∞). So the approximate random intensity function is

$$Z_{\text{Pg}}(\xi | \Phi^{\text{trunc}}, \omega) = \sum_{(c,\gamma) \in \Phi^{\text{trunc}}} \gamma \exp\left(-\|\xi - c\|^2/(2\omega^2)\right)/(2\pi\omega^2), \quad \xi \in W,$$

and the joint density of X_W and Φ^{trunc} with respect to (X_1, X_2) is

$$p_\theta(x, \phi) = p(x | Z_{\text{Pg}}(\cdot | \phi, \omega)) \times$$
$$\prod_{(c,\gamma) \in \phi} \frac{\chi_{(\kappa,\tau)}(\gamma)}{\tilde{\chi}(\gamma)} \exp\left(-|B_{\text{ext}}| \int_\epsilon^\infty \left(\chi_{(\kappa,\tau)}(\gamma) - \tilde{\chi}(\gamma)\right) d\gamma\right) \quad (10.22)$$

where $\chi_{(\kappa,\tau)}(\gamma) = \kappa \exp(-\tau\gamma)/\gamma$, cf. (5.16) with $\alpha = 0$. For estimation of the likelihood $L(\theta) = \mathbb{E} p_\theta(x, X_2)$ we use the methods in Section 8.6 with $X = X_W$ and $Y = \Phi^{\text{trunc}}$. For MCMC simulation under a fixed value $\theta_0 = (\kappa_0, \tau_0, \omega_0)$, we take $\tilde{\chi} = \chi_{(\kappa_0, \tau_0)}$.

For the Poisson-gamma process we use the same approach as for the Thomas process in Section 10.3.1 and compute profile likelihood estimates $(\hat{\kappa}_l, \hat{\tau}_l)$ for varying ω values ω_l, $l = 1, \ldots, K$. The derivatives of $\log p_\theta(x, \phi)$ with respect to κ and τ are

$$d \log p_\theta(x, \phi)/d\kappa = \left(n(\phi) - q_{(\kappa,\tau)}(\epsilon)\right)/\kappa$$

and
$$d\log p_\theta(x,\phi)/d\tau = \kappa|B_{\text{ext}}|\exp(-\tau\epsilon)/\tau - \sum_{(c,\gamma)\in\phi}\gamma$$
where
$$q_{(\kappa,\tau)}(\epsilon) = \kappa|B_{\text{ext}}|\int_\epsilon^\infty \exp(-\tau\gamma)/\gamma \, d\gamma.$$

Using a notation as in Section 10.3.1 and letting $\sum_{\Phi^{\text{trunc}}}$ denote summation over $(c,\gamma)\in\Phi^{\text{trunc}}$, the profile score function becomes

$$u_l(\kappa,\tau)$$
$$= \left((\mathbb{E}_{\theta_l,x}n(\Phi^{\text{trunc}}) - q_{(\kappa,\tau)}(\epsilon))/\kappa, \kappa|B_{\text{ext}}|\exp(-\tau\epsilon)/\tau - \mathbb{E}_{\theta_l,x}\sum_{\Phi^{\text{trunc}}}\gamma\right).$$

The profile observed information is

$$j_l(\kappa,\tau)$$
$$= \begin{pmatrix} \mathbb{E}_{\theta_l,x}n(\Phi^{\text{trunc}})/\kappa^2 & -|B_{\text{ext}}|\exp(-\tau\epsilon)/\tau \\ -|B_{\text{ext}}|\exp(-\tau\epsilon)/\tau & \exp(-\tau\epsilon)|B_{\text{ext}}|\kappa(1+\epsilon\tau)/\tau^2 \end{pmatrix} +$$
$$\begin{pmatrix} -\text{Var}_{\theta_l,x}n(\Phi^{\text{trunc}})/\kappa^2 & \text{Cov}_{\theta_l,x}(n(\Phi^{\text{trunc}}),\sum_{\Phi^{\text{trunc}}}\gamma)/\kappa \\ \text{Cov}_{\theta_l,x}(n(\Phi^{\text{trunc}}),\sum_{\Phi^{\text{trunc}}}\gamma)/\kappa & -\text{Var}_{\theta_l,x}\sum_{\Phi^{\text{trunc}}}\gamma \end{pmatrix}.$$

The path sampling and bridge sampling formulae are as for the Thomas process in Section 10.3.1, except that α is replaced by τ in (10.19) and (10.20).

The EM equations for the $(m+1)$th update become

$$\mathbb{E}_{(\tilde{\kappa}_l^{(m)},\tilde{\tau}_l^{(m)},\omega_l),x}n(\Phi^{\text{trunc}}) = q_{(\tilde{\kappa}_l^{(m+1)},\tilde{\tau}_l^{(m+1)})}(\epsilon)$$
$$\tilde{\kappa}_l^{(m+1)}|B_{\text{ext}}|\exp(-\tilde{\tau}_l^{(m+1)}\epsilon)/\tilde{\tau}_l^{(m+1)} = \mathbb{E}_{(\tilde{\kappa}_l^{(m)},\tilde{\tau}_l^{(m)},\omega_l),x}\sum_{(c,\gamma)\in\Phi^{\text{trunc}}}\gamma$$

which are not easily solved with respect to $\tilde{\kappa}_l^{(m+1)}$ and $\tilde{\tau}_l^{(m+1)}$ even when the conditional expectations are replaced by Monte Carlo estimates.

EXAMPLE 10.5 (*Likelihood estimation for the hickory trees*) For the hickory tree data we consider the same five values $\omega_l = 1+l$, $l=1,\ldots,5$, for ω as for the Thomas process in Example 10.4. In analogy with Example 10.4 we find profile likelihood estimates $\hat{\kappa}_l$ and $\hat{\tau}_l$, using Newton-Raphson optimisation with initial values given by the minimum contrast estimates of κ and τ from Example 10.1, and compute bridge sampling estimates of $L(\hat{\kappa}_{l+1},\hat{\tau}_{l+1},\omega_{l+1})/L(\hat{\kappa}_l,\hat{\tau}_l,\omega_l)$, $l=1,\ldots,K-1$. The Monte Carlo sample sizes and burn-in periods are similar to those used in Example 10.4. Monte Carlo maximum profile likelihood estimates and log likelihood ratios are given in Table 10.3. The maximal likelihood is obtained for $\hat{\theta}_3$ which is rather close to the minimum contrast estimate

MAXIMUM LIKELIHOOD INFERENCE

from Example 10.1. Figure 10.5 shows the path sampling estimate of the log profile likelihood for (κ, τ) in the case $l = 3$.

l	1	2	3	4	5
ω_l	2	3	4	5	6
$\hat{\kappa}_l$	0.0033	0.0020	0.0015	0.0012	0.0009
$\hat{\tau}_l$	0.54	0.32	0.23	0.18	0.12
$\log(L(\hat{\theta}_l)/L(\hat{\theta}_1))$	0	8.3	10.1	9.3	7.6

Table 10.3 *Monte Carlo maximum profile likelihood estimates of κ and τ for five ω values and Monte Carlo estimates of log likelihood ratios.*

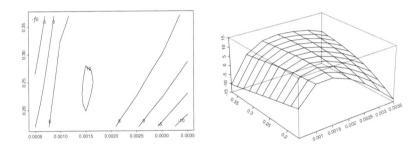

Figure 10.5 *Contour and perspective plot of* $\log L_3(\kappa, \tau)$.

Wolpert & Ickstadt (1998) considered a Bayesian analysis of the hickory data. They obtained posterior means 0.0072, 1.785, and 3.3 for κ/τ (intensity), $1/\tau$ (measures over dispersion), and ω, respectively. For comparison we obtain $\hat{\kappa}_3/\hat{\tau}_3 = 0.006$ and $1/\hat{\tau}_3 = 1.9$ for the Poisson-gamma model, and $\hat{\kappa}_3 \hat{\alpha}_3 = 0.0065$ for the intensity under the Thomas model.

10.3.3 Likelihood inference for LGCPs

For an LGCP we assume that a parametric model m_β is specified for the mean function of the Gaussian field and that the covariance function is given in terms of (5.21) and (5.22) so that the parameter vector is $\theta = (\beta, \sigma^2, \alpha, \delta)$. The likelihood is thus

$$L(\theta) = \mathbb{E}_\theta[p(x|(\exp(Y_W))]$$

which we approximate by

$$\tilde{L}(\theta) = \mathbb{E} p(x| \exp(y^{\text{step}}(\Gamma, \theta))) \qquad (10.23)$$

where using the notation from Section 10.2.3, $y^{\text{step}}(\Gamma, \theta)$ is the step function with step heights given by $\tilde{y}(\Gamma, \theta) = \Gamma Q + m_{\beta,\text{ext}}$ with $m_{\beta,\text{ext}}(\eta) = $

$m_\beta(\eta)$ if $\eta \in I$, and $m_{\beta,\text{ext}}(\eta) = 0$ if $\eta \in I_{\text{ext}} \setminus I$. Note that Q depends on $(\sigma^2, \alpha, \delta)$. The importance sampling approximation based on (10.23) is

$$\tilde{L}(\theta)/\tilde{L}(\theta_0) \approx \frac{1}{m} \sum_{m=0}^{n-1} \frac{p(x|\exp(y^{\text{step}}(\Gamma_m, \theta)))}{p(x|\exp(y^{\text{step}}(\Gamma_m, \theta_0)))}$$

where $\Gamma_0, \Gamma_1, \ldots$ is an MCMC sample with invariant density

$$p_{\theta_0}(\cdot|x) \propto p(x|\exp(y^{\text{step}}(\cdot, \theta_0))) \exp(-\|\cdot\|^2)/2.$$

The MCMC sample can e.g. be generated using the Langevin-Hastings algorithm (Algorithm 10.1) or its truncated version.

The path sampling formula becomes

$\log(\tilde{L}(\theta)/\tilde{L}(\theta_0))$

$$= \int_0^1 \mathbb{E}[\mathrm{d}\log p(x|\exp(y^{\text{step}}(\Gamma, \theta(s))))/\mathrm{d}\theta(s)|X = x] \cdot \theta'(s)\mathrm{d}s.$$

The derivatives of $\log p(x|\exp(y^{\text{step}}(\Gamma, \theta(s))))$ with respect to α and δ are fairly complex, so as in the previous sections, a profile likelihood approach may be useful, where maximum profile likelihood estimates for β and σ^2 are found for a range of values of (α, δ).

We do not pursue likelihood inference for LGCPs further, but return to LGCPs in Example 10.6 below where a fully Bayesian analysis is carried out.

10.4 Bayesian inference

Bayesian analysis of Poisson-gamma processes is considered in e.g. Wolpert & Ickstadt (1998) and Best et al. (2000). In the latter paper the Poisson-gamma process is a model for unexplained spatial variation in a marked point process regression model, where cases of child respiratory diseases are linked to various individual specific and spatial risk factors. Bayesian inference for LGCPs is considered in Benes et al. (2002), Møller & Waagepetersen (2003) and Waagepetersen (2002). In connection with MCMC computation of posterior distributions, it is necessary to discretise the Gaussian field as described in Sections 5.7 and 10.2.3. Since the posterior for a discretised LGCP is formally equivalent to a posterior for a spatial GLMM, some guidance for choosing priors can be found in the literature on Bayesian analysis of GLMMs, see e.g. Christensen & Waagepetersen (2002). Waagepetersen (2002) discusses convergence of the posterior for discretised LGCPs when the cell sizes tends to zero. Using certain approximations from geostatistics, Kelsall & Wakefield (2002) consider a Bayesian analysis for an aggregated LGCP where the data consists of counts of disease cases.

BAYESIAN INFERENCE

The algorithms in Section 10.2 can straightforwardly be extended to accommodate posterior simulation in fully Bayesian analyses with priors also on the model parameters as in the following example.

EXAMPLE 10.6 (*Bayesian analysis for hickory data*) A Bayesian analysis for the hickory tree data is carried out in Wolpert & Ickstadt (1998) who use a Poisson-gamma process as in Example 10.5. Here we consider alternatively the LGCP from Example 10.1 with $\theta = (\beta, \sigma, \alpha)$. Using the notation in Section 10.2.3, let $B = W = [10, 130] \times [10, 130]$ be the observation window for the hickory trees, and consider a discretisation where B is partitioned into 100^2 square cells each of side length 1.2 and centred in the locations in the lattice I from Example 10.3.

In the MCMC computations we reparametrise α by $\kappa = \log(\alpha)$. A priori we let β, σ, and κ be independent with marginal priors

$$p_1(\beta) \propto 1, \ \beta \in \mathbb{R},$$
$$p_2(\sigma) \propto \exp(-10^{-5}/\sigma)/\sigma, \ \sigma > 0,$$
$$p_3(\kappa) \propto 1, \ -2 < \kappa < 4.$$

For the discretised LGCP, we can exploit that the conditional density (10.11) is formally equivalent to the conditional density of random effects Γ given observations n_η, $\eta \in I$, in a generalised linear mixed model (GLMM) with log link and Poisson error distribution (see e.g. Christensen, Diggle & Ribeiro 2003) and check as in Christensen & Waagepetersen (2002) that these priors yield a proper posterior. However, strictly speaking we do not know whether a proper posterior is also obtained for the original LGCP. The improper prior p_1 is completely flat, and the improper p_2 yields an essentially flat prior for $\log \sigma$ on $(0, \infty)$. The choice of prior for κ is a delicate matter for two reasons: First, the posterior distribution of κ is highly sensitive to the prior distribution — it can be verified that the posterior support will always contain the prior support. So using a noninformative improper prior for κ is not possible and it is on the other hand difficult to elicit an informative prior. Second, for numerical reasons we cannot use an unbounded prior for κ. Specifically, we use the circulant embedding technique in Appendix E, and this will typically fail when κ exceeds a certain maximal value depending on the choice of M_{ext} and N_{ext} (the extended grid I_{ext} is of size $M_{\text{ext}} \times N_{\text{ext}}$, and we chose $M_{\text{ext}} = N_{\text{ext}} = 512$).

The posterior density for (Γ, θ) using the discretised LGCP is

$$\pi(\gamma, \beta, \sigma, \kappa | x) \propto p_1(\beta) p_2(\sigma) p_3(\kappa)$$
$$\times \exp\left(-\|\gamma\|^2/2 + \sum_{\eta \in I}(\tilde{y}_\eta n_\eta - A_\eta \exp(\tilde{y}_\eta))\right), \quad (10.24)$$

cf. (10.11). Note that $\tilde{y}$ is a function of $(\gamma, \beta, \sigma, \kappa)$. As in Christensen & Waagepetersen (2002), Christensen et al. (2001), and Beneš et al. (2002), we use a hybrid MCMC algorithm with a systematic updating scheme: In each iteration of the algorithm, γ, β, σ, and κ are updated in turn, using a truncated Langevin-Hastings update for γ and one-dimensional Metropolis-Hastings updates for β, σ, and κ (for further details, see the above-mentioned references). Due to sensitivity to the initial value for γ, we generate first 5000 iterations with a small value of the proposal variance h for the truncated Langevin-Hastings update to let the algorithm settle in equilibrium. Subsequently, we generate 250,000 additional iterations with a value of h chosen to obtain an acceptance rate close to 0.57 for γ. The first 5000 iterations are discarded and we retain each 10th of the last 250,000 iterations in the Monte Carlo sample.

Posterior distributions for β, σ, and κ are shown in Figure 10.6 where also the shape of the prior densities is shown. The posterior distribution for κ is strongly affected by the upper limit of the prior support. Posterior means for β, σ, and κ are -5.82 (0.03), 1.85 (0.01), and 3.12 (0.02), respectively (Monte Carlo standard errors for the posterior means are given in parenthesis). The posterior mean of $\alpha = \exp(\kappa)$ is 25 which yields a considerably stronger correlation than the minimum contrast estimate of $\alpha = 8.1$ in Example 10.1. Figure 10.6 also shows the posterior expectation of $\exp(Y^{\text{step}})$ where Y^{step} is the step function approximation of Y_B, see Section 10.2.3. The minimal and maximal values are 0.0006 and 0.24, respectively. For comparison with Table 10.1, Table 10.4 shows the posterior expectations for $Y^{\text{step}}(\eta)$ at the four locations $\eta \in I$ considered in Table 10.1 and the posterior mean of the Riemann approximation $IZ = \sum_{\eta \in I} A_\eta \exp(\tilde{Y}_\eta)$ of $\int_B \exp(Y(s))\mathrm{d}s$. The values obtained are, except for the location (117.6,123.6), rather close to the values given in Table 10.1 for the Poisson-gamma process.

	(33.6,92.4)	(81.6,42)	(117.6,123.6)	(42.0,51.6)	IZ
LGCP	0.005	0.001	0.20	0.009	84.9
	(0.0002)	(0.00007)	(0.004)	(0.0002)	(0.06)

Table 10.4 *Estimated posterior expectations and Monte Carlo standard errors (in parentheses) for the random intensity step function* $\exp(Y^{\text{step}})$ *at four different locations and for the integrated random intensity step function.*

Figure 10.7 is similar to the lower plots in Figure 10.1. It shows the estimated g and F-functions, but now with envelopes calculated from simulations under the posterior predictive distribution (see e.g. Gelfand 1996) for the LGCP. That is, 39 approximately independent posterior realisations of (Γ, θ) are obtained by subsampling from the MCMC sample, and

BAYESIAN INFERENCE

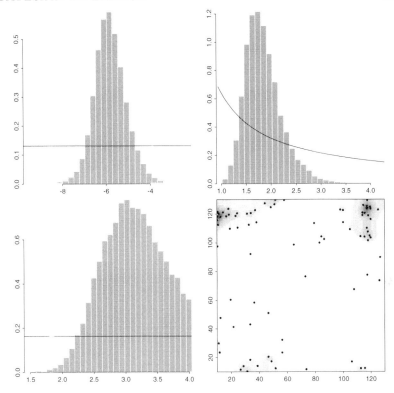

Figure 10.6 *Posterior distributions for β (upper left), σ (upper right), κ (lower left), and posterior mean of $\exp(Y^{\text{step}})$ (lower right). The solid lines show the shapes of the prior densities.*

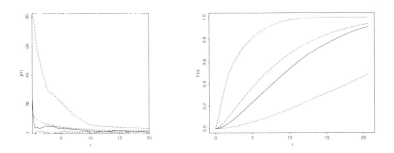

Figure 10.7 *Left: $\hat{g}(r)$ for the hickory data (solid line) and envelopes and average calculated from 39 posterior predictive simulations (dashed line). Right: as left but for $\hat{F}$.*

conditional on each such realisation a simulation of the Poisson process X_W is drawn. Due to the posterior variability of the parameters and since the posterior distribution favours large values of α compared with the minimum contrast estimate, much wider envelopes are obtained in Figure 10.7 than in Figure 10.1.

10.4.1 Bayesian inference for cluster processes

So far we have focused on estimation of parameters and prediction of the intensity function for various parametric Cox process models. Alternatively, if the aggregation is due to some kind of cluster mechanism (e.g. like in an SNCP), we may consider the unobserved cluster centres as the quantity of primary interest, see van Lieshout & Baddeley (2002), McKeague & Loizeaux (2002), and the references therein. In these papers repulsive Markov point process priors with known parameters are imposed on the unobserved pattern of cluster centres. For example, van Lieshout & Baddeley (2002) model the redwood seedlings data (Strauss 1975, Ripley 1977) by a Cox process model with random intensity function given by

$$Z(\xi|\tilde{C}, \epsilon, \mu) = \epsilon + \mu \sum_{c \in \tilde{C}} \mathbf{1}[\|c - \xi\| \leq R]/(\pi R^2). \qquad (10.25)$$

Here $\epsilon > 0$ and $\mu > 0$ are parameters, $\tilde{C}$ is a priori a hard core Markov point process, and $R > 0$ is the radius in the uniform kernel as for the Matérn process in Example 5.3 (Section 5.3). The Cox process can be viewed as a superposition of a "clutter" Poisson process with intensity ϵ and a Cox process with random intensity function given by the last term of the right hand side in (10.25), where the two processes are independent. Both R and the hard core distance in the model for $\tilde{C}$ are assumed to be known, so the only unknown parameters are ϵ and μ. An empirical Bayesian approach is considered in van Lieshout & Baddeley (2002), where μ and ϵ are estimated from the data using the Monte Carlo MLE.

CHAPTER 11

Spatial birth-death processes and perfect simulation

One of the most exciting recent developments in stochastic simulation is perfect (or exact) simulation. Following the seminal work by Propp & Wilson (1996), many perfect samplers have been developed in recent years, particularly for spatial point processes. We study such algorithms in Section 11.2, including the Propp-Wilson algorithm, Wilson's read-once algorithm, and perfect simulation based on spatial birth-death processes.

Section 11.1 discusses how to use spatial birth-death processes for simulation and estimation of mean values. For a long time in the spatial statistical community, spatial birth-death processes were considered as the simulation method for spatial point processes (see Kelly & Ripley 1976, Ripley 1977, Baddeley & Møller 1989), although Metropolis-Hastings algorithms as studied in Section 7.1 were known long before in statistical physics (e.g. Metropolis et al. 1953 and Norman & Filinov 1969). Today the birth-death type Metropolis-Hastings algorithms in Section 7.1.2 seem more popular for MCMC applications, possibly because of simplicity and efficiency (Clifford & Nicholls 1994, Geyer & Møller 1994). On the other hand, spatial birth-death processes seem more applicable for perfect simulation (Kendall & Møller 2000).

11.1 Simulation based on spatial birth-death processes

Preston (1977) provides a detailed mathematical study of finite spatial birth-death processes, and shows among other things that under suitable conditions, (approximate) realisations of a finite point process can be obtained by running a spatial birth-death process for a long enough time; this point was also taken by Kelly & Ripley (1976) and Ripley (1977). Spatial birth-death processes have also been used as statistical models for geological data (Fiksel 1984b, Stoyan et al. 1995) and sand dunes (Møller & Sørensen 1994), but the focus in this chapter is on their close relationship to finite point processes with a density and the simulation aspect.

Sections 11.1.1–11.1.2 describe how spatial birth-death processes can be constructed, and Section 11.1.3 discusses how to simulate a finite spatial point process with an unnormalised density. Section 11.1.4 considers the most common case, i.e. the locally stable and constant death rate case where a useful coupling construction can be obtained. Section 11.1.5 shows that for Monte Carlo estimation, instead of generating a spatial birth-death process, we only need to generate the corresponding discrete time jump chain. For clarity of presentation, many of the mathematical details are given in Appendix G.

We use a notation as in Section 7.1.2 and restrict attention to the case of spatial birth-death processes and point processes with points in a given set $B \subset \mathbb{R}^d$ with $|B| < \infty$. However, everything in this section and in Appendix G easily extends to finite spatial birth-death processes and finite point processes defined on an "arbitrary" state space B (Preston 1977, Møller 1989).

11.1.1 General definition and description of spatial birth-death processes

The distribution of a *spatial birth-death process* $Y_t \in N_f$, $t \in [0, \infty)$, may be specified in terms of the *birth rate* b and *death rate* d which are nonnegative functions defined on $N_f \times B$. Specifically, for $x \in N_f$, let

$$\beta(x) = \int_B b(x, \xi) \mathrm{d}\xi, \quad \delta(\emptyset) = 0, \quad \delta(x) = \sum_{\xi \in x} d(x \setminus \xi, \xi) \text{ if } x \neq \emptyset,$$

and

$$\alpha(x) = \beta(x) + \delta(x). \qquad (11.1)$$

We assume that $\beta(x) < \infty$ for all $x \in N_f$. For $\beta(x) > 0$, define a density $\bar{b}(x, \cdot)$ on B by $\bar{b}(x, \xi) = b(x, \xi)/\beta(x)$, and for $\delta(x) > 0$, a discrete density $\bar{d}(x, \cdot)$ on x by $\bar{d}(x, \eta) = d(x \setminus \eta, \eta)/\delta(x)$. The process Y_t, $t \geq 0$, is right-continuous and piecewise constant except at the *jump times* $T_1 < T_2 < \ldots$: Given an initial state $x_0 \in N_f$, set $T_0 = 0$, $Y_0 = x_0$, and $J_m = Y_{T_m}$, $m = 0, 1, \ldots$, and suppose we condition on $J_0, T_0, \ldots, J_m, T_m$, where $J_m = x$ with $\alpha(x) > 0$. Then

$$T_{m+1} - T_m \sim \mathrm{Exp}(\alpha(x)). \qquad (11.2)$$

If we furthermore condition on T_{m+1}, then either a birth or a death occurs at time T_{m+1}, where

a birth occurs with probability $\beta(x)/\alpha(x)$, in which case $J_{m+1} = x \cup \xi_m$ where the newborn point ξ_m has density $\bar{b}(x, \cdot)$,
$$\qquad (11.3)$$

SPATIAL BIRTH-DEATH PROCESSES

and

a death occurs with probability $\delta(x)/\alpha(x)$, in which case $J_{m+1} = x \setminus \eta_m$ where the point η_m to be deleted from x has density $\bar{d}(x, \cdot)$. (11.4)

It remains to specify what happens if $\alpha(x) = 0$. Then *absorption* occurs, i.e. we set $Y_s = x$ for $s \geq T_m$, and $T_k = \infty$ for $k > m$. We shall mainly exclude this case in the following, since it is not of interest for the purpose of simulation of spatial point processes.

Clearly the *jump chain* $J_0, J_1, J_2, \ldots$, is a Markov chain (when we later in Section 11.2.6 generate the process Y_t backwards in time $t \leq 0$, $J_0 = Y_0$ is the state of the jump which happens just before time 0). Since the exponential distribution has no memory (see Remark 3.3 in Section 3.1.3), we have that Y_t, $t \geq 0$, is a *continuous time Markov process*, meaning that for any time $t \geq 0$, conditional on $Y_t = x$ and the past history Y_s, $s < t$, the waiting time to the first jump in Y_s, $s > t$, is $\text{Exp}(\alpha(x))$-distributed, and the distribution of the jump is still given by (11.3)–(11.4).

The process is said to *explode* if an infinite number of jumps in a finite time interval can occur with a positive probability. We assume henceforth that explosion does not occur; sufficient conditions are given in Section G.2.

11.1.2 General algorithms

Let the situation be as in the previous section. Formally, the construction of a spatial birth-death process is given by the following algorithm.

ALGORITHM 11.1 (*Spatial birth-death process algorithm: I*) For $m = 0, 1, \ldots$, given that $(T_m, J_m) = (t, x)$ with $\alpha(x) > 0$, generate (T_{m+1}, J_{m+1}) as follows:

(i) draw $R'_m \sim \text{Uniform}([0, 1])$ and $R''_m \sim \text{Uniform}([0, 1])$;

(ii) set $T_{m+1} = t + \log(-R'_m)/\alpha(x)$;

(iii) if $R''_m \leq \beta(x)/\alpha(x)$, generate $\xi_m \sim \bar{b}(x, \cdot)$ and set $J_{m+1} = x \cup \xi_m$;

(iv) else generate $\eta_m \sim \bar{d}(x, \cdot)$ and set $J_{m+1} = x \setminus \eta_m$.

Conditional on the random variables used for the generation of $(T_0, J_0, \ldots, T_m, J_m)$, it is assumed that R'_m, R''_m, and ξ_m or η_m are mutually independent.

Algorithm 11.1 can be too slow, since it is needed to calculate $\beta(x)$ and, in case of (iii), to make a simulation from $\bar{b}(x, \cdot)$. However, we can often solve this problem by a kind of *rejection sampling*: Suppose that

$b(x,\xi) \le b'(x,\xi)$ where $\beta'(x) = \int_B b'(x,\xi)\mathrm{d}\xi < \infty$ is easy to calculate and it is easy to simulate from the density $\bar{b}'(x,\cdot) = b'(x,\cdot)/\beta'(x)$ for $\beta'(x) > 0$. For $a > 0$, set $a/0 = \infty$. Then we proceed as follows.

ALGORITHM 11.2 (*Spatial birth-death process algorithm: II*) For $m = 0, 1, \ldots$, given that $T_m = t$ and $J_m = x = \{x_1, \ldots, x_n\}$ with $n \ge 0$ and $\alpha(x) > 0$, generate (T_{m+1}, J_{m+1}) as follows:

(i) set $j = 1$ and if $n > 0$ then for $i = 1, \ldots, n$, draw $R_m^{(i)} \sim$ Uniform $([0,1])$ and set $\tau_i = \log(-R_m^{(i)})/d(x \setminus x_i, x_i)$;

(ii) draw $R_{jm} \sim$ Uniform$([0,1])$ and set $\tau_0 = \log(-R_{jm})/\beta'(x)$ and $\tau = \min\{\tau_0, \ldots, \tau_n\}$;

(iii) if $\tau = \tau_i$ with $i > 0$ then return $T_{m+1} = t + \tau_i$ and $J_{m+1} = x \setminus x_i$;

(iv) else

  (a) generate $\xi_{jm} \sim \bar{b}'(x, \cdot)$ and $R'_{jm} \sim$ Uniform$([0,1])$;
  (b) if $R'_{jm} \le b(x, \xi_{jm})/b'(x, \xi_{jm})$ then return $T_{m+1} = t + \tau_0$ and $J_{m+1} = x \cup \xi_{jm}$;
  (c) else $j \leftarrow j + 1$ and $t \leftarrow t + \tau_0$, and repeat (ii)–(iv).

Conditional on the random variables used for the generation of $(T_0, J_0, \ldots, T_m, J_m)$, it is assumed that $R_m^{(1)}, \ldots, R_m^{(n)}, R_{jm}, R'_{jm}, \xi_{jm}, j = 1, 2, \ldots$, are mutually independent.

PROPOSITION 11.1 *Algorithm 11.2 produces a spatial birth-death process with birth rate b and death rate d.*

Proof. Assume that $\beta'(x) > 0$, and imagine that we repeat all of the steps (ii)–(iv) in Algorithm 11.2 for all $j = 1, 2, \ldots$ (that is, (iv) is performed irrespective of whether $\tau = \tau_i$ with $i > 0$ in (iii) or not, and (c) is performed irrespective of whether the condition in (b) is satisfied or not). Let $Z+1$ denote the value of j the first time the inequality in (b) is satisfied, and let $T = \sum_{j=1}^{Z+1} \log(-R'_{jm})/\beta'(x)$ be the corresponding waiting time. Then $Z \sim \text{Geo}(q)$ with

$$q = \int_B \bar{b}'(x,\xi)b(x,\xi)/b'(x,\xi)\mathrm{d}\xi = \beta(x)/\beta'(x),$$

and the point born at time $T_m + T$ has density

$$[\bar{b}'(x,\xi)b(x,\xi)/b'(x,\xi)]/q = \bar{b}(x,\xi).$$

Further, for $j \in \mathbb{N}_0$, $T|Z = j$ follows a gamma distribution $\Gamma(j+1, \beta'(x))$ with shape parameter $j+1$ and inverse scale parameter $\beta'(x)$. Hence by Lemma 11.1 below, $T \sim \text{Exp}(\beta(x))$. Furthermore, by Lemma 11.2 below and the independence assumptions in Algorithm 11.2, $T_{m+1} - T_m$ is

SPATIAL BIRTH-DEATH PROCESSES   209

distributed as $\tau = \min\{T, \tau_1, \ldots, \tau_n\} \sim \mathrm{Exp}(\alpha(x))$, and a death $x \to x \setminus x_i$ happens with probability $P(\tau = \tau_i) = \delta(x \setminus x_i, x_i)/\alpha(x)$. Thereby (11.2)–(11.4) are verified.

The case $\beta'(x) = 0$ is verified in a similar way, using that $\tau_0 = \infty$ and $\tau_i < \infty$ for some $i > 0$, since it assumed that $\alpha(x) > 0$. □

LEMMA 11.1 *Assume that for any $n \in \mathbb{N}_0$, $T|Z = n \sim \Gamma(n+1, a)$ where $a > 0$ and $Z \sim \mathrm{Geo}(q)$ with $0 < q < 1$. Then $T \sim \mathrm{Exp}(aq)$.*

Proof. By the law of total probability, T has density

$$f_T(t) = \sum_{n=0}^{\infty} q(1-q)^n \frac{a^{n+1}}{n!} t^n e^{-at} = aqe^{-at} \sum_{n=0}^{\infty} \frac{[(1-q)at]^n}{n!} = aqe^{-aqt}$$

for $t > 0$. □

LEMMA 11.2 *Let $\tau = \min\{\tau_1, \ldots, \tau_n\}$ where $\tau_1 \sim \mathrm{Exp}(\alpha_1), \ldots, \tau_n \sim \mathrm{Exp}(\alpha_n)$ are mutually independent with $\alpha_i \geq 0$, $i = 1, \ldots, n$, so that $\alpha = \alpha_1 + \ldots + \alpha_n$ is positive (where $\mathrm{Exp}(0)$ denotes the distribution concentrated at ∞). Then $\tau \sim \mathrm{Exp}(\alpha)$ and $P(\tau = \tau_i) = \alpha_i/\alpha$, $i = 1, \ldots, n$.*

Proof. For $t > 0$,

$$P(\tau > t) = \prod_{i=1}^{n} P(\tau_i > t) = \prod_{i=1}^{n} e^{-\alpha_i t} = e^{-\alpha t},$$

and using the law of total probability by conditioning on e.g. τ_1, we obtain

$$P(\tau = \tau_1) = \int_0^{\infty} \alpha_1 e^{-\alpha_1 t} \prod_{i=2}^{n} e^{-\alpha_i t} dt = \alpha_1/\alpha.$$

□

11.1.3 Simulation of spatial point processes with a density

We now discuss how Algorithms 11.1–11.2 can be used for simulation of a point process X with unnormalised density h with respect to Poisson $(B, 1)$.

Suppose that for all $x \in N_{\mathrm{f}}$ and $\xi \in B \setminus x$,

$$h(x)b(x, \xi) = h(x \cup \xi)d(x, \xi). \tag{11.5}$$

This is a *detailed balance condition* which with the additional assumption $\mathbb{E}\beta(X) < \infty$ implies that the spatial birth-death process is reversible with respect to h, cf. Proposition G.3 (Section G.3). The detailed balance condition also implies that with probability one,

$$Y_0 \in E \Rightarrow Y_t \in E, \ t \geq 0,$$

where E is defined by (7.8).

LEMMA 11.3 *If we in addition to (11.5) assume*

$$x \cup \xi \in E \Rightarrow d(x, \xi) > 0, \tag{11.6}$$

then h is hereditary, $\alpha(\emptyset) = 0$ if $E = \{\emptyset\}$, and $\alpha(x) > 0$ for all $x \in E$ if $E \neq \{\emptyset\}$.

Proof. From (11.5)–(11.6) follow immediately that h is hereditary. Since h is a hereditary unnormalised density, $h(\emptyset) > 0$, and so by (11.5),

$$\alpha(\emptyset) = \int_B b(\emptyset, \xi) \mathrm{d}\xi = \frac{1}{h(\emptyset)} \int_B h(\xi) d(\emptyset, \xi) \mathrm{d}\xi.$$

Thus $\alpha(\emptyset) = 0$ if $E = \{\emptyset\}$, and $\alpha(\emptyset) > 0$ if $E \neq \{\emptyset\}$ because of (11.6) and since $\int_B h(\xi) \mathrm{d}\xi > 0$ when h is hereditary and $E \neq \{\emptyset\}$. Suppose that $x \in E \setminus \{\emptyset\}$. By (11.6), $d(x \setminus \xi, \xi) > 0$ for $\xi \in x$, so $\alpha(x) \geq \delta(x) > 0$. $\square$

We also obtain irreducibility under (11.5)–(11.6), since the process can move from any state in E to $\emptyset$, and from $\emptyset$ to any state in E. Under weak conditions, Y_t converges in distribution towards X (Propositions G.4–G.7 in Section G.4). Properties of MCMC estimates are discussed in Sections 11.1.5 and G.4.

In practice we are given h and construct b and d such that (11.5) and (11.6) hold. When $h(\cdot) > 0$ the computationally simplest case is to let $b(\cdot, \cdot) = 1$, and hence by detailed balance, $d(x, \xi) = 1/\lambda^*(x, \xi)$. Empirically, unless h is close to the density for a homogeneous Poisson process, the spatial birth-death process tends to delete a newborn point immediately; compare with Remark 7.2 in Section 7.1.1. The commonly used case in the point process literature is the other extreme where

$$b(x, \xi) = \lambda^*(x, \xi), \quad d(\cdot, \cdot) = 1. \tag{11.7}$$

We refer to this as the *constant death rate case*. If also h is ϕ^*-locally stable, we refer to (11.7) as the *locally stable and constant death rate case*. In that case,

$$\mathbb{E}\beta(X) \leq \int_B \phi^*(\xi) \mathrm{d}\xi = c^*$$

is finite.

Algorithm 11.2 applies in the ϕ^*-locally stable and constant death rate case, as we can take $b'(x, \xi) = \phi^*(\xi)$. Then $\beta'(x) = c^*$ does not depend on x, and a slightly more efficient modification of Algorithm 11.2 is given

SPATIAL BIRTH-DEATH PROCESSES   211

as follows. Instead of generating the random variables $R_m^{(1)}, \ldots, R_m^{(n)}$ in step (i) of Algorithm 11.2, we generate a life time $\tau(\xi) \sim \mathrm{Exp}(1)$ for each newborn point ξ. These life times are assumed to be mutually independent and independent of $R_{jm}, R'_{jm}, \xi_{jm}, m = 0, 1, \ldots, j = 1, 2, \ldots$. For $i = 1, \ldots, n$ in step (i) of Algorithm 11.2, we then set $\tau_i = T(x_i) + \tau(x_i) - t$ where $T(x_i)$ denotes the time at which x_i was born. This works because $d(\cdot, \cdot) = 1$ and the exponential distribution has no memory (see Remark 3.3, Section 3.1.3).

Finally, suppose that
$$\tilde{h}(x) = \alpha(x)h(x) \qquad (11.8)$$
is integrable with respect to $\mathrm{Poisson}(B, 1)$. If (11.5) holds, then the jump chain is reversible with respect to $\tilde{h}$ (Proposition G.6, Section G.4); or if Y_t has invariant unnormalised density h, then the jump chain has invariant unnormalised density $\tilde{h}$ (Remark G.2, Section G.4). Proposition G.7 in Section G.4 implies that in the locally stable and constant death rate case, the jump chain is ergodic.

11.1.4 A useful coupling construction in the locally stable and constant death rate case

In the locally stable and constant death rate case a useful *coupling construction* can be obtained (Kendall 1998, Kendall & Møller 2000). We shall use this construction several times and in particular in Sections 11.2.6–11.2.7 for making perfect simulation. The coupling construction is given by a bivariate process (Y_t, D_t) where both Y_t and D_t are spatial birth-death processes, with birth and death rate for Y_t given by (11.7). A formal description of the coupling construction is given in Algorithm 11.3 below.

Recall that by ϕ^*-local stability, $\phi^*(\xi) \geq \lambda^*(\cdot, \xi)$. We construct D_t, $t \geq 0$, as a spatial birth-death process with birth and death rate
$$b_D(x, \xi) = \phi^*(\xi), \quad d_D(\cdot, \cdot) = 1. \qquad (11.9)$$
Write D_{t-} for the state just before time t, i.e. a jump occurs at time t if and only if $D_t \neq D_{t-}$. In the coupling construction we assume that
$$Y_0 \in E, \quad Y_0 \subseteq D_0 \qquad (11.10)$$
(e.g. take $Y_0 = \emptyset$). The construction is so that Y_t only jumps when D_t jumps:
$$D_t = D_{t-} \Rightarrow Y_t = Y_{t-}. \qquad (11.11)$$
Further,
$$D_t = D_{t-} \setminus \eta_t \Rightarrow Y_t = Y_{t-} \setminus \eta_t, \qquad (11.12)$$
i.e. a deletion of a point in D_{t-} implies the deletion of the same point

212 BIRTH-DEATH PROCESSES AND PERFECT SIMULATION

in Y_{t-} (provided the point is in Y_{t-}). Finally, if a point ξ_t is added to D_{t-}, then it is added to Y_{t-} with probability $\lambda^*(Y_{t-},\xi_t)/\phi^*(\xi_t)$, i.e.

$$D_t = D_{t-} \cup \xi_t \Rightarrow Y_t = Y_{t-} \cup \xi_t \text{ with probability } \lambda^*(Y_{t-},\xi_t)/\phi^*(\xi_t), \tag{11.13}$$

and otherwise we retain $Y_t = Y_{t-}$.

Note that $Y_t \in E$ for all $t \geq 0$, since h is hereditary and (11.10) holds. Further, D_t *dominates* Y_t in the sense that

$$Y_t \subseteq D_t, \quad t \geq 0. \tag{11.14}$$

This follows by induction from (11.10)–(11.13).

In Algorithm 11.3 we let $T'_0 = 0 < T'_1 < T'_2 < \ldots$ denote the jump times and $J'_0 = D_0, J'_1, J'_2, \ldots$ the jump chain for the dominating process D_t, $t \geq 0$. The density for newborn points in D_t is denoted $\bar{b}'(\xi) = \phi^*(\xi)/c^*$. Furthermore, the jump chain of Y_t is given by the jumps in $\tilde{J}_m = Y_{T'_m}$, $m = 0, 1, \ldots$.

ALGORITHM 11.3 (*Coupling construction in the locally stable and constant death rate case*) For $m = 0, 1, \ldots$, given that $(T'_m, J'_m, \tilde{J}_m) = (t, x, y)$ with $y \in E$, $x \in N_f$, and $y \subseteq x$, generate $(T'_{m+1}, J'_{m+1}, \tilde{J}_{m+1})$ as follows:

(i) draw $R'_m \sim \text{Uniform}([0, 1])$ and $R''_m \sim \text{Uniform}([0, 1])$;

(ii) set $T'_{m+1} = t + \log(-R'_m)/(c^* + n(x))$;

(iii) if $R''_m \leq c^*/(c^*+n(x))$, generate $\xi_m \sim \bar{b}'$ and $R'''_m \sim \text{Uniform}([0, 1])$, and set $J'_{m+1} = x \cup \xi_m$ and

$$\tilde{J}_{m+1} = \begin{cases} y \cup \xi_m & \text{if } R'''_m \leq \lambda^*(y,\xi_m)/\phi^*(\xi_m) \\ y & \text{otherwise;} \end{cases}$$

(iv) else generate $\eta_m \sim \text{Uniform}(x)$ and set $J'_{m+1} = x \setminus \eta_m$ and $\tilde{J}_{m+1} = y \setminus \eta_m$.

Conditional on the random variables used for the generation of $(T'_0, J'_0, \tilde{J}_0, \ldots, T'_m, J'_m, \tilde{J}_m)$, it is assumed that $R'_m, R''_m, R'''_m, \xi_m, \eta_m$ are mutually independent.

REMARK 11.1 The (Y_t, D_t) process can be constructed using another approach which is stochastically equivalent to the construction in Algorithm 11.3 (i.e. the two constructions generate processes which are identically distributed). Consider a marked Poisson process with points $\tilde{T}_m$ and marks $(\xi_m, \tau_m, \tilde{R}_m)$, $m = 1, 2, \ldots$, where $\tilde{T}_0 = 0 < \tilde{T}_1 < \tilde{T}_2 < \ldots$, and where the random variables

$$\tilde{T}_m - \tilde{T}_{m-1} \sim \text{Exp}(c^*), \quad \xi_m \sim \bar{b}', \quad \tau_m \sim \text{Exp}(1), \quad \tilde{R}_m \sim \text{Uniform}([0, 1]), \tag{11.15}$$

SPATIAL BIRTH-DEATH PROCESSES

$m = 1, 2, \ldots$, are mutually independent. We interpret $\tilde{T}_m$ as the birth time and $\tilde{T}_m + \tau_m$ as the death time of the point ξ_m in D_t. The jump chain of (Y_t, D_t), $t \geq 0$, is then given by

$$D_{\tilde{T}_m} = D_{\tilde{T}_m-} \cup \xi_m$$

$$Y_{\tilde{T}_m} = \begin{cases} Y_{\tilde{T}_m-} \cup \xi_m & \text{if } \tilde{R}_m \leq \lambda^*(Y_{\tilde{T}_m-}, \xi_m)/\phi^*(\xi_m) \\ Y_{\tilde{T}_m-} & \text{otherwise} \end{cases}$$

$$D_{\tilde{T}_m+\tau_m} = D_{\tilde{T}_m+\tau_m-} \setminus \xi_m$$

$$Y_{\tilde{T}_m+\tau_m} = Y_{\tilde{T}_m+\tau_m-} \setminus \xi_m$$

for $m = 1, 2, \ldots$. We verify in Proposition G.1 in Section G.2 that this construction is stochastic equivalent to that in Algorithm 11.3, and that Y_t and D_t individually become spatial birth-death processes with the correct birth and death rates.

REMARK 11.2 From (11.5) and (11.9) follows that the process D_t is reversible with respect to h_D given by

$$h_D(x) = \prod_{\xi \in x} \phi^*(\xi), \qquad (11.16)$$

that is, an unnormalised density for Poisson(B, ϕ^*), cf. (6.2). Recall that the process Y_t is reversible with respect to h. However, as noted in Berthelsen & Møller (2002a), since in general we have obtained Y_t by a *dependent thinning* of D_t, the process (Y_t, D_t) is not reversible, and it seems complicated to obtain a closed form expression for its limit distribution as $t \to \infty$ (existence of the limit distribution is given by Proposition G.7 in Section G.4).

To see this, suppose for simplicity that $h(\cdot) > 0$, and let $f(\cdot, \cdot)$ denote a density with respect to two independent standard Poisson processes on B. Reversibility of the process (Y_t, Z_t) with respect to f is essentially equivalent to the following detailed balance equations:

$$f(y, z)\lambda^*(y, \xi) = f(y \cup \xi, z) \qquad (11.17)$$

and

$$f(y, z)(\phi^*(\xi) - \lambda^*(y, \xi)) = f(y, z \cup \xi) \qquad (11.18)$$

(this follows by similar arguments as in the proof of Proposition G.3 in Section G.3). We know that Y_t is reversible with respect to the marginal density $f_Y(y) \propto h(y)$ obtained by integrating $f(y, z)$ over z (with respect to a standard Poisson process). By (11.17),

$$f(z|y) = \frac{f(y, z)}{f_Y(y)} = \frac{f(y, z)\lambda^*(y, \xi)}{f_Y(y \cup \xi)} = \frac{f(y \cup \xi, z)}{f_Y(y \cup \xi)} = f(z|y \cup \xi),$$

and so by induction $f(z|y) = f_Z(z)$ does not depend on $y \in N_f$. Hence by (11.18),

$$\phi^*(\xi) - \lambda^*(y,\xi) = f(y, z \cup \xi)/f(y,z) = f(z \cup \xi|y)/f(z|y) = f_Z(z \cup \xi)/f_Z(z)$$

(taking $0/0 = 0$) does not depend on $y \in N_f$. Consequently, $\lambda^*(y,\xi) = \rho(\xi)$ is the intensity function for X, so the process (Y_t, D_t) is not reversible except in the following trivial case: if we condition on (Y_0, Z_0), the processes Y_t, $t > 0$, and Z_t, $t > 0$, are independent spatial birth-death processes, both with death rates 1, and birth rate ρ, respectively, $\phi^* - \rho$. This is in accordance with Proposition 3.7 (Section 3.2.2), since we are then obtaining Y_t by an independent thinning of D_t.

11.1.5 Ergodic averages for spatial birth-death processes

For Monte Carlo computations based on spatial birth-death processes we argue now why a certain ergodic average based on the jump chain should be used.

Consider again a spatial birth-death process Y_t, $t \geq 0$, with invariant unnormalised density h, jump times $T_0 = 0 < T_1 < T_2 \ldots$, and jump chain $J_0 = Y_0, J_1, J_2, \ldots$. Suppose we wish to estimate $\mathbb{E}k(X)$ where k is a real function and $X \sim h$, and we decide to generate the spatial birth-death process for times $t \leq T_n$. Under suitable conditions, an obvious and consistent estimate is given by

$$F_n(k) = \frac{\sum_{m=0}^{n-1} k(J_m)(T_{m+1} - T_m)}{\sum_{m=0}^{n-1} (T_{m+1} - T_m)} \quad (11.19)$$

By a Rao-Blackwellisation argument we show now that a better estimate is given by

$$G_n(k) = \frac{\sum_{m=0}^{n-1} k(J_m)/\alpha(J_m)}{\sum_{m=0}^{n-1} 1/\alpha(J_m)}. \quad (11.20)$$

PROPOSITION 11.2 *Under suitable conditions, with probability one, both $F_n(k) \to \mathbb{E}k(X)$ and $G_n(k) \to \mathbb{E}k(X)$ as $n \to \infty$, and $G_n(k)$ is asymptotically more efficient than $F_n(k)$.*

Proof. We only give a sketch proof which indicates what is meant by "suitable conditions".

Suppose that $\tilde{h}$ given by (11.8) is integrable with respect to Poisson$(B, 1)$. Then, if $Y_0 \sim h$ and hence $Y_t \sim h$, $t \geq 0$,

$$\mathbb{E}[k(J_m)(T_{m+1} - T_m)] = \mathbb{E}[k(J_m)\mathbb{E}(T_{m+1} - T_m|J_m)] = \mathbb{E}[k(J_m)/\alpha(J_m)]$$
$$= (c/\tilde{c})\mathbb{E}k(X)$$

where c and $\tilde{c}$ are the normalising constants for h and $\tilde{h}$. Hence, if a law of

SPATIAL BIRTH-DEATH PROCESSES   215

large numbers applies for the numerator and denumerator in both (11.19) and (11.20), with probability one, both $F_n(k)$ and $G_n(k)$ converge to

$$(c/\tilde{c})\mathbb{E}k(X)/(c/\tilde{c}) = \mathbb{E}k(X) \quad \text{as } n \to \infty.$$

To establish the other assertion, consider

$$\sqrt{n}(F_n(k) - \mathbb{E}k(X)) = \frac{\frac{1}{\sqrt{n}}\sum_{m=0}^{n-1}[k(J_m) - \mathbb{E}k(X)][T_{m+1} - T_m]}{\frac{1}{n}\sum_{m=0}^{n-1}(T_{m+1} - T_m)}$$

and

$$\sqrt{n}(G_n(k) - \mathbb{E}k(X)) = \frac{\frac{1}{\sqrt{n}}\sum_{m=0}^{n-1}[k(J_m) - \mathbb{E}k(X)]/\alpha(J_m)}{\frac{1}{n}\sum_{m=0}^{n-1}1/\alpha(J_m)}.$$

Arguing as above, the denumerators converge with probability one toward the same limit, namely $c/\tilde{c}$. Hence, if a central limit theorem applies for the numerators, it also applies for $\sqrt{n}(F_n(k) - \mathbb{E}k(X))c/\tilde{c}$ and $\sqrt{n}(G_n(k) - \mathbb{E}k(X))c/\tilde{c}$, which are asymptotically normally distributed with mean 0 and variances given by the asymptotic variances of the numerators. Conditional on $J_0, \ldots, J_{n-1}$, we know that $T_{m+1} - T_m \sim \text{Exp}(\alpha(J_m))$, $m = 0, \ldots, n-1$, are mutually independent. By Rao-Blackwellisation (i.e. using the well-known equation $\text{Var}(Z) = \mathbb{E}\text{Var}(Z|Y) + \text{Var}\mathbb{E}(Z|Y)$ for random variables Y and Z with $E(Z^2) < \infty$),

$$\text{Var}\left(\sum_{m=0}^{n-1}[k(J_m) - \mathbb{E}k(X)][T_{m+1} - T_m]\right)$$

$$\geq \text{Var}\left(\mathbb{E}\left(\sum_{m=0}^{n-1}[k(J_m) - \mathbb{E}k(X)][T_{m+1} - T_m]\,\bigg|\,J_0, \ldots, J_{n-1}\right)\right)$$

$$= \text{Var}\left(\sum_{m=0}^{n-1}[k(J_m) - \mathbb{E}k(X)]/\alpha(J_m)\right)$$

(provided the variances exist). Consequently, the asymptotic variance of $\sqrt{n}(F_n(k) - \mathbb{E}k(X))$ is larger than that of $\sqrt{n}(G_n(k) - \mathbb{E}k(X))$. □

We can rewrite $G_n(k)$ as a ratio of ergodic averages by dividing both the numerator and denumerator in (11.20) by n. However, as in Section 11.1, one problem is that we usually cannot easily calculate $\alpha(J_m)$. Fortunately, in the locally stable and constant death rate case, we can derive an alternative estimate of $\mathbb{E}k(X)$ as follows.

Recall the notation and the coupling construction in Section 11.1.4. Proposition G.7 in Section G.4 gives the following. With probability one, if $k \in \mathcal{L}(h)$,

$$G'_n(k) \equiv \frac{\sum_{m=0}^{n-1} k(\tilde{J}_m)/(c^* + n(J'_m))}{\sum_{m=0}^{n-1} 1/(c^* + n(J'_m))} \to \mathbb{E}k(X) \quad \text{as } n \to \infty. \quad (11.21)$$

Furthermore, if e.g. $k \in \mathcal{L}^{2+\epsilon}(h)$ for some $\epsilon > 0$, then $\sqrt{n}(G'_n(k) - \mathbb{E}k(X))$ converges in distribution towards $N(0, \bar{\sigma}^2)$ where $\bar{\sigma}^2$ is the asymptotic variance (see Proposition G.7). Finally, by a similar Rao-Blackwellisation argument as above, $G'_n(k)$ is an asymptotically more efficient estimate of $\mathbb{E}k(X)$ than the obvious estimate

$$F'_n(k) \equiv \frac{\sum_{m=0}^{n-1} k(\tilde{J}_m)(T'_{m+1} - T'_m)}{\sum_{m=0}^{n-1} (T'_{m+1} - T'_m)}.$$

11.2 Perfect simulation

A Markov chain algorithm is said to be *exact* if it returns exact draws from a given target distribution when the algorithm completes; usually the algorithm running time is then random but still finite. For several reasons simulations cannot be "exact" in the precise sense: in practice random number generators always have defects, and there is the possibility that the algorithm fails to deliver an answer within practical constraints of time (however, see the discussion in Section 11.2.9 on Fill's algorithm). It is therefore preferable to use the term *perfect simulation* for these "exact" simulation methods (Kendall 1998, Kendall & Møller 2000).

Perfect simulation is obviously appealing and potentially very useful for many reasons: The problem of assessing an "appropriate" burn-in, which sometimes can be a difficult task, is completely eliminated. Further, i.i.d. sampling is available by perfect simulation, so that e.g. asymptotic variances of Monte Carlo estimates can be calculated very easily. Furthermore, one can assess the approximation error incurred by a "non-perfect" algorithm via comparison with an algorithm which is a perfect variation of the original (a useful point if the perfect version is very costly in computational terms), see Berthelsen & Møller (2003) for an example. Although perfect simulation can be very time consuming, a careful output analysis can be time consuming as well, and in contrast to perfect simulation it is not automatic.

In this section we describe different versions of perfect simulation, starting in Section 11.2.1 with a general setting for *coupling from the past (CFTP)* algorithms. Sections 11.2.2–11.2.3 introduce the innovative CFTP algorithm by Propp & Wilson (1996). Section 11.2.4 illustrates how the algorithm applies for continuum Ising models. Section 11.2.5 deals with the read-once algorithm by Wilson (2000a). However, as

PERFECT SIMULATION   217

pointed out in Section 11.2.6, the algorithms in Sections 11.2.2–11.2.5 do not easily apply for most spatial point process models. Sections 11.2.6–11.2.7 therefore describe other versions of perfect samplers based on spatial birth-death processes, which apply for locally stable point processes. Section 11.2.8 discusses some empirical findings for such algorithms. Finally, Section 11.2.9 discusses some other perfect samplers for spatial point processes.

David Wilson's web site (http://dimacs.rutgers.edu/~dbwilson/exact) provides a comprehensive list of references and the history of perfect simulation. Surveys on perfect simulation for spatial point processes are given in Møller (2001) and Berthelsen & Møller (2002a).

11.2.1 General CFTP algorithms

We consider first a general setting with a discrete time Markov chain $\{Y_m\}_{m=0}^{\infty}$ with invariant distribution Π defined on a state space Ω. Under mild conditions, the chain can be constructed by a so-called *stochastic recursive sequence (SRS)*,

$$Y_{m+1} = \varphi(Y_m, R_m), \quad m = 0, 1, 2, \ldots, \qquad (11.22)$$

where the R_m are i.i.d. random variables and φ is a deterministic function, called the *updating function*; see Foss & Tweedie (1998) and the references therein. In practice, φ and R_m are given by the computer code with R_m generated by a vector of pseudo-random numbers $V_m = (V_{m,1}, \ldots, V_{m,N_m})$, where $N_m \in \mathbb{N}$ is either a constant or yet another pseudo-random number (the latter situation will be exemplified in Section 11.2.4; in the descriptions of algorithms to follow, it is convenient to let R_m be a function of V_m, though from a mathematical point of view we may let $R_m = V_m$). Moreover, we include negative times and let for $x \in E$ and integers $m \leq n$,

$$Y_m^n(x) = x \quad \text{if } m = n$$

and

$$Y_m^n(x) = \varphi\Big(\cdots\varphi\big(\varphi(x, R_m), R_{m+1}\big)\cdots, R_{n-1}\Big) \quad \text{if } m < n \qquad (11.23)$$

denote the state of a chain at time n when it is started in x at time m. Notice that the $Y_m^n(x)$ for $x \in \Omega$ are *coupled* via the common "random bits" $R_m, R_{m+1}, \ldots, R_{n-1}$.

EXAMPLE 11.1 (*Random walk*) Consider a random walk on $\Omega = \{0, 1, \ldots, k\}$, where $k \in \mathbb{N}$, the R_m are i.i.d. and uniformly distributed on $\{-1, 1\}$, and the updating function is given by

$$\varphi(x, 1) = \min\{x + 1, k\}, \quad \varphi(x, -1) = \max\{x - 1, 0\}. \qquad (11.24)$$

218 BIRTH-DEATH PROCESSES AND PERFECT SIMULATION

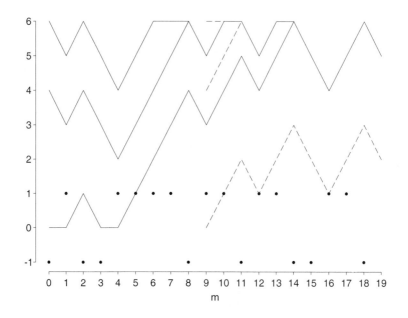

Figure 11.1 For x equal to $0, 4,$ or 6, the solid lines connects $Y_0^n(x)$, $n = 0, \ldots, 19$, and the dashed lines connects $Y_9^n(x)$, $n = 9, \ldots, 19$. The dots show R_m, $m = 0, \ldots, 18$.

Figure 11.1 shows how the paths $Y_m^n(x)$, $n = m, m+1, \ldots$, are coupled by the random bits in the case $k = 6$.

The following proposition has some similarity to Kendall & Møller (2000, Theorem 2.1) and covers Propp & Wilson (1996, Theorem 3). It is stated so that it applies for both discrete and continuous time Markov chains (the latter case is needed in Sections 11.2.6 11.2.7). Let $\mathbb{T}$ denote either discrete time $\mathbb{T} = \mathbb{Z}$ (the set of integers) or continuous time $\mathbb{T} = \mathbb{R}$. We extend (11.23) by assuming that for any $s \in \mathbb{T}$ and a given $\hat{x} \in \Omega$, there is defined a stochastic process $Y_s^t(\hat{x})$, $t \in \mathbb{T} \cap [s, \infty)$, on Ω by $Y_s^s(\hat{x}) = \hat{x}$ and

$$Y_s^t(\hat{x}) = \varphi(\hat{x}, (Z_u; u \in [s, t) \cap \mathbb{T})), \quad t \in \mathbb{T} \cap (s, \infty), \qquad (11.25)$$

for some deterministic function φ and a stationary stochastic process Z_t, $t \in \mathbb{T}$ (i.e. for any $s \in \mathbb{T}$, the processes $(Z_t; t \in \mathbb{T})$ and $(Z_{s+t}; t \in \mathbb{T})$ are identically distributed). This is clearly satisfied in the discrete time case (11.23) where the $Z_t = R_t$ are i.i.d. Note that for $r, s, t \in \mathbb{T}$ with $r < s < t$, $Y_r^t(\hat{x}) = Y_s^t(Y_r^s(\hat{x}))$ and $Y_s^t(\hat{x})$ are *coupled* via the common

PERFECT SIMULATION   219

path $(Z_u; u \in [s,t) \cap \mathbb{T})$ and $Y_r^s(\hat{x})$ (which in turn is determined by the path $(Z_u; u \in [r,s) \cap \mathbb{T}))$. Finally, we say that a random variable T with values in $\mathbb{T} \cap [0, \infty]$ is a *stopping time* if for any $s \in \mathbb{T} \cap (0, \infty)$, the $\{-T \geq -s\}$ is determined by $(Z_{-t}; -t \in \mathbb{T} \cap [-s, 0))$. Notice that knowing $(Z_{-t}; -t \in \mathbb{T} \cap [-s, 0))$ means that we know the path $(Y_{-s}^{-t}(\hat{x}); -t \in \mathbb{T} \cap [-s, 0])$.

PROPOSITION 11.3 Assume that
(i) the process Z_t, $t \in \mathbb{T}$, is stationary,
(ii) there exists a state $\hat{x} \in \Omega$ so that for any $F \subseteq \Omega$,
$$P\left(Y_0^t(\hat{x}) \in F\right) \to \Pi(F) \quad \text{as } t \to \infty,$$
(iii) there is a stopping time $T \geq 0$ such that
$$P(T < \infty) = 1$$
and
$$Y_{-t}^0(\hat{x}) = Y_{-T}^0(\hat{x}) \quad \text{whenever } -t \leq -T. \tag{11.26}$$
Then $Y_{-T}^0(\hat{x}) \sim \Pi$.

Proof. The result is intuitively clear if we imagine a process started in the infinite past (conceptually at time $-\infty$) in $\hat{x}$, and then employ (i)–(iii). Formally, by (iii), $Y_{-T}^0(\hat{x}) = \lim_{t \to \infty} Y_{-t}^0(\hat{x})$, so
$$P\left(Y_{-T}^0(\hat{x}) \in F\right) = \lim_{t \to \infty} P\left(Y_{-t}^0(\hat{x}) \in F\right)$$
$$= \lim_{t \to \infty} P\left(Y_0^t(\hat{x}) \in F\right) = \Pi(F)$$
where respectively the dominated convergence theorem, (i), and (ii) are used for obtaining the first, second, and third equality. □

The conditions (i) (ii) are naturally satisfied for MCMC algorithms, cf. Chapter 7, and for spatial birth-death process algorithms as discussed in Section 11.2.6. The art in practice is therefore to find an SRS, or more generally a coupling construction as in (11.25), and a stopping time T so that (iii) is satisfied and T is not too large.

REMARK 11.3 Generally speaking, by a *CFTP algorithm* we understand a way of determining a coupling construction as in (11.25) and a stopping time T so that the conditions in Proposition 11.3 are satisfied, whereby we can return a perfect simulation $Y_{-T}^0(\hat{x}) \sim \Pi$. If (11.26) holds for all $\hat{x} \in \Omega$, we talk about a *vertical CFTP* algorithm and $-T$ as a *vertical backward coupling time*; otherwise, if T depends on a fixed $\hat{x}$, we refer to a *horizontal CFTP* algorithm and to $-T$ as a *horizontal backward coupling time*.

11.2.2 Propp-Wilson's CFTP algorithm

The Propp & Wilson (1996) algorithm (Algorithm 11.4 below) is a vertical CFTP algorithm for an SRS as in (11.22) with invariant distribution Π, where $-T = -T_{\mathrm{PW}}$ is the smallest vertical backward coupling time, that is, the first time before time 0 for coalescence of all possible chains:

$$T_{\mathrm{PW}} = \inf\left\{m \in \mathbb{N}_0 : Y^0_{-m}(x) = Y^0_{-m}(y) \text{ for all } x, y \in \Omega\right\}. \quad (11.27)$$

Note that by stationarity, if

$$T_{\mathrm{for}} = \inf\{m \in \mathbb{N}_0 : Y^m_0(x) = Y^m_0(y) \text{ for all } x, y \in \Omega\}$$

denotes the *forward coupling time*, then

$$T_{\mathrm{for}} \sim T_{\mathrm{PW}}. \quad (11.28)$$

The following proposition states when Algorithm 11.4 works.

PROPOSITION 11.4 *For $t \in \mathbb{N}$, define the event*

$$C_t = \{Y^t_0(x) = Y^t_0(y) \text{ for all } x, y \in \Omega\}.$$

Then

$$P(T_{\mathrm{PW}} < \infty) = 1 \Leftrightarrow \exists t \in \mathbb{N}: P(C_t) > 0. \quad (11.29)$$

If $P(T_{\mathrm{PW}} < \infty) = 1$, then the chain is uniformly ergodic, and for any $x \in \Omega$, $Y^0_{-T_{\mathrm{PW}}}(x) \sim \Pi$.

Proof. The "only if part" in (11.29) follows immediately from (11.28), while the "if part" follows by considering the independent and equiprobable events $\{Y^{it}_{(i-1)t}(x) = Y^{it}_{(i-1)t}(y) \text{ for all } x, y \in \Omega\}$, $i = 1, 2, \ldots$.

Suppose that $P(T_{\mathrm{PW}} < \infty) = 1$. By (11.29) we obtain uniform ergodicity as follows: Let $\epsilon = P(C_t) > 0$ and set $m = t + 1$ and $M(F) = \epsilon P(\varphi(Z, R_t) \in F \mid C_t)$ where Z denotes the common value of $Y^t_0(x)$, $x \in \Omega$, when the event C_t happens to occur. Then for any $x \in \Omega$,

$$P^m(x, F) = P(Y^m_0(x) \in F) \geq M(F),$$

i.e. the chain is (m, M)-small and thus uniformly ergodic (Proposition 7.8, Section 7.2.3). To see that $Y^0_{-T_{\mathrm{PW}}}(x) \sim \Pi$, we check (i)–(iii) in Proposition 11.3. We have already noticed that (i) is satisfied, and (ii) is implied by uniform ergodicity. Finally, by the definition (11.27), (iii) is obviously satisfied. $\square$

EXAMPLE 11.2 (*Random walk (continued)*) Since (11.28) holds, a natural question is why we cannot return $Y^{T_{\mathrm{for}}}_0$ as a perfect simulation. A counterexample is provided by the random walk introduced in Example 11.1. The invariant distribution for the random walk is $\Pi = \mathrm{Uniform}(\Omega)$, while $Y^{T_{\mathrm{for}}}_0 \sim \mathrm{Uniform}(\{0, k\})$ (see Figure 11.1). These distributions are different for $k \geq 2$.

PERFECT SIMULATION

Let now $0 < T_1 < T_2 < \ldots$ denote any increasing sequence of nonnegative integers, and suppose that $P(T_{\text{PW}} < \infty) = 1$.

ALGORITHM 11.4 (*Propp-Wilson's CFTP algorithm*) For $j = 1, 2, \ldots$, generate $Y_{-T_j}^{-T_j+1}(x) = \varphi(x, R_{-T_j}), \ldots, Y_{-T_j}^{0}(x) = \varphi(Y_{-T_j}^{-1}(x), R_{-1})$ for all $x \in \Omega$, until the $Y_{-T_j}^{0}(x)$ all agree,

and then return any of those $Y_{-T_j}^{0}(x)$.

The algorithm produces a perfect simulation by Proposition 11.4 and the definition of T_{PW}. One essential point here is that T_{PW} is required to be finite with probability one, which implies uniform ergodicity. This limits the applicability of the Propp-Wilson CFTP algorithm, as many MCMC algorithms are not uniformly ergodic. Though mainly of theoretical interest, it is interesting to notice that uniform ergodicity implies the existence of an SRS construction so that $P(T_{\text{PW}} < \infty) = 1$ (Foss & Tweedie 1998).

REMARK 11.4 As argued in Propp & Wilson (1996), instead of using the sequence $T_j = j$, $j = 1, 2, \ldots$, it is usually more efficient to use a *doubling scheme* $T_j = 2^{j-1}$, $j = 1, 2, \ldots$: If $T_j = j$ for all j, the total number of iterations,

$$1 + \ldots + T_{\text{PW}} = T_{\text{PW}}(T_{\text{PW}} + 1)/2$$

is quadratic in T_{PW}. If instead $T_j = 2^{j-1}$ for all j and $2^{m-1} < T_{\text{PW}} \leq 2^m$, the total number of iterations,

$$1 + 2 + \ldots + 2^m = 2^{m+1} - 1 < 4T_{\text{PW}}$$

is linear in T_{PW}. See also the discussion in Wilson (2000b).

11.2.3 Propp-Wilson's monotone CFTP algorithm

Usually the difficult step in Algorithm 11.4 is to check if all paths started at time $-T_j$ coalesce before time 0. For large state space it may be too time consuming or even impossible to consider all possible paths. However, as pointed out in Propp & Wilson (1996), this problem may be overcome if there is a partial ordering $\prec$ on Ω such that the following conditions hold.

Assume that the updating function is *monotone* in its first argument,

$$\varphi(x, \cdot) \prec \varphi(y, \cdot) \quad \text{whenever } x \prec y, \tag{11.30}$$

and that there exist a unique minimum $\hat{0} \in \Omega$ and a unique maximum $\hat{1} \in \Omega$ such that

$$\forall x \in \Omega : \hat{0} \prec x \prec \hat{1}.$$

For $m \in \mathbb{Z}$, let $L_m^n = Y_m^n(\hat{0})$ and $U_m^n = Y_m^n(\hat{1})$, $n = m, m+1, \ldots$. Then

$$L_m^n \prec Y_m^n(x) \prec U_m^n \quad \text{for all } x \in \Omega \text{ and integers } m \leq n, \qquad (11.31)$$

i.e. any chain started at time m in x sandwiches between the *lower chain* L_m^n and the *upper chain* U_m^n. Thereby T_{PW} is determined by only two paths "started in the infinite past", since

$$T_{\text{PW}} = \inf \left\{ m \in \mathbb{N}_0 : L_{-m}^0 = U_{-m}^0 \right\}. \qquad (11.32)$$

EXAMPLE 11.3 (*Random walk (continued)*) For the random walk (Example 11.1), $\prec$ could be given by the usual partial ordering $\leq$ of the integers. Then $\hat{0} = 0$ and $\hat{1} = k$, and the updating function (11.24) is obviously monotone in its first argument. Hence for the Propp-Wilson CFTP algorithm it suffices to generate the paths started in 0 and k, see Figure 11.1.

Suppose the monotonicity property (11.30) is replaced by the *anti-monotonicity* property,

$$\varphi(y, \cdot) \prec \varphi(x, \cdot) \quad \text{whenever } x \prec y$$

(such cases are considered in Section 11.2.4; other examples are given in Häggström & Nelander 1998 and Møller 1999). Using the *cross-over trick* introduced in Kendall (1998), we redefine L_m^n and U_m^n as follows: set $L_m^m = \hat{0}$, $U_m^m = \hat{1}$, and

$$L_m^{n+1} = \varphi\left(U_m^n, R_m\right), \quad U_m^{n+1} = \varphi\left(L_m^n, R_m\right), \quad n = m, m+1, \ldots, \qquad (11.33)$$

whereby (11.31)–(11.32) remain true. Note that these *lower* and *upper processes* are not individually Markov chains.

In both the monotone and the anti-monotone case, in addition to the *sandwiching* property (11.31), we have the *funnelling* property,

$$L_m^n \prec L_{m'}^n \prec U_{m'}^n \prec U_m^n \quad \text{for integers } m' \leq m \leq n, \qquad (11.34)$$

and the *coalescence* property,

$$L_m^n = U_m^n \;\Rightarrow\; L_m^{n'} = U_m^{n'} \quad \text{for integers } m \leq n \leq n', \qquad (11.35)$$

see Figure 11.2. So putting things together with Algorithm 11.4, we obtain the following algorithm, where we now both assume that $P(T_{\text{PW}} < \infty) = 1$ and that there is a partial ordering with unique minimal and maximal states.

ALGORITHM 11.5 (*Propp-Wilson's monotone CFTP algorithm*) In both the monotone and anti-monotone case, for $j = 1, 2, \ldots$,

$$\text{generate } (L_{-T_j}^{-T_j}, U_{-T_j}^{-T_j}) \ldots, (L_{-T_j}^0, U_{-T_j}^0)$$
$$\text{until } L_{-T_j}^0 = U_{-T_j}^0,$$

PERFECT SIMULATION

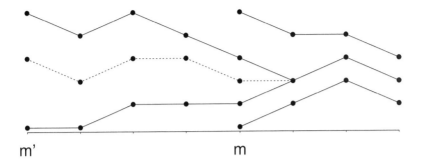

Figure 11.2 *The plot illustrates the sandwiching, funnelling, and coalescence properties for upper and lower processes (solid lines) started at times $m' < m$ in $\hat{1}$ and $\hat{0}$, respectively, and the chain $Y_{m'}^n(x)$ (dashed line) started at time m' in some state x.*

and then return $L_{-T_j}^0$.

11.2.4 Perfect simulation of continuum Ising models

We now show how the Propp-Wilson monotone CFTP algorithm applies for a continuum Ising model Π with $k \geq 2$ components and density (7.12), using the Gibbs sampler in Algorithm 7.6 (Section 7.1.3).

The Gibbs sampler amounts to use a SRS with $R_m = (R_{m,1}, \ldots, R_{m,k})$ where the

$$R_{m,j} \sim \text{Poisson}(T \times [0,1], \beta_j), \quad m \in \mathbb{Z},\ j \in \{1, \ldots, k\},$$

are mutually independent. We view $R_{m,j}$ as a marked point process where the points are given by a Poisson(T, β_j)-process and the marks are i.i.d. and Uniform$([0,1])$-distributed. For finite point configurations $x_1, \ldots, x_k \subset T$ and finite marked point configurations $r_1, \ldots, r_k \subset T \times [0,1]$, let

$$(y_1, \ldots, y_k) = \varphi((x_1, \ldots, x_k), (r_1, \ldots, r_k))$$

denote the update after one cycle of the Gibbs sampler, where each of the y_j are obtained by independent thinning of the r_j as follows. If (ξ, v) is a marked point in r_1, then ξ is included in y_1 if

$$v \leq \prod_{\eta_2 \in x_2} \cdots \prod_{\eta_k \in x_k} \phi(\xi, \eta_2, \ldots, \eta_k), \tag{11.36}$$

and deleted otherwise. Similarly, we obtain $y_2, \ldots, y_k$; e.g., for a marked point $(\xi, v) \in r_k$, we include ξ in y_k if

$$v \leq \prod_{\eta_1 \in y_1} \cdots \prod_{\eta_{k-1} \in y_{k-1}} \phi(\eta_1, \ldots, \eta_{k-1}, \xi), \qquad (11.37)$$

and delete it otherwise.

In the special case of a Widom-Rowlinson penetrable spheres mixture model (Example 6.8, Section 6.6.3), i.e. when $\phi(\xi_1, \ldots, \xi_k) = \mathbf{1}[\|\xi_i - \xi_j\| > R$, for all $i \neq j]$, there is no need for generating the marks for the $R_{m,j}$, since (11.36)–(11.37) reduce to

retain ξ if its distance to $x_2 \cup \ldots \cup x_k$ is larger than R

and

retain ξ if its distance to $y_1 \cup \ldots \cup y_{k-1}$ is larger than R.

The natural state space is given by the support of Π. However, as we shall see in a moment, it is convenient to extend this to a state space

$$\Omega = \{(x_1, \ldots, x_k) : \text{each } x_i \subseteq T \text{ is either finite or } x_i = T\}.$$

A natural partial ordering is given by set-inclusion with respect to the k types of points: for $(x_1, \ldots, x_k), (y_1, \ldots, y_k) \in \Omega$, define

$$(x_1, \ldots, x_k) \prec (y_1, \ldots, y_k) \Leftrightarrow x_i \subseteq y_i, \; i = 1, \ldots, k. \qquad (11.38)$$

Then

$$\hat{0} = (\emptyset, \ldots, \emptyset), \quad \hat{1} = (T, \ldots, T),$$

are unique minima and maxima, and the SRS construction for the Gibbs sampler is seen to be anti-monotone, since $\phi \leq 1$. Hence Algorithm 11.5 easily applies. Note that $\hat{0}$ belongs to the support of Π (it is even an atom), while $\hat{1}$ does not (it is at this point the definition of Ω becomes convenient). However, with probability one, even if the Gibbs sampler is started in $\hat{1}$, the states after one or more cycles of the Gibbs sampler all consist of k-tuples of finite point configurations contained in the support of Π.

There is another partial ordering which makes the two-component Gibbs sampler monotone: suppose that $k = 2$ and

$$(x_1, x_2) \prec (y_1, y_2) \Leftrightarrow x_1 \subseteq y_1, \; x_2 \supseteq y_2,$$

in which case

$$\hat{0} = (\emptyset, T), \quad \hat{1} = (T, \emptyset), \qquad (11.39)$$

are unique minima and maxima. Note that now neither $\hat{0}$ nor $\hat{1}$ is in the support of Π. The monotone case does not extend to the case of $k > 2$ components.

The point patterns in Figure 6.2 (Section 6.6.3) are produced by perfect simulation, using the monotone version; for example, in the case

PERFECT SIMULATION

$\beta = 200$ and $R = 0.1$ (the lower right plot in Figure 6.2), using a doubling scheme $T_j = 2^j$, the upper and lower processes need to be started about $2^{10} = 1024$ iterations back in time in order to obtain coalescence at time 0. Further empirical results for the Widom-Rowlinson penetrable spheres mixture model are reported in Häggström et al. (1999) (but with (11.39) replaced by "quasi-minimal and quasi-maximal" states). Georgii (2000) shows perfect simulations of another continuum Ising model.

It would be very nice if the Swendsen-Wang type algorithm for a symmetric Widom-Rowlinson penetrable spheres mixture model (Algorithm 7.7, Section 7.1.3) could be used for perfect simulation, but unfortunately it seems very difficult to find any partial ordering for which the algorithm has a useful monotonicity property.

11.2.5 Wilson's read-once algorithm

Since Algorithms 11.4–11.5 are reusing the random bits, there is the potential danger of running into storage problems when T_{PW} is large (in practice this problem may be handled by keeping the random seeds used in the pseudo-random generator). This section describes Wilson's (2000a) read-once algorithm, which runs forwards in time, starting at time 0, and reading the R_m only once. As we shall see, it works precisely when the Propp-Wilson CFTP algorithm works, and it can be naturally used for producing i.i.d. perfect samples.

Figure 11.3 illustrates the meaning of the following notation. For a

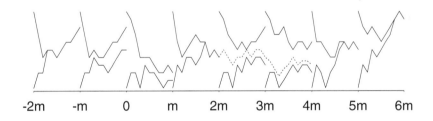

Figure 11.3 *Illustration of notation for the read-once algorithm. In the plot $K_0 = -2$, $K_1 = 1$, $K_2 = 4$, and $K_3 = 5$. The waiting times τ_0, τ_1, and τ_2 are 1, 2, and 0, respectively. The value of G_1 (i.e. the first perfect simulation) is given by the dashed curve at time $4m$.*

given $m \in \mathbb{N}$, set

$$F_i(x) = Y_{im}^{(i+1)m}(x) \quad \text{for } i \in \mathbb{Z} \text{ and } x \in \Omega.$$

We can view the F_i as i.i.d. random maps. By (11.29), $P(T_{\mathrm{PW}} < \infty) = 1$

if and only if

$$p_m \equiv P(\text{range}(F_0) \text{ is a singleton}) > 0 \quad \text{for } m \text{ sufficiently large.}$$

Set
$$K_0 = \sup\{k < 0 : \text{range}(F_k) \text{ is a singleton}\},$$
$$K_1 = \inf\{k \geq 0 : \text{range}(F_k) \text{ is a singleton}\},$$
and define recursively,
$$K_i = \inf\{k > K_{i-1} : \text{range}(F_k) \text{ is a singleton}\}, \quad i = 1, 2, \ldots.$$

Finally, set
$$G_0 = F_{-1} \circ F_{-2} \circ \ldots \circ F_{K_0}$$
where $\circ$ denotes composition of mappings, and set
$$G_i = F_{K_{i+1}-1} \circ \ldots \circ F_{K_i}, \quad i = 1, 2, \ldots,$$
$$\tau_0 = -1 - K_0,$$
and
$$\tau_i = K_{i+1} - 1 - K_i, \quad i = 1, 2, \ldots.$$

The maps G_i, $i \in \mathbb{N}$, correspond to i.i.d. "cycles" starting at times mk where F_k is a singleton and ending at the next time $m(k'-1)$ where $k' > k$ and $F_{k'}$ is a singleton. The $m\tau_i$ are waiting times between occurrences of singletons. The "cycle" for the map G_0 starts at the last singleton before time 0 and ends at the fixed time point 0 after the waiting time $m\tau_0$. However, all the (G_i, τ_i), $i \in \mathbb{N}_0$, are i.i.d. as shown in the following proposition.

PROPOSITION 11.5 If $p_m > 0$ then the (G_i, τ_i), $i \in \mathbb{N}_0$, are i.i.d. with $G_i \sim \Pi$ and $\tau_i \sim \text{Geo}(p_m)$.

Proof. By Proposition 11.4, $G_0 \sim \Pi$. Let Π_+ denote the conditional distribution of F_0 given that range(F_0) is a singleton, and let Π_- denote the conditional distribution of F_0 given that range(F_0) is not a singleton. Using that the random maps are i.i.d., we immediately obtain the following properties. The τ_i, $i \in \mathbb{N}_0$, are i.i.d., and each $\tau_i \sim \text{Geo}(p_m)$. Conditional on the τ_i, the random maps $F_{K_0}, F_{K_0+1}, \ldots$, are mutually independent, where $F_{K_i} \sim \Pi_+$ for $i \in \mathbb{N}_0$, and $F_j \sim \Pi_-$ for $j \in \{K_0+1, K_0+2, \ldots\} \setminus \{K_1, K_2, \ldots\}$. Thus $(F_{K_0}, \ldots, F_0, \tau_0)$, $(F_{K_1}, \ldots, F_{K_2-1}), \tau_1)$, $(F_{K_2}, \ldots, F_{K_3-1}), \tau_2)$, …, are i.i.d., and so the (G_i, τ_i) are i.i.d. □

REMARK 11.5 Somewhat counterintuitively, the time interval $K_1 - 1 - K_0$ is not Geo(p_m) (it is the sum of τ_0 and K_1 which are independent and both Geo(p_m)). This is analogous to the discussion in Remark 3.3 (Section 3.1.3).

Given $m \in \mathbb{N}$ so that the conditions of Proposition 11.5 are fulfilled, Wilson's *read-once algorithm* for making n i.i.d. perfect simulations simply consists in generating $G_1, \ldots, G_n$. Note that $G_1, \ldots, G_n$ are given by generating the path starting in the unique element of range(F_{K_1}) at time $m(K_1 + 1)$ and sampling at times $mK_2, \ldots, mK_{n+1}$. If there is a partial ordering with unique minimal and maximal states, then in both the monotone and anti-monotone case, we exploit that only the lower and upper processes started at the minimal and maximal states at time im and followed up to time $(i+1)m$ are needed for determining range$(F_i) = \{F_i(\hat{0}), F_i(\hat{1})\}$, cf. Figure 11.3. Some experimentation may be needed in order to determine an "appropriate" value of p_m; Wilson (2000a) recommends that m should be chosen so that $p_m > 1/2$ or equivalently $\mathbb{E}\tau_i < 1$.

11.2.6 Perfect simulation for locally stable point processes using dominated CFTP

We now turn to the issue of perfect simulation of spatial point processes, using for specificity the same setting as in Section 11.1. So N_f denotes the set of finite point configurations contained in $B \subset \mathbb{R}^d$ with $|B| < \infty$. For an extension to an "arbitrary" state space B, see Kendall & Møller (2000).

A natural partial ordering on N_f is given by set inclusion,

$$x \prec y \ \Leftrightarrow \ x \subseteq y$$

for $x, y \in N_\mathrm{f}$. Then

$$\hat{0} = \emptyset$$

is the unique minimum, but there is no maximal element. Crucially, uniform ergodicity of the Metropolis-Hastings algorithms in Section 7.1 is the exception rather than the rule, so the Propp-Wilson CFTP algorithm needs modification. Following the ideas introduced in Kendall (1998), Kendall & Møller (2000) show how a special "dominating process" which acts as a kind of stochastic maximum can be used for making perfect simulation based on either spatial birth-death processes or Metropolis-Hasting sampling: the resulting *dominated CFTP* algorithm provides a way to establish when *horizontal CFTP* has taken place.[†]

In the sequel we concentrate on the approach based on spatial birth-death processes, using the coupling construction in Section 11.1.4 for the locally stable and constant death rate case, with birth rate $b(x, \xi) =$

[†] Propp-Wilson's monotone CFTP algorithm has sometimes also been called dominated CFTP, but the difference is that $\hat{1}$ is now replaced by a process which dominates "target" processes started in $\emptyset$, whereby Proposition 11.3 with $\hat{x} = \emptyset$ applies (the details are given later).

$\lambda^*(x,\xi) \le \phi^*(\xi)$ and death rate $d(\cdot,\cdot) = 1$. This approach is much simpler than that based on Metropolis-Hastings sampling, cf. Kendall & Møller (2000).

The dominating process is given by the spatial birth-death process with birth rate $b_D(x,\xi) = \phi^*(\xi)$ and death rate $d_D(\cdot,\cdot) = 1$. Since this is reversible with invariant distribution Poisson(B, ϕ^*), cf. (11.16), we can easily generate the dominating process both forwards and backwards in time, starting with $D_0 \sim$ Poisson(B, ϕ^*) (using the methods in Section 3.2.3), and so we obtain a stationary process D_t, $t \in \mathbb{R}$. Moreover, whenever a forward birth happens we associate a Uniform$([0,1])$-distributed mark, where the marks are mutually independent and independent of the process D_t.

For the coupling construction (11.25), let $\hat{x} = \emptyset$, and let Z_t denote D_t with associated marks for births used in the same way as the R_m''' in step (iii) of Algorithm 11.3 so that $Y_s^t(\emptyset)$, $t \ge s$, becomes a spatial birth-death process with birth rate $b(x,\xi) = \lambda^*(x,\xi)$, death rate $d(\cdot,\cdot) = 1$, and started in $\emptyset$ at time $s \in \mathbb{R}$. Conditions (i)–(ii) in Proposition 11.3 are then obviously satisfied. As for (iii) in Proposition 11.3, since

$$D_t = \emptyset \;\Rightarrow\; Y_s^t(\emptyset) = \emptyset \quad \text{whenever } s \le t,$$

one possible choice of a horizontal stopping time is

$$-T_\emptyset = \sup\{-t \le 0 : D_{-t} = \emptyset\}$$

(more efficient choices will be discussed later). As observed in Section G.4, $\emptyset$ is an ergodic atom, which implies that $P(T_\emptyset < \infty) = 1$. Hence, by Proposition 11.3, $Y^0_{-T_\emptyset}$ follows the target density.

To generate $Y^0_{-T_\emptyset}(\emptyset)$, we need only first to generate D_0, then until time $-T_\emptyset$, the jump chain of D_{-t}, $-t \le 0$, generated backwards in time together with associated marks for deaths (corresponding to forward births), and finally to construct the jump chain of $Y^{-t}_{-T_\emptyset}(\emptyset)$, $-T_\emptyset \le -t \le 0$, forwards in time in the same way as in Algorithm 11.3. Specifically, using a notation like in Algorithm 11.3, let $T'_{-1} > T'_{-2} > \ldots$ denote the jump times and $J'_{-1}, J'_{-2}, \ldots$ the jump chain of D_{-t}, $-t < 0$, considered backwards in time, set $J'_0 = D_0$ and $T'_0 = 0$, let $N'_\emptyset$ denote the number of jumps for the dominating process in the time interval $[-T_\emptyset, 0)$, and set $\tilde{J}_{-m} = Y^{T'_{-m}}_{-T_\emptyset}(\emptyset)$, $m = 0, \ldots, N'_\emptyset$. Note that (J_0, J'_0) is the last jump before time 0 in the process (Y_t, D_t), and $\tilde{J}_0 = Y^0_{-T_\emptyset}(\emptyset)$ follows the target distribution.

ALGORITHM 11.6 (*Simple dominated CFTP algorithm*) Generate $J'_0 \sim$ Poisson(B, ϕ^*). If $J'_0 = \emptyset$, then stop and return $\tilde{J}_0 = \emptyset$, else

(i) generate backwards $J'_{-1}, \ldots, J'_{-N'_\emptyset}$ (i.e. until the first time $-m$ with $J'_{-m} = \emptyset$): for $m = 0, \ldots, N'_\emptyset - 1$, given that $J'_{-m} = x$,

PERFECT SIMULATION

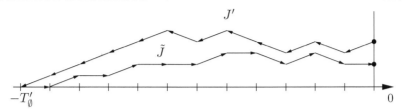

Figure 11.4 *Illustration of the simple dominated CFTP algorithm.*

(a) draw $R''_{-m-1} \sim \text{Uniform}([0,1])$;

(b) if $R''_{-m-1} \geq c^*/(c^* + n(x))$, generate $\xi_{-m-1} \sim \text{Uniform}(x)$ and $R'''_{-m-1} \sim \text{Uniform}([0,1])$, and set $J'_{-m-1} = x \setminus \xi_{-m-1}$;

(c) else generate $\eta_{-m-1} \sim \bar{b}'$ and set $J'_{-m-1} = x \cup \eta_{-m-1}$;

here, conditional on the random variables used for the generation of $J'_0, \ldots, J'_{-m}$ and the associated marks in (b), it is assumed that R''_{-m-1}, R'''_{-m-1}, and ξ_{-m-1} or η_{-m-1} are mutually independent;

(ii) set $\tilde{J}_{-T'_\emptyset} = \emptyset$ and construct forwards $\tilde{J}_{-T'_\emptyset+1}, \ldots, \tilde{J}_0$: for $-m = -N'_\emptyset, \ldots, -1$,

(a) if $J'_{-m+1} = J'_{-m} \cup \xi_{-m}$, set
$$\tilde{J}_{-m+1} = \begin{cases} \tilde{J}_{-m} \cup \xi_{-m} & \text{if } R'''_{-m} \leq \lambda^*(\tilde{J}_{-m}, \xi_{-m})/\phi^*(\xi_{-m}) \\ \tilde{J}_{-m} & \text{otherwise;} \end{cases}$$

(b) else set $\tilde{J}_{-m+1} = \tilde{J}_{-m} \setminus \eta_{-m}$.

Return $\tilde{J}_0$.

The backwards-forwards construction used in Algorithm 11.6 is illustrated in Figure 11.4.

REMARK 11.6 The facts that $\emptyset$ is an ergodic atom and that the coupled process (Y_t, D_t) regenerates each time $D_t = \emptyset$ are used for theoretical considerations in Section G.4. As verified in Berthelsen & Møller (2002a),

$$\mathrm{E}T'_\emptyset \geq \exp(c^*) - 1/2. \tag{11.40}$$

For instance, for a Strauss process (Example 6.1, Section 6.2.2) on the unit square, $c^* = \beta$, so $\mathrm{E}T'_\emptyset$ is at least exponentially growing in β, and if e.g. $\beta = 100$ then $\mathrm{E}T'_\emptyset \geq e^{100} - 1/2 \approx 2.7 \times 10^{43}$. Consequently, the simple dominated CFTP algorithm is infeasible for applications of real interest. However, Algorithm 11.6 becomes useful for understanding Algorithms 11.7–11.8 below.

A much faster perfect simulation algorithm is given in Kendall (1998) and Kendall & Møller (2000), using upper processes U_m^n, $n = m, m + 1, \ldots$, started at times $m \in \mathbb{Z}$ in the states of the dominating jump chain, and using lower processes L_m^n, $n = m, m+1, \ldots$, started in the minimum $\emptyset$. Specifically, for $m, n \in \mathbb{Z}$ with $n \geq m$, set $U_m^m = J'_m$ and $L_m^m = \emptyset$, and if $U_m^n = u$, $L_m^n = l$, and $J'_n = x$, then

$$J'_{n+1} = x \setminus \eta_n \quad \Rightarrow \quad U_m^{n+1} = u \setminus \eta_n \quad \text{and} \quad L_m^{n+1} = l \setminus \eta_n,$$

$$J'_{n+1} = x \cup \xi_n \quad \Rightarrow \quad U_m^{n+1} = \begin{cases} u \cup \xi_n & \text{if } R'''_n \leq \alpha_{\max}(u, l, \xi_n) \\ u & \text{otherwise} \end{cases}$$

$$\text{and} \quad L_m^{n+1} = \begin{cases} l \cup \xi_n & \text{if } R'''_n \leq \alpha_{\min}(u, l, \xi_n) \\ l & \text{otherwise.} \end{cases}$$

Here again the R'''_m denote the i.i.d. and Uniform$([0, 1])$-distributed marks associated to the births in the dominating jump chain, and for $\phi^*(\xi) > 0$,

$$\alpha_{\max}(u, l, \xi) = \max\{\lambda^*(y, \xi)/\phi^*(\xi) : l \subseteq y \subseteq u\} \tag{11.41}$$

and

$$\alpha_{\min}(u, l, \xi) = \min\{\lambda^*(y, \xi)/\phi^*(\xi) : l \subseteq y \subseteq u\}. \tag{11.42}$$

Hence, by construction, the sandwiching, funnelling, and coalescence properties (11.31)–(11.35) also hold in the present situation, with a slight modification for the sandwiching property:

$$L_m^n \subseteq \tilde{J}_n \subseteq U_m^n \quad \text{for integers } -N'_\emptyset \leq m \leq n \leq 0.$$

The definitions (11.41)–(11.42) seem natural as they provide the minimal upper and maximal lower processes so that the sandwiching, funnelling, and coalescence properties are satisfied for all possible realisations of the marks R'''_m. The calculation of $\alpha_{\max}$ and $\alpha_{\min}$ is particular simple in the attractive or repulsive case, since in the attractive case (6.7),

$$\alpha_{\max}(u, l, \xi) = \lambda^*(u, \xi)/\phi^*(\xi) \quad \text{and} \quad \alpha_{\min}(u, l, \xi) = \lambda^*(l, \xi)/\phi^*(\xi),$$

while in the repulsive case (6.8),

$$\alpha_{\max}(u, l, \xi) = \lambda^*(l, \xi)/\phi^*(\xi) \quad \text{and} \quad \alpha_{\min}(u, l, \xi) = \lambda^*(u, \xi)/\phi^*(\xi). \tag{11.43}$$

For a given sequence $0 = T_0 < T_1 < T_2 < \ldots$, used for specifying the starting times of the upper and lower processes, the algorithm proceeds as follows.

ALGORITHM 11.7 (*Dominated CFTP algorithm using upper and lower processes*) Generate $J'_0 \sim \text{Poisson}(B, \phi^*)$. If $J'_0 = \emptyset$, then stop and return $\tilde{J}_0 = \emptyset$, else for $j = 1, 2, \ldots$,

PERFECT SIMULATION

(i) along similar lines as in Algorithm 11.6, generate backwards $J'_{-T_{j-1}-1}$, ..., J'_{-T_j}, including the associated marks for backward deaths;

(ii) set $U^{-T_j}_{-T_j} = J'_{-T_j}$ and $L^{-T_j}_{-T_j} = \emptyset$, and for $-m = -T_j, \ldots, -1$, using the same notation as in Algorithm 11.6,

(a) if $J'_{-m+1} = J'_{-m} \cup \xi_{-m}$, set

$$U^{-m+1}_{-T_j} = \begin{cases} U^{-m}_{-T_j} \cup \xi_{-m} & \text{if } R'''_{-m} \leq \alpha_{\max}(U^{-m}_{-T_j}, L^{-m}_{-T_j}, \xi_{-m}) \\ U^{-m}_{-T_j} & \text{otherwise} \end{cases}$$

and

$$L^{-m+1}_{-T_j} = \begin{cases} L^{-m}_{-T_j} \cup \xi_{-m} & \text{if } R'''_{-m} \leq \alpha_{\min}(U^{-m}_{-T_j}, L^{-m}_{-T_j}, \xi_{-m}) \\ L^{-m}_{-T_j} & \text{otherwise;} \end{cases}$$

(b) else set $U^{-m+1}_{-T_j} = U^{-m}_{-T_j} \setminus \eta_{-m}$ and $L^{-m+1}_{-T_j} = L^{-m}_{-T_j} \setminus \eta_{-m}$;

until $U^0_{-T_j} = L^0_{-T_j}$, and then return $\tilde{J}_0 = U^0_{-T_j}$.

Because of the sandwiching, funnelling, and coalescence properties, $\tilde{J}_0 = U^0_{-T_j}$ is a perfect simulation when Algorithm 11.7 terminates. The perfect simulations of the Strauss process shown in Figure 6.1 (Section 6.2.2) are produced by Algorithm 11.7. Note that the two point patterns for $\gamma = 0.5$ and $\gamma = 0$ are included in the point pattern for $\gamma = 1$ (which is just a realisation of the dominating Poisson process), but the point pattern for $\gamma = 0$ is not included in that for $\gamma = 0.5$ (this is in fact a consequence of (11.43)).

REMARK 11.7 The running time of the backwards-forwards construction in Algorithm 11.7 can be defined by 1 if $J'_0 = \emptyset$, and

$$RT = T_M + (T_1 + \ldots + T_M) \qquad (11.44)$$

otherwise, where

$$T_M = \inf\{T_j : U^0_{-T_j} = L^0_{-T_j}\}$$

is the number of steps needed backwards in time. The first T_M in (11.44) appears because of the generation of the dominating jump chain.

It is advantageous to notice that the strictly increasing sequence T_j may be random. Berthelsen & Møller (2002a) consider both the case of a usual *doubling scheme* $T_j = 2^{j-1}n$ where $n \in \mathbb{N}$ is chosen by the user, and a *refined doubling scheme* where n is replaced by the random variable

$$T_{\min} = \inf\{m \in \mathbb{N} : J'_{-m} \cap J'_0 = \emptyset\},$$

that is, the number of jumps backwards in time needed until all points in J'_0 are deleted. If T^n_M and $T^{\min}_M$ denote T_M for these two doubling schemes, then RT in (11.44) is given by

$$RT_n = 3T^n_M - n \quad \text{if } T_j = 2^{j-1}n$$

and

$$RT_{\min} = 3T^{\min}_M - T_{\min} \quad \text{if } T_j = 2^{j-1}T_{\min}.$$

Empirical results for RT_1 and $RT_{\min}$ are shown in Figure 11.7, Section 11.2.8. Based on this and other results (not shown in this book) with varying values of n, Berthelsen & Møller (2002a) conclude that all things considered the refined doubling scheme seems the best choice.

11.2.7 Perfect simulation for locally stable point processes using clans of ancestors

We consider now an alternative algorithm due to Fernández, Ferrari & Garcia (2002). For simplicity we assume that λ^* is defined in terms of an interaction function with finite range of interaction, i.e. $\lambda^*(x, \xi) = \lambda^*(x \cap b(\xi, R), \xi)$ where $R < \infty$ (see Definition 6.1 and Remark 6.4 in Section 6.3.1).

In order to understand the following definitions it may be useful to consider Figure 11.5 and to keep in mind how the simple dominated CFTP algorithm (Algorithm 11.6) works. For $\xi \in J'_{\leq 0}$ where $J'_{\leq 0} = \cup_{m=0}^{T'_\emptyset} J'_{-m}$, let $-I(\xi) \leq 0$ be the time at which ξ was born in Algorithm 11.6, i.e. if $\xi = \xi_{-m}$ then $-I(\xi_{-m}) = -m+1$. We call

$$\mathrm{an}^1(\xi) = J'_{-I(\xi)-1} \cap b(\xi, R)$$

the first generation of ancestors of ξ. Since the range of interaction is given by R, we need only to know $\mathrm{an}^1(\xi_{-m})$ to see whether ξ_{-m} is in $\tilde{J}_{-m+1}$ or not in step (ii)(b) of Algorithm 11.6. Define recursively the jth generation of ancestors of ξ by

$$\mathrm{an}^j(\xi) = \cup_{\eta \in \mathrm{an}^{j-1}(\xi)} \mathrm{an}^1(\eta), \quad j = 2, 3, \ldots,$$

and call $\mathrm{an}(\xi) = \cup_{j=1}^\infty \mathrm{an}^j(\xi)$ the ancestors of ξ. Then $\tilde{J}_0$ depends only on $J'_{\leq 0}$ through $J'_C = C \cup J'_0$ where

$$C = \cup_{\xi \in J'_0} \mathrm{an}(\xi)$$

is called the *clan of ancestors* of J'_0 (taking $C = \emptyset$ if $J'_0 = \emptyset$). Finally, let

$$T_C = \inf\{m \in \mathbb{N}_0 : J'_{-m} \cap J'_C = \emptyset\}$$

specify the time interval in which the points in J'_C are living. Then $T_C \leq T'_\emptyset$, and J'_0 is unaffected if we set $J'_{-T_C} = \emptyset$ and generate J'_{-m} forwards in time $-m \geq -T_C$ as usual, but considering only the transitions for the points in $J'_{-T_C} \cap J'_C, \ldots, J'_0 \cap J'_C$.

PERFECT SIMULATION

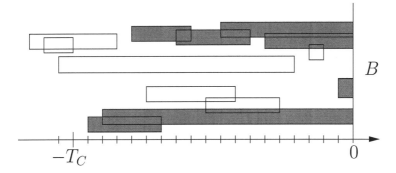

Figure 11.5 *Example of a clan of ancestors when B is a line segment. The points in the dominating jump process are given by the midpoints of the vertical edges of the rectangles. Each horizontal edge of a rectangle shows the life time of the corresponding point in the dominating jump process. The vertical edges are all of length R. Shaded rectangles represent members of the clan.*

ALGORITHM 11.8 (*Dominated CFTP algorithm based on the clan of ancestors*) Generate $J'_0 \sim \text{Poisson}(B, \phi^*)$. If $J'_0 = \emptyset$, then stop and return $\tilde{J}_0 = \emptyset$, else

(i) generate backwards $J'_0, \ldots, J'_{-T_C}$ as in step (i) of Algorithm 11.6, where at each backwards transition $J'_{-m} \to J'_{-m-1}$ the first generation of ancestors of the points in $\cup_{\xi \in J'_0} \text{an}^m(\xi)$ are determined in order to construct C and to check if $T_C = m + 1$;

(ii) set $\tilde{J}_{-T_C} = \emptyset$ and generate forwards $\tilde{J}_{-T_C+1}, \ldots, \tilde{J}_0$ along similar lines as in step (ii) of Algorithm 11.6, but so that $\tilde{J}_{-m+1} = \tilde{J}_{-m}$ is unchanged whenever $J'_{-m+1} \cap J'_C = J'_{-m} \cap J'_C$ is unchanged.

Return $\tilde{J}_0$.

REMARK 11.8 Since T_C depends only on ϕ^* and R, no monotonicity properties such as attractivity (6.7) or repulsivety (6.8) are required. The algorithm can easily be modified to perfect Metropolis-Hastings simulation of locally stable point processes (Kendall & Møller 2000), and to perfect Gibbs sampling of continuum Ising models where the special case of the Widom-Rowlinson penetrable spheres mixture model turns out to be particular simple.

11.2.8 Empirical findings

This section presents some empirical findings for the dominated CFTP algorithm based on upper and lower processes (Algorithm 11.7) and the clan algorithm (Algorithm 11.8). The algorithms are applied on a Strauss process defined on the unit square, with $\beta = 100$ and varying values of $R > 0$ (Example 6.1, Section 6.2.2).

We start by comparing Algorithm 11.7 with the simplest form of perfect simulation, namely *rejection sampling* which works as follows. The unnormalised density for the Strauss process given by (6.14) is dominated by $\beta^{n(x)}$, that is, the density for a homogeneous Poisson process X on the unit square with intensity β. Given a simulation $X = x$ and $U = u$ where $U \sim \text{Uniform}([0,1])$ is independent of X, if $u \leq \gamma^{s_R(x)}$ then return x, else repeat generating a new simulation of (X, U) (independently of previously generated realisations) until a simulation is accepted. As in Proposition 7.1 (Section 7.1.1), the output follows a Strauss process with parameter $\theta = (\beta, \gamma, R)$.

The two first plots in Figure 11.6 show mean CPU times for rejection sampling and Algorithm 11.7 based on a refined doubling scheme. For each algorithm and each value of θ, the mean CPU time is given by the average obtained from 10000 independent perfect simulations. As expected rejection sampling performs well when the interaction is weak, i.e. when R is sufficiently small or γ is sufficiently close to 1. But even for rather small values of R or modest values of γ, dominated CFTP is much more efficient. The last plot in Figure 11.6 shows that as R grows and $\gamma = 0.1$ is fixed, $s_R(X)$ is much larger for the dominating Poisson process than for the Strauss process, whereby rejection sampling becomes infeasible.

We consider next the running times RT_1 and $RT_{\min}$ for Algorithm 11.7 (see Remark 11.7). The left plot in Figure 11.7 shows how $\mathbb{E}T_1$ and $\mathbb{E}T_{\min}$ depend on R when $\gamma = 0$ (a hard core process). Each mean is estimated by the empirical average based on 500 independent perfect simulations. For all values of R in the plot, $\mathbb{E}T_{\min} < \mathbb{E}T_1$, but the difference decreases as R increases.

We next compare Algorithms 11.7 and 11.8. As these algorithms are not immediately comparable, it makes little sense to compare the number of steps involved in the backwards-forwards construction in the two algorithms. Instead we consider $\mathbb{E}T_M^{\min}$ (see Remark 11.7) and $\mathbb{E}T_C$ when either $\gamma = 0$ or $\gamma = 0.5$, though this of course is not telling the whole story about which algorithm is the fastest.

The right plot in Figure 11.7 shows $\mathbb{E}T_M^{\min}$ and $\mathbb{E}T_C$ versus R when $\gamma = 0$ and $\gamma = 0.5$, respectively, where each mean is estimated by the empirical average based on 500 independent perfect simulations. Recall that T_C does not depend on γ, cf. Remark 11.8. As expected the mean running times agree as R tends to 0, and $\mathbb{E}T_M^{\min}$ decreases as γ increases. For both $\gamma = 0$ and $\gamma = 0.5$, it is only for rather small values of R that $\mathbb{E}T_M^{\min}$ is larger than $\mathbb{E}T_C$. The situation changes as γ tends to 1, since in

PERFECT SIMULATION

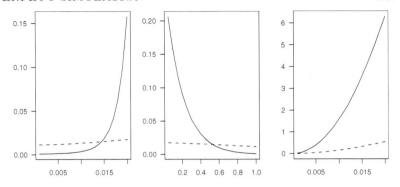

Figure 11.6 *Comparison of rejection sampling and dominated CFTP for a Strauss process on the unit square with $\beta = 100$ and varying values of γ and R. Left plot: mean CPU time per sample (in seconds) versus R for the Strauss process with $\gamma = 0.1$ (rejection sampling: solid line; dominated CFTP: dashed line). Centre plot: mean CPU time per sample versus γ for the Strauss process with $R = 0.02$ (rejection sampling: solid line; dominated CFTP: dashed line). Right plot: $\mathbb{E} s_R(X)$ versus R for the Strauss process with $\gamma = 0.1$ (dashed line) and a homogeneous Poisson process on the unit square with intensity $\beta = 100$ (solid line). (Kasper K. Berthelsen is gratefully acknowledged for providing the plots.)*

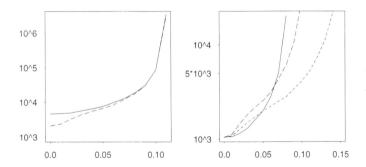

Figure 11.7 *Various mean values related to the dominated CFTP Algorithms 11.7 and 11.8, where each mean is estimated from 500 independent runs. Left plot: $\mathbb{E} T_1$ (full line) and $\mathbb{E} T_{\min}$ (dashed line) versus R. Right plot: $\mathbb{E} T_C$ (full line) and $\mathbb{E} T_{M}^{\min}$ when $\gamma = 0$ (long dashed line) and $\gamma = 0.5$ (short dashed line) versus R.*

the limit $\mathbb{E}T_M^{\min}$ agrees with $T_{\min}$ which is smaller than T_C. Furthermore, as R increases, $\mathbb{E}T_M^{\min}$ becomes much smaller than $\mathbb{E}T_C$.

Berthelsen & Møller (2002a) also report that similar conclusions as above are obtained for varying values of β, and they conclude that all things considered the dominated CFTP algorithm using upper and lower processes and a refined doubling scheme seems to be the best choice. Further empirical results for perfect simulation of the Strauss process and other locally stable point processes can be found in Berthelsen & Møller (2002b).

11.2.9 Other perfect simulation algorithms

We conclude with a brief discussion on some other perfect simulation algorithms for spatial point processes.

Fill (1998) introduces a clever form of rejection sampling, assuming a finite state space and a monotone setting with unique minimal and maximal states. Applications and extensions of *Fill's algorithm* can be found in Fill, Machida, Murdoch & Rosenthal (2000) and the references therein. The advantage of Fill's algorithm compared to CFTP is that it is interruptible in the sense that the output is independent of the running time (like in any rejection sampler). The disadvantages may be problems with storage and that it seems more limited for applications than CFTP. Regarding applications of Fill's algorithm for spatial point processes, Thönnes (1999) considers the special case of the Widom-Rowlinson penetrable spheres mixture model in Example 6.8 in Section 6.6.3 (using the anti-monotone setting, this can be extended to perfect simulation of continuum Ising models with k-components as considered in Example 6.9). Furthermore, Møller & Schladitz (1999) consider more general spatial point processes approximated by lattice processes.

Horizontal CFTP algorithms based on Metropolis-Hastings algorithms for locally stable point processes are studied in Kendall & Møller (2000), while the use of spatial jump processes for more general classes of spatial point processes is studied in Berthelsen & Møller (2002b). It is noticed in Kendall & Møller (2000) that the considered Metropolis-Hastings algorithm is geometrically ergodic, but in general it is not uniformly ergodic — recall that uniform ergodicity is a necessary condition for vertical CFTP to work. However, inspired by the "multigamma" coupler in Murdoch & Green (1998) which applies for uniformly ergodic chains, Wilson (2000a) shows how his read-once algorithm applies in connection to spatial birth-death processes.

Often one considers point processes with infinitely many points contained in an "infinite volume" such as $\mathbb{R}^d$ (see e.g. Section 6.4). In order to avoid edge effects, a perfect sample within a bounded region may sometimes be achieved by extending simulations both backwards in time and in space; this may be possible if the interaction radius is sufficiently weak (Kendall 1997, Fernández et al. 2002). Such coupling constructions may be of great theoretical interest, but in our opinion they remain so far unpractical for applications of real interest.

APPENDIX A

History, bibliography, and software

A.1 Brief history

Daley & Vere-Jones (1988) provide a nice review of the history of point processes, particularly the probabilistic aspects. The history of statistics for spatial point processes goes back at least to Student (1907) (incidentally one of the earliest applications of the χ^2-test for goodness of fit). Statistical aspects were treated in the seminal work by Cox (1955), Matérn (1960), and Bartlett (1963, 1964). Substantial progress was initiated with the introduction of complex stochastic models such as Markov point processes (Ripley & Kelly 1977, Ripley 1977) and MCMC methods in the 1970s, and the access to fast and cheap computers during the last decades; Clifford (1990) surveys the history of Markov random fields, including Markov point processes. Today spatial point processes are a main topic in spatial statistics, cf. Ripley (1981, 1988) and Cressie (1993), and in stochastic geometry, cf. Matheron (1975) and Stoyan et al. (1995).

The history of Monte Carlo methods stretches back to Laplace's method of estimating the number π using Buffon's (1777) needle, cf. Newman & Barkema (2000) (incidentally, Buffon's needle is considered as the starting point of stochastic geometry, cf. Stoyan et al. 1995). MCMC methods originate from statistical physics; for a brief history, see Newman & Barkema (2000). Many important MCMC algorithms have first been applied on spatial point processes, including the seminal work by Metropolis et al. (1953) where the Metropolis algorithm was introduced and applied on a hard core point process. Also the Gibbs sampler (Geman & Geman 1984), which is now widely used in many areas of statistics, was first introduced in spatial statistics and statistical physics for the simulation of random fields and spatial point processes (Suomela 1976, Ripley 1977, Ripley 1979, Creutz 1979). Today MCMC techniques play a major role in statistics, particularly in Bayesian statistics and spatial statistics, cf. Møller (2003c) and the references therein.

A.2 Brief bibliography

The area of spatial point processes is well established, with a number of books on this subject and on the associated subjects of spatial statistics and stochastic geometry. Monographs with a focus on the probabilistic aspects include Kerstan, Matthes & Mecke (1974), Kallenberg (1975), Daley & Vere-Jones (1988), Karr (1991), and Reiss (1993). Textbooks and review articles with a view to statistical applications include Matérn (1960) (reprinted as Matérn 1986), Bartlett (1975), Ripley (1977, 1981, 1988), Cox & Isham (1980), Diggle (1983), Penttinen (1984), Baddeley & Møller (1989), Mecke, Schneider, Stoyan & Weil (1990), Cressie (1993), Stoyan & Stoyan (1994), Stoyan et al. (1995), Geyer (1999), Møller (1999a), van Lieshout (2000), Ohser & Mücklich (2000), and Møller & Waagepetersen (2003).

General textbooks on MCMC and statistical applications include Gilks, Richardson & Spiegelhalter (1996), Gamerman (1997), Robert & Casella (1999), Chen et al. (2000), and Liu (2001), but they contain very little material associated to spatial point processes. MCMC methods specific to spatial point processes are reviewed in the survey papers by Geyer (1999) and Møller & Waagepetersen (2003) and in the book by van Lieshout (2000).

A.3 Software

The computations for the data examples in this book were done using our own programmes in C and Splus, and using A. J. Baddeley and R. Turner's R-package spatstat for estimation of F, G, and J-functions and for computation of MPLEs. Methods for Monte Carlo maximum likelihood estimation will probably be added to spatstat in the near future. Other software for statistics for spatial point patterns include B. S. Rowlingson and P. J. Diggle's Splancs (in Splus/R) and the module spatial in W. N. Venables and B. D. Ripley's package MASS. For references to these and other sources of code, see http://www.maths.uwa.edu.au/~adrian/spatlinks.html.

Pseudo random numbers were obtained using the Mersenne Twister algorithm (Matsumoto & Nishimura 1998). The right plots in Figure 5.2 were produced using the R-package scatterplot3d by U. Ligges.

APPENDIX B

Measure theoretical details

This appendix gives the rigorous framework for point processes defined on a metric space S with metric $d(\cdot,\cdot)$ and taking values in the space N_{lf} of locally finite point configurations in S, cf. the notation introduced in Chapter 2. A more detailed treatment can be found in Daley & Vere-Jones (1988) and the other references mentioned below. For background material on measure theory, see e.g. Billingsley (1995).

B.1 Preliminaries

Let $(\Omega, \mathcal{F}, \mu)$ be a given measure space. A set $\mathcal{A}$ of subsets $A \subseteq \Omega$ is called a *paving*. We write $\sigma(\mathcal{A})$ for the σ-algebra generated by $\mathcal{A}$ (i.e. the smallest σ-algebra for Ω containing $\mathcal{A}$). In order to simplify the exposition we work under the following conventions.

When we say that the measure μ is defined on Ω, we more precisely mean that μ is defined on the σ-algebra $\mathcal{F}$. When we refer to the measure $\mu(A)$ of a set $A \subseteq \Omega$ and to the integral $\int f \mathrm{d}\mu$ of a function $f : \Omega \to \mathbb{R}$ (where $\mathbb{R}$ is equipped with the usual Borel σ-algebra), it will be understood that $A \in \mathcal{F}$ and f are measurable.

If $\Omega = \Omega_1 \times \cdots \times \Omega_k$ is a *product space* and each space Ω_i is equipped with a σ-algebra $\mathcal{F}_i$, we let $\mathcal{F} = \mathcal{F}_1 \otimes \cdots \otimes \mathcal{F}_k$ be the corresponding product σ-algebra (i.e. the σ-algebra generated by the paving of product sets $A_1 \times \cdots \times A_k$ with $A_i \in \mathcal{F}_i$, $i = 1, \ldots, k$).

We shall often make use of the following well-known result (a slight extension of Theorem 3.3 in Billingsley 1995).

LEMMA B.1 Let μ_1 and μ_2 be two measures defined on a space Ω equipped with a σ-algebra $\mathcal{F} = \sigma(\mathcal{A})$ for a paving $\mathcal{A}$ which is closed under intersection. If $\mu_1(\Omega) = \mu_2(\Omega) < \infty$ and $\mu_1 = \mu_2$ on $\mathcal{A}$, then $\mu_1 = \mu_2$ on $\mathcal{F}$.

B.2 Formal definition of point processes

We now give a precise definition of a point process defined on S. We equip S with the Borel σ-algebra $\mathcal{B}$ (the σ-algebra generated by the open sets), and $\mathcal{B}_0$ denotes the class of bounded Borel sets. We equip

$\mathcal{N}_{lf}$ with the σ-algebra

$$\mathcal{N}_{lf} = \sigma(\{x \in N_{lf} : n(x_B) = m\} : B \in \mathcal{B}_0, \, m \in \mathbb{N}_0). \tag{B.1}$$

DEFINITION B.1 A *point process* X defined on S is a measurable mapping defined on some probability space $(\Omega, \mathcal{F}, \mathcal{P})$ and taking values in $(N_{lf}, \mathcal{N}_{lf})$. The *distribution* P_X of X is given by $P_X(F) = P(\{\omega \in \Omega : X(\omega) \in F\})$ for $F \in \mathcal{N}_{lf}$. We shall sometimes identify X and P_X, and call them both a point process.

Thus measurability of X is equivalent to that the count $N(B)$ is a random variable for any $B \in \mathcal{B}_0$. This implies that $N(B)$ is a random variable for any $B \in \mathcal{B}$, since S is a metric space and hence a countable union of bounded Borel sets.

The result in the following Lemma B.2 is often seen in textbooks as the definition of the distribution of a point process. Lemma B.3 gives a more simple characterisation.

LEMMA B.2 The distribution of a point process X is determined by the finite dimensional distributions of its count function, i.e. the joint distribution of $N(B_1), \ldots, N(B_m)$ for any $B_1, \ldots, B_m \in \mathcal{B}_0$ and $m \in \mathbb{N}_0$.

Proof. Follows immediately from Lemma B.1 and the definition (B.1), since the paving

$$\mathcal{A} = \{\{x \in N_{lf} : n(x_{B_i}) = n_i, \, i = 1, \ldots, m\} : \\ n_i \in \mathbb{N}_0, \, B_i \in \mathcal{B}_0, \, i = 1, \ldots, m, \, m \in \mathbb{N}_0\}$$

is closed under intersection and $\mathcal{N}_{lf} = \sigma(\mathcal{A})$. □

We may identify a point process X with a locally finite *random counting measure* Φ on S where $\Phi(B) = n(X_B)$. In a more general framework we may consider all locally finite random counting measures as point processes. In this book we restrict attention to so-called *simple point processes* for which the corresponding random measure is one or zero on singletons, i.e. $\Phi(\{\xi\}) \leq 1$ for $\xi \in S$. In fact many of the concepts and results below may be extended to a general setting for random counting measures, see e.g. Carter & Prenter (1972) and Daley & Vere-Jones (1988). However, our setting for simple locally finite point processes covers almost all cases of practical importance, and point processes with multiple points may in fact be viewed as a marked point process, where the mark m_ξ is the multiplicity of the point ξ.

B.3 Some useful conditions and results

Henceforth we assume that S is *separable*, i.e. S contains a countable dense set. Thereby the following useful lemma and theorem can be established.

LEMMA B.3 We have that
$$\mathcal{N}_{\mathrm{lf}} = \sigma(\mathcal{N}_{\mathrm{lf}}^0) \tag{B.2}$$
where
$$\mathcal{N}_{\mathrm{lf}}^0 = \{\{x \in N_{\mathrm{lf}} : n(x_B) = 0\} : B \in \mathcal{B}_0\} \tag{B.3}$$
is the class of *void events*.

Proof. Clearly, $\sigma(\mathcal{N}_{\mathrm{lf}}^0) \subseteq \mathcal{N}_{\mathrm{lf}}$, so we only need to show that $\{x \in N_{\mathrm{lf}} : n(x_B) = m\} \in \sigma(\mathcal{N}_{\mathrm{lf}}^0)$ for $B \in \mathcal{B}_0$ and $m \in \mathbb{N}$.

Since S is a separable metric space it contains a so-called dissecting system (Proposition A2.1.V in Daley & Vere-Jones 1988), i.e. a sequence $\mathcal{F}_n = \{A_{n,i} \in \mathcal{B} : i = 1, \ldots, k_n\}$, $n \in \mathbb{N}$, of partitions of S (i.e. S is a disjoint union of the sets in each $\mathcal{F}_n$), so that the $\mathcal{F}_n$ are nested (i.e. $A_{n-1,j} \cap A_{n,j}$ is either $\emptyset$ or $A_{n,j}$) and separate points in S (i.e. for each pair of distinct points $\xi, \eta \in S$ we have that $\xi \in A_{n,i}$ implies $\eta \notin A_{n,i}$ for all sufficiently large n). Hence $n(x_B) = m$ if and only if there exists an n_0 with $k_{n_0} \geq m$ such that for all $n \geq n_0$ there exists $\{j_1, \ldots, j_m\} \subseteq \{1, 2, \ldots, k_n\}$ such that $n(x \cap B \cap A_{n,j_i}) > 0$, $i = 1, \ldots, m$, and $n(x \cap (B \setminus \cup_{i=1}^m A_{n,j_i})) = 0$. In other words

$$\{x \in N_{\mathrm{lf}} : n(x_B) = m\} =$$
$$\bigcup_{n_0 : k_{n_0} \geq m} \bigcap_{n \geq n_0} \bigcup_{1 \leq j_1 < \ldots < j_m \leq k_n} A(n, m, B, \{j_1, \ldots, j_m\})$$

where each of the events
$$A(n, m, B, \{j_1, \ldots, j_m\}) =$$
$$\bigcap_{i=1}^m \{x \in N_{\mathrm{lf}} : n(x \cap B \cap A_{n,j_i}) > 0\}$$
$$\bigcap \{x \in N_{\mathrm{lf}} : n(x \cap (B \setminus \cup_{i=1}^m A_{n,j_i})) = 0\}$$

belongs to $\sigma(\mathcal{N}_{\mathrm{lf}}^0)$. Consequently, $\{x \in N_{\mathrm{lf}} : n(x_B) = m\} \in \sigma(\mathcal{N}_{\mathrm{lf}}^0)$. $\square$

Often (B.3) can be replaced by a smaller paving (see e.g. Daley & Vere-Jones 1988 and Kingman 1993), but this will not be needed for our purpose.

The following characterisation of the distribution of a point process is extremely useful, cf. Chapters 3 and 5.

THEOREM B.1 A point process is uniquely determined by its void probabilities.

Proof. This follows from Lemmas B.1 and B.3, since the paving (B.3) is closed under intersection. □

If S is a *Polish space*, i.e. if S is both a separable and complete metric space (completeness means that any Cauchy sequence is convergent), then the following result concerning separability of $\mathcal{N}_{lf}$ can be established.

PROPOSITION B.1 If S is a Polish space, then $\mathcal{N}_{lf}$ is separable (i.e. countably generated).

Proof. By Theorem A2.6.III in Daley & Vere-Jones (1988), N_{lf} is a Polish space, and $\mathcal{N}_{lf}$ agrees with the Borel sets in N_{lf}. Hence by Lemma A2.1.III in Daley & Vere-Jones (1988), $\mathcal{N}_{lf}$ is separable. □

The use of separability of $\mathcal{N}_{lf}$ is briefly commented in a footnote at the beginning of Section 7.2. In all our examples of applications we use a metric so that S becomes a Polish space. When $S \subseteq \mathbb{R}^d$ we usually take $d(\xi, \eta) = \|\xi - \eta\|$, the usual Euclidian distance between the points $\xi, \eta \in S$. If e.g. S is the $(d-1)$-dimensional unit sphere, a more natural metric is given by geodesic distance (i.e. $d(\xi, \eta)$ is the length of the shortest arc on the unit sphere between the two points).

For marked point process we use different metrics depending on the context. Recall that we can view a marked point process with points ξ in T and associated marks m_ξ in M as a point process with points (ξ, m_ξ) in $S = T \times M$, cf. Section 2.3. If we assume that T and M are Polish spaces with metrics d_T, d_M and Borel σ-algebras $\mathcal{B}_T, \mathcal{B}_M$, then S is a Polish space with metric

$$d((\xi_1, m_{\xi_1}), (\xi_2, m_{\xi_2})) = \max\{d_T(\xi_1, \xi_2), d_M(m_{\xi_1}, m_{\xi_1})\}.$$

This is a natural metric in the sense that the Borel σ-algebra $\mathcal{B}_S$ for S is given by the product σ-algebra $\mathcal{B}_S = \mathcal{B}_T \otimes \mathcal{B}_M$. For a multitype point process with $M = \{1, \ldots, k\}$, we obtain for any metric d_M that M is a Polish space where $\mathcal{B}_M = \{B \subseteq \{1, \ldots, k\}\}$ does not depend on d_M. If $T \subseteq \mathbb{R}^d$ and the marks are geometrical objects, we may use the formalism of stochastic geometry, considering M as the space of compact subsets of $\mathbb{R}^d$ equipped with the *Hausdorff metric* defined by

$$d_M(A, B) = \inf\{r \geq 0 : A \subset B \oplus b(0, r), B \subset A \oplus b(0, r)\}$$

where $\oplus$ denotes Minkowski-addition, i.e. $A \oplus B = \{a+b : a \in A, b \in B\}$. Then M is a Polish space (Matheron 1975). For example, for discs $A = b(0, r_1)$ and $B = b(0, r_2)$ with $0 < r_1 < r_2 < \infty$, $d_M(A, B) = r_2 - r_1$.

SOME USEFUL CONDITIONS AND RESULTS

We conclude this appendix with a useful specialisation of Lemma B.1.

LEMMA B.4
(i) Let μ and ν be measures on $S \times N_{\mathrm{lf}}$. Then $\mu = \nu$ provided $\mu(B \times N_{\mathrm{lf}}) = \nu(B \times N_{\mathrm{lf}}) < \infty$ and $\mu(B \times F) = \nu(B \times F)$ for all bounded $B \subseteq S$ and all $F \in \mathcal{N}_{\mathrm{lf}}^0$.
(ii) Let μ and ν be measures on $N_{\mathrm{lf}} \times N_{\mathrm{lf}}$. Then $\mu = \nu$ provided $\mu(N_{\mathrm{lf}} \times N_{\mathrm{lf}}) = \nu(N_{\mathrm{lf}} \times N_{\mathrm{lf}}) < \infty$ and $\mu(F_1 \times F_2) = \nu(F_1 \times F_2)$ for all $F_1, F_2 \in \mathcal{N}_{\mathrm{lf}}^0$.

Proof. (i) Fix an arbitrary $C \in \mathcal{B}_0$. The paving

$$\mathcal{A}_C = \{B \times F : F \in \mathcal{N}_{\mathrm{lf}}^0, B \in \mathcal{B}_0, B \subseteq C\}$$

is closed under intersection. By Lemma B.3, $\mathcal{A}_C$ generates the restriction of the product σ-algebra $\mathcal{B} \otimes \mathcal{N}_{\mathrm{lf}}$ to $C \times N_{\mathrm{lf}}$. Then by Lemma B.1, $\mu(F) = \nu(F)$ for $F \in \sigma(\mathcal{A}_C)$. Hence, since S is a countable union of disjoint sets $C \in \mathcal{B}_0$, we have $\mu(F) = \nu(F)$ for any $F \in \mathcal{B} \otimes \mathcal{N}_{\mathrm{lf}}$.

(ii) The paving given by sets of the form $F_1 \times F_2$ with $F_1, F_2 \in \mathcal{N}_{\mathrm{lf}}^0$ generates the product σ-algebra $\mathcal{N}_{\mathrm{lf}} \otimes \mathcal{N}_{\mathrm{lf}}$, cf. Lemma B.3, and the paving is closed under intersection. The result then follows directly from Lemma B.1. □

APPENDIX C

Moment measures and Palm distributions

Appendices C.1 and C.2 provide some background material on moment and Palm measures. In Appendices C.1.1 and C.2.1 we consider a point process X on a general metric space S (see Appendix B).

C.1 Moment measures

C.1.1 Moment measures in a general setting

The first and higher order moments of the counts $N(B)$, $B \subseteq S$, can be expressed by the following measures.

DEFINITION C.1 *For a point process X on S and each $n \in \mathbb{N}$, define the nth order moment measure $\mu^{(n)}$ on S^n by*

$$\mu^{(n)}(D) = \mathbb{E} \sum_{\xi_1,\ldots,\xi_n \in X} \mathbf{1}[(\xi_1,\ldots,\xi_n) \in D], \quad D \subseteq S^n,$$

and the nth order reduced moment measure $\alpha^{(n)}$ on S^n by

$$\alpha^{(n)}(D) = \mathbb{E} \sum_{\xi_1,\ldots,\xi_n \in X}^{\neq} \mathbf{1}[(\xi_1,\ldots,\xi_n) \in D], \quad D \subseteq S^n,$$

where the $\neq$ over the summation sign means that the n points $\xi_1,\ldots,\xi_n$ are pairwise distinct. In particular, $\mu = \mu^{(1)} = \alpha^{(1)}$ is called the intensity measure.

The nth order moment measure $\mu^{(n)}$ determines the nth order moments of the count variables $N(B)$, $B \subseteq S$, since

$$\mu^{(n)}(B_1 \times \cdots \times B_n) = \mathbb{E} \prod_{i=1}^{n} N(B_i) \quad \text{for } B_i \subseteq S.$$

For any $n \in \mathbb{N}$, there is a one-to-one correspondence between $(\mu^{(1)},\ldots,\mu^{(n)})$ and $(\alpha^{(1)},\ldots,\alpha^{(n)})$. It is often more convenient to work with the reduced moment measures.

Definition C.1 immediately extends to that

$$\mathbb{E} \sum_{\xi_1,\ldots,\xi_n \in X}^{\neq} h(\xi_1,\ldots,\xi_n) = \int \cdots \int h(\xi_1,\ldots,\xi_n) \mathrm{d}\alpha^{(n)}(\xi_1,\ldots,\xi_n) \quad (\text{C.1})$$

for nonnegative functions h. Hence by the extended Slivnyak-Mecke Theorem 3.3 (Section 3.2) extended to the present general setting, we obtain the following result.

PROPOSITION C.1 If $X \sim \text{Poisson}(S, \mu)$ (see Remark 3.1 in Section 3.1.1), then $\alpha^{(n)} = \mu^n$ (the product measure of μ n-times).

C.1.2 The second order reduced moment measure

In the case $S = \mathbb{R}^d$, we can express $\alpha^{(2)}$ in terms of the intensity function ρ and the second order reduced moment measure $\mathcal{K}$ (provided $\mathcal{K}$ is well defined, see Definition 4.5 in Section 4.1.2). By (4.4) and (C.1),

$$\int h(\xi, \eta - \xi) \mathrm{d}\alpha^{(2)}(\xi, \eta) = \int\int h(\xi, \eta) \rho(\xi) \rho(\eta) \mathrm{d}\mathcal{K}(\eta) \mathrm{d}\xi \quad (\text{C.2})$$

for nonnegative functions h.

C.2 Campbell measures and Palm distributions

C.2.1 Campbell measures and Palm distributions in a general setting

Throughout this section we assume that μ is σ-finite, i.e. $\mu(B_i) < \infty$ for a countable partition $\{B_i\}$ of S (this is e.g. satisfied if μ is locally finite).

We start by defining the so-called reduced Campbell measure, which in turn is used for defining so-called reduced Palm distributions. As shown in the following, the first and second order properties of a point process are closely related to these concepts.

DEFINITION C.2 For a point process X on S, define the *reduced Campbell measure* $C^!$ on $S \times N_{\text{lf}}$ by

$$C^!(D) = \mathbb{E} \sum_{\xi \in X} \mathbf{1}[(\xi, X \setminus \xi) \in D], \quad D \subseteq S \times N_{\text{lf}}.$$

The *Campbell measure* is defined by $C(D) = \mathbb{E} \sum_{\xi \in X} \mathbf{1}[(\xi, X) \in D]$; hence the term reduced Campbell measure for $C^!$. We have that

$$\int h(\xi, x) \mathrm{d}C^!(\xi, x) = \mathbb{E} \sum_{\xi \in X} h(\xi, X \setminus \xi)$$

CAMPBELL MEASURES AND PALM DISTRIBUTIONS

for nonnegative functions h. Note that $C^!$ determines $(\mu, \alpha^{(2)})$, since $\mu(\cdot) = C^!(\cdot \times \mathcal{N}_{\mathrm{lf}})$ and

$$\alpha^{(2)}(B_1 \times B_2) = \int 1[\xi \in B_1] n(x_{B_2}) \mathrm{d}C^!(\xi, x).$$

For each $F \in \mathcal{N}_{\mathrm{lf}}$, $C^!(\cdot \times F) \leq \mu(\cdot)$, so $C^!(\cdot \times F)$ is absolutely continuous with respect to μ. Then by the Radon-Nikodym theorem, there exists a μ-almost surely unique density $\xi \to P_\xi^!(F)$ so that

$$C^!(B \times F) = \int_B P_\xi^!(F) \mathrm{d}\mu(\xi) \tag{C.3}$$

where $P_\xi^!(\cdot)$ is a probability measure for each $\xi \in S$; for details, see e.g. Daley & Vere-Jones (1988, page 455).

DEFINITION C.3 *The probability measure $P_\xi^!(\cdot)$ on $\mathcal{N}_{\mathrm{lf}}$ is called a reduced Palm distribution at the point ξ.*

REMARK C.1 We can interpret $P_\xi^!$ as the conditional distribution of $X \setminus \xi$ given that $\xi \in X$: heuristically, for a very small ball $B = b(\xi, \epsilon)$, it is unlikely that $N(B) > 1$, so $\mu(B) \approx P(N(B) > 0)$ and $C^!(B \times F) \approx P(X \setminus \xi \in F, N(B) > 0)$, whereby

$$P_\xi^!(F) \approx P(X \setminus \xi \in F | N(b(\xi, \epsilon)) > 0).$$

Therefore, $P_\xi^!$ is often interpreted as the conditional distribution of $X \setminus \xi$ given that $\xi \in X$.

From (C.3) we obtain the so-called *Campbell-Mecke theorem*:

$$\mathbb{E} \sum_{\xi \in X} h(\xi, X \setminus \xi) = \int \int h(\xi, x) \mathrm{d}P_\xi^!(x) \mathrm{d}\mu(\xi) \tag{C.4}$$

for nonnegative functions h. For later use, note that by Lemma B.4 (Section B.3) and (C.4), the reduced Palm distributions are for μ-almost all ξ uniquely characterised by the equations

$$\mathbb{E} \sum_{\xi \in X} 1[\xi \in A, (X \setminus \xi) \cap B = \emptyset] = \int \int 1[\xi \in A, x \cap B = \emptyset] \mathrm{d}P_\xi^!(x) \mathrm{d}\mu(\xi) \tag{C.5}$$

for bounded $A, B \subseteq S$.

In the Poisson case, $P_\xi^!$ is just the process itself:

PROPOSITION C.2 *If $X \sim \mathrm{Poisson}(S, \mu)$, then we can take $P_\xi^! = \mathrm{Poisson}(S, \mu)$ for all $\xi \in S$.*

Proof. Combining Theorem 3.2 (Section 3.2) and (C.4) we obtain that $P^!_\xi = \text{Poisson}(S, \mu)$ for μ-almost all $\xi \in S$. □

EXAMPLE C.1 (*Differential characterisation of Gibbs point processes*) Let the situation be as in Section 6.1, i.e. X is a finite point process on $S \subset \mathbb{R}^d$ with density f with respect to the standard Poisson process $\text{Poisson}(S, 1)$ where $|S| < \infty$. Suppose that X has intensity function $\rho(\xi) > 0$ for all $\xi \in S$, and f is hereditary. Let λ^* denote the Papangelou conditional intensity, and let $Y \sim \text{Poisson}(S, 1)$. Then

$$C^!(B \times F) = \mathbb{E} \sum_{\xi \in Y} \mathbf{1}[\xi \in B, Y \setminus \xi \in F] f(Y)$$

$$= \int_B \mathbb{E} \mathbf{1}[Y \in F] f(Y \cup \xi) \mathrm{d}\xi$$

$$= \int_B \mathbb{E}\big(\mathbf{1}[X \in F] \lambda^*(X, \xi)\big) \mathrm{d}\xi$$

where the second identity follows from the Slivnyak-Mecke Theorem 3.2 (Section 3.2), and the last from the hereditary condition. So as the right hand side of (C.3) is given by

$$\int_B P^!_\xi(F) \rho(\xi) \mathrm{d}\xi$$

we can take

$$P^!_\xi(F) = \mathbb{E}\big(\mathbf{1}[X \in F] \lambda^*(X, \xi)\big) / \rho(\xi). \tag{C.6}$$

Hence $P^!_\xi$ is absolutely continuous with respect to the distribution P of X, with density

$$\frac{\mathrm{d}P^!_\xi}{\mathrm{d}P}(x) = \frac{\lambda^*(x, \xi)}{\rho(\xi)}. \tag{C.7}$$

Let next the situation be as in Section 6.4.1, where $S = \mathbb{R}^d$ and λ^* is defined in terms of a given interaction function, which is Ruelle stable and of bounded range of interaction. Suppose that X is a point process on $\mathbb{R}^d$ with intensity function ρ. Then X is a Gibbs point process if and only if for Lebesgue almost all ξ with $\rho(\xi) > 0$, $P^!_\xi$ is absolutely continuous with respect to the distribution P of X, with density (C.7), cf. Theorem 3.5 in Georgii (1976).

Appendix C.2.2 considers in more detail the reduced Palm distribution for stationary point processes. Palm distributions for shot noise Cox processes are studied in Møller (2003b).

C.2.2 Palm distributions in the stationary case

Assume that X is a stationary point process on $\mathbb{R}^d$ with intensity $0 < \rho < \infty$. Then as shown in the following theorem, $P_\xi^!$ can be expressed in terms of $P_0^!$, and there is an explicit expression for $P_0^!$. For $\xi \in \mathbb{R}^d$, $x \in N_{\mathrm{lf}}$, and $F \subseteq N_{\mathrm{lf}}$, we let $x + \xi = \{\eta + \xi : \eta \in x\}$ denote the translation of the point configuration x by ξ, and $F + \xi = \{x + \xi : x \in F\}$ the translation of the event F by ξ.

THEOREM C.1 In the stationary case we can take

$$P_0^!(F) = \mathbb{E} \sum_{\xi \in X_B} 1[X \setminus \xi \in F + \xi]/(\rho|B|), \quad F \subseteq N_{\mathrm{lf}}, \tag{C.8}$$

for an arbitrary set $B \subseteq S$ with $0 < |B| < \infty$, and

$$P_\xi^!(F) = P_0^!(F - \xi), \quad F \subseteq N_{\mathrm{lf}}, \ \xi \in \mathbb{R}^d. \tag{C.9}$$

Moreover,

$$\mathbb{E} \sum_{\xi \in X} h(\xi, X \setminus \xi) = \rho \int \int h(\xi, x + \xi) \mathrm{d} P_0^!(x) \mathrm{d}\xi \tag{C.10}$$

for nonnegative functions h.

Proof. For fixed $F \subseteq N_{\mathrm{lf}}$, define

$$\kappa_F(B) = \mathbb{E} \sum_{\xi \in X_B} 1[X \setminus \xi \in F + \xi], \quad B \subseteq S.$$

Since X is stationary, κ_F is a translation invariant measure on S, so κ_F is proportional to Lebesgue measure. Hence the right hand side in (C.8) does not depend on the choice of B. Further, $P_0^!$ as defined by (C.8) is easily seen to be a probability measure. By the standard proof, (C.8) implies (C.10). Defining $P_\xi^!$ by (C.8) and (C.9), each $P_\xi^!$ is a probability measure. It can be verified that $\xi \to P_\xi^!(F)$ is measurable (we omit the details). Combining (C.9) and (C.10), we immediately obtain (C.4), so the $P_\xi^!$ are reduced Palm distributions. $\square$

REMARK C.2 • Letting $X_\xi \sim P_\xi^!$, (C.9) states that X_ξ is distributed as $X_0 + \xi$.

• The relations (C.8) and (C.9) further support the heuristic interpretation of $P_\xi^!$ (Remark C.1) as the conditional distribution of $X \setminus \xi$ given $\xi \in X$. Consider the problem of estimating $P_0^!(F)$ for some $F \subseteq S$. Heuristically, since X is stationary we may for a bounded B regard $X - \xi$, $\xi \in X$, as representing observations of X conditional on $0 \in X$.

Thus a natural estimator of $P_0^!(F)$ is the empirical average

$$\frac{1}{N(B)} \sum_{\xi \in X_B} \mathbf{1}[X \setminus \xi \in F + \xi].$$

If B is large, we may expect that $\rho \approx N(B)/|B|$, and so we obtain the estimator

$$\frac{1}{\rho|B|} \sum_{\xi \in X_B} \mathbf{1}[X \setminus \xi \in F + \xi].$$

This is exactly the unbiased estimator of $P_0^!(F)$ obtained from (C.8).

- Another common interpretation based on (C.8) and (C.9) is that $P_0^!$ is the *distribution of the further points of X given a 'typical' point of X*.
- Notice that if X is isotropic, then $P_0^!$ is isotropic, cf. (C.8).

C.2.3 Interpretation of $\mathcal{K}$ and G as Palm expectations

The second order reduced moment measure $\mathcal{K}$ has an interpretation as a Palm expectation, since

$$\mathcal{K}(B) = \int \sum_{\eta \in x} \mathbf{1}[\eta - \xi \in B]/\rho(\eta) \, dP_\xi^!(x) \tag{C.11}$$

for Lebesgue almost all $\xi \in \mathbb{R}^d$. This follows from the Campbell-Mecke theorem (C.4). In the stationary case, by (C.10),

$$\mathcal{K}(B) = \mathbb{E}_0^! N(B)/\rho, \quad B \subseteq \mathbb{R}^d, \tag{C.12}$$

where $\mathbb{E}_0^!$ denotes expectation with respect to $P_0^!$. Note that if $P_0^!$ is isotropic, then $\mathcal{K}$ is isotropic.

For the nearest neighbour function G defined by (4.9) we obtain from (C.10) that

$$G(r) = P_0^!(N(b(0,r)) > 0), \quad r > 0.$$

APPENDIX D

Simulation of SNCPs without edge effects and truncation

Throughout this appendix we use a notation as in Section 5.4. As in Section 5.5 we consider simulation of the restriction X_B of a SNCP X to a bounded set $B \subset \mathbb{R}^d$. Further, we consider X as a cluster process, cf. Section 5.4.1. The basic problem is that the process Φ of mother points (c,γ) is infinite. However, given Φ, most of the clusters $X_{(c,\gamma)}$ typically do not contribute with points to X_B either because γ is very small so that $X_{(c,\gamma)}$ is empty with high probability, or because c is far from B so that $k(c,\xi)$ is close to zero for $\xi \in B$. We follow Brix & Kendall (2002) and Møller (2003b) in showing how we can simulate the typically finite number of clusters which intersect B whereby edge effects and truncation are avoided.

The simulation method is a consequence of the following lemma, where

$$\Phi_B = \{(c,\gamma) \in \Phi : X_{(c,\gamma)} \cap B \neq \emptyset\},$$

$$p_B(c,\gamma) = 1 - \exp\left(-\gamma \int_B k(c,\xi)\mathrm{d}\xi\right),$$

and

$$\zeta_B(c,\gamma) = p_B(c,\gamma)\zeta(c,\gamma).$$

LEMMA D.1 We have that

(i) conditional on Φ, $p_B(c,\gamma)$ is the probability that a cluster associated to $(c,\gamma) \in \Phi$ has at least one point falling in B;

(ii) Φ_B and $\Phi \setminus \Phi_B$ are Poisson processes with intensity functions ζ_B and $\zeta - \zeta_B$, respectively, and $(\Phi_B, \{X_{(c,\gamma)} : (c,\gamma) \in \Phi_B\})$ and $(\Phi \setminus \Phi_B, \{X_{(c,\gamma)} : (c,\gamma) \in \Phi \setminus \Phi_B\})$ are independent;

(iii) conditional on Φ_B, the clusters $X_{(c,\gamma)}$ with $(c,\gamma) \in \Phi_B$ are independent, and $X_{(c,\gamma)}$ is distributed as a conditional Poisson process with intensity function $\xi \to \gamma k(c,\xi)$ where we have conditioned on that the Poisson process has at least one point in B;

(iv) conditional on $\Phi \setminus \Phi_B$, the clusters $X_{(c,\gamma)}$ with $(c,\gamma) \in \Phi \setminus \Phi_B$ are independent, and $X_{(c,\gamma)}$ is a Poisson process with intensity function $\xi \to \gamma k(c,\xi)\mathbf{1}[\xi \notin B]$.

Proof. (i) follows immediately from viewing X as a cluster process. Since Φ_B is obtained by independent thinning of Φ with retention probabilities $p_B(c,\gamma)$, (ii) follows from Proposition 3.7 in Section 3.2.2. Regarding (iii) and (iv), notice that to condition on $(\Phi_B, \Phi \setminus \Phi_B)$ is equivalent to condition on first Φ and second the information concerning which clusters have points or not in B. Thus conditional on $(\Phi_B, \Phi \setminus \Phi_B)$, the clusters $X_{(c,\gamma)}$, $(c,\gamma) \in \Phi$, are independent point processes, and $X_{(c,\gamma)}$ follows a conditional Poisson process with intensity function $\gamma k(c,\cdot)$ where we condition on that $X_{(c,\gamma)} \cap B \neq \emptyset$ if $(c,\gamma) \in \Phi_B$, and $X_{(c,\gamma)} \cap B = \emptyset$ if $(c,\gamma) \notin \Phi_B$. In the latter case, the conditional distribution coincides with a Poisson process with intensity function $\xi \to \gamma k(c,\xi)\mathbf{1}[\xi \notin B]$. Hence, (iii) and (iv) follow. □

Assume that Φ_B is finite, i.e.

$$\int_{\mathbb{R}^d} \int_0^\infty p_B(c,\gamma)\zeta(c,\gamma)\mathrm{d}c\mathrm{d}\gamma < \infty.$$

According to Lemma D.1, we can obtain a perfect simulation of X_B by first simulating Φ_B, cf. (ii), second simulate independent realisations $X_{(c,\gamma)}$ of the clusters given that they intersect B, cf. (iii), and third return $B \cap (\cup_{(c,\gamma) \in \Phi_B} X_{(c,\gamma)})$.

In practice, however, it may be difficult to compute $p_B(c,\gamma)$ and then it may be advantageous to use a thinning procedure. Suppose that a kernel k_B^{dom} satisfies (C1)–(C2) in Section 5.5. Given Φ, let $X_{(c,\gamma)}^{\mathrm{dom}}$, $(c,\gamma) \in \Phi$, denote independent Poisson processes with intensity functions $\gamma k^{\mathrm{dom}}(c,\cdot)$. Then we can obtain a realisation of $X_{(c,\gamma)} \cap B$ by independent thinning of $X_{(c,\gamma)}^{\mathrm{dom}}$, using retention probabilities $k(c,\xi)/k^{\mathrm{dom}}(c,\xi)$, $\xi \in X_{(c,\gamma)}^{\mathrm{dom}}$. Let

$$\Phi_B^{\mathrm{dom}} = \{(c,\gamma) \in \Phi : X_{(c,\gamma)}^{\mathrm{dom}} \neq \emptyset\}.$$

By (i)-(ii) in Lemma D.1, Φ_B^{dom} is a Poisson process on $\mathbb{R}^d \times (0,\infty)$ with intensity function

$$\zeta_B^{\mathrm{dom}}(c,\gamma) = p_B^{\mathrm{dom}}(c,\gamma)\zeta(c,\gamma)$$

where

$$p_B^{\mathrm{dom}}(c,\gamma) = 1 - \exp\left(-\gamma \int_B k_B^{\mathrm{dom}}(c,\xi)\mathrm{d}\xi\right) = 1 - \exp(-\gamma a_B^{\mathrm{dom}}(c)).$$

We assume that with probability one

(C3) Φ_B^{dom} is finite, i.e.

$$\beta_B^{\mathrm{dom}} \equiv \int_{\mathbb{R}^d} \int_0^\infty p_B^{\mathrm{dom}}(c,\gamma)\zeta(c,\gamma)\mathrm{d}c\mathrm{d}\gamma < \infty.$$

In many cases (see Brix & Kendall 2002 and Example D.1 below) we can

PERFECT SIMULATION OF SNCPS 255

choose a k_B^{dom} for which it is easy to calculate β_B^{dom} (at least by numerical methods) and where using Lemma D.1 it is feasible to generate a perfect simulation of Φ_B^{dom} and the $X_{(c,\gamma)}^{\text{dom}}$ with $(c,\gamma) \in \Phi_B^{\text{dom}}$. We thus arrive at the following algorithm.

ALGORITHM D.1 (*Perfect simulation algorithm for SNCPs*)
(a) Generate the Poisson process Φ_B^{dom}.
(b) (i) Generate independent processes $X_{(c,\gamma)}^{\text{dom}}$, $(c,\gamma) \in \Phi_B^{\text{dom}}$, distributed as conditional Poisson processes with intensity functions $\gamma k_B^{\text{dom}}(c,\cdot)$ and given that they are nonempty;
  (ii) For each $(c,\gamma) \in \Phi_B^{\text{dom}}$, obtain $X_{(c,\gamma)}$ by an independent thinning of $X_{(c,\gamma)}^{\text{dom}}$ with retention probabilities $k(c,\xi)/k_B^{\text{dom}}(c,\xi)$, $\xi \in X_{(c,\gamma)}^{\text{dom}}$;
(c) return $B \cap (\cup_{(c,\gamma) \in \Phi_B^{\text{dom}}} X_{(c,\gamma)})$ as a perfect simulation of X_B.

Examples of simulated point patterns using Algorithm D.1 are given in Brix & Kendall (2002). Steps (a) and (b)(i) are easily performed by the methods in Section 3.2.3, where in case (b)(i) we may use rejection sampling, i.e. we repeat to simulate a Poisson process $X_{(c,\gamma)}^{\text{dom}}$ with intensity function $\gamma k_B^{\text{dom}}(c,\cdot)$ until $X_{(c,\gamma)}^{\text{dom}} \neq \emptyset$. In Example D.1 below, we let $k_B^{\text{dom}}(c,\xi) = \sup_{\eta \in B} k(c,\eta) 1[\xi \in B]$ be constant for $\xi \in B$, whereby the generation of $X_{(c,\gamma)}^{\text{dom}}$ in Algorithm D.1 becomes straightforward.

EXAMPLE D.1 Let the situation be as in Example 5.8 (Section 5.5), so $B = b(0,R)$, and recall the cases (I) and (II) introduced in Example 5.3 (Section 5.3).

In case (I) the centres of Φ_B^{dom} are contained in $b(0, R+r)$,
$$p_B^{\text{dom}}(c,\gamma) = \left(1 - \exp(-\gamma(R/r)^d)\right) 1[\|c\| \leq R+r]$$
(see Example 5.8), and
$$\beta_B^{\text{dom}} = \omega_d (R+r)^d \int_0^\infty [1 - \exp(-\gamma(R/r)^d)] \chi(\gamma) \mathrm{d}\gamma.$$

In case (II) also $p_B^{\text{dom}}(c,\gamma) = 1 - \exp(-\gamma a_{II}(\|c\|))$ is a radially symmetric function of c which decays fast to zero (where a_{II} is defined in Example 5.8), and β_B^{dom} is equal to
$$\sigma_d \int_0^\infty \int_0^\infty s^{d-1}\left[1 - \exp\left(-\frac{\gamma \omega_d R^d}{(2\pi\omega^2)^{d/2}} \exp\left(-\frac{1[s>R]}{2\omega^2}(s-R)^2\right)\right)\right] \chi(\gamma) \mathrm{d}s \mathrm{d}\gamma.$$

As in Example 5.4 (Section 5.4.3), if χ is concentrated at α and $\kappa = \chi(\alpha)$, we have a Matérn cluster process in case (I) and a Thomas process in case (II). For the Matérn cluster process, the centres in Φ_B^{dom} form a homogeneous Poisson process on $b(0, R+r)$ with rate

$$\beta_B^{\mathrm{dom}} = \kappa \omega_d (R+r)^d \big[1 - \exp\big(-\alpha(R/r)^d\big)\big].$$

For the Thomas process,

$$\beta_B^{\mathrm{dom}} = \kappa\sigma_d \int_0^\infty s^{d-1}\bigg[1 - \exp\bigg(-\frac{\alpha\omega_d R^d}{(2\pi\omega^2)^{d/2}}\exp\bigg(-\frac{\mathbf{1}[s>R]}{2\omega^2}(s-R)^2\bigg)\bigg)\bigg]ds$$

is finite and easily determined by numerical integration. We can first generate $N \sim \mathrm{po}(\beta_B^{\mathrm{dom}})$ and next the N i.i.d. centres c_j, $j = 1, \ldots, N$, where the direction of c_j is uniformly distributed and independent of $s_j = \|c_j\|$, and s_j has a density proportional to

$$s^{d-1}\bigg[1 - \exp\bigg(-\frac{\alpha\omega_d R^d}{(2\pi\omega^2)^{d/2}}\exp\bigg(-\frac{\mathbf{1}[s>R]}{2\omega^2}(s-R)^2\bigg)\bigg)\bigg], \quad s > 0$$

(each s_j can be generated by e.g. rejection sampling).

Consider a shot noise G Cox process with parameter (α, τ, κ) as in Example 5.6 (Section 5.4.3). Then in case (I),

$$\beta_B^{\mathrm{dom}} = \kappa \omega_d (R+r)^d[(\tau + (R/r)^d)^\alpha - \tau^\alpha]/\alpha$$

if $\alpha < 0$ and

$$\beta_B^{\mathrm{dom}} = \frac{\kappa\omega_d(R+r)^d}{\Gamma(1-\alpha)}\int_0^\infty \big[1 - \exp\big(-(R/r)^d\gamma\big)\big]\gamma^{-\alpha-1}\exp(-\tau\gamma)d\gamma$$

if $0 \leq \alpha < 1$, while in case (II),

$$\beta_B^{\mathrm{dom}} = \frac{\kappa\sigma_d}{\Gamma(1-\alpha)}$$

$$\int_0^\infty \int_0^\infty \bigg[1 - \exp\bigg(-\frac{\gamma\omega_d R^d}{(2\pi\omega^2)^{d/2}}\exp\bigg(-\frac{\mathbf{1}[s>R]}{2\omega^2}(s-R)^2\bigg)\bigg)\bigg]$$
$$s^{d-1}\gamma^{-\alpha-1}\exp(-\tau\gamma)ds d\gamma.$$

In all cases β_B^{dom} is finite, and it is known or can be determined by numerical integration. Finally, Φ_B^{dom} is a finite inhomogeneous Poisson process with intensity function $(c, \gamma) \to (1 - \exp(-\gamma a_B^{\mathrm{dom}}(c)))\chi(\gamma)$, and we can generate this along similar lines as above for the Thomas process.

APPENDIX E

Simulation of Gaussian fields

Suppose that $Y = \{Y(\xi) : \xi \in \mathbb{R}^2\}$ is a Gaussian random field with zero mean and translation invariant covariance function c. Consider $\tilde{Y} = (Y(\xi))_{\xi \in I}$ where

$$I = \{(\Delta_1 m, \Delta_2 n) : m = 0, \ldots, M-1, n = 0, \ldots, N-1\}$$

is a regular lattice with spacings $\Delta_1 > 0$ and $\Delta_2 > 0$. In the following we describe an efficient method for simulation of $\tilde{Y}$ using *circulant embedding*. For a review of other methods, see Schlater (1999).

Let M_{ext} and N_{ext} be integers with $M_{\text{ext}} \geq 2(M-1)$ and $N_{\text{ext}} \geq 2(N-1)$, and let

$$d_{1il} = \Delta_1 \min\{|i-l|, M_{\text{ext}} - |i-l|\}, \quad i,l = 0, \ldots, M_{\text{ext}} - 1,$$

and

$$d_{2jk} = \Delta_2 \min\{|j-k|, N_{\text{ext}} - |j-k|\}, \quad j,k = 0, \ldots, N_{\text{ext}} - 1.$$

Define the symmetric matrix $C = (C_{ij,lk})$ by

$$C_{ij,lk} = c((d_{1il}, d_{2jk})), \quad i,l = 0, \ldots, M_{\text{ext}} - 1, j, k = 0, \ldots, N_{\text{ext}} - 1.$$

Suppose for a moment that C is positive semidefinite. Then C is the covariance matrix of a circulant stationary random field $\tilde{Y}_{\text{ext}} = (Y_{\text{ext}}(\xi))_{\xi \in I_{\text{ext}}}$ on

$$I_{\text{ext}} = \{(\Delta_1 m, \Delta_2 n) : m = 0, \ldots, M_{\text{ext}} - 1, n = 0, \ldots, N_{\text{ext}} - 1\}.$$

The construction of C corresponds to wrapping I_{ext} on a torus and using the shortest distances d_{1il} and d_{2lk} on the torus when defining the covariance structure of $\tilde{Y}_{\text{ext}}$ in terms of c. The field $\tilde{Y}$ is embedded in $\tilde{Y}_{\text{ext}}$ in the sense that the marginal distribution of $(Y_{\text{ext}}(\xi))_{\xi \in I}$ coincides with that of $\tilde{Y}$. Simulations of $\tilde{Y}$ can therefore be obtained from simulations of $\tilde{Y}_{\text{ext}}$.

By construction C is a block-circulant matrix with circulant $N_{\text{ext}} \times N_{\text{ext}}$ blocks $C_0, \ldots, C_{M_{\text{ext}}-1}$, say, see Davis (1979). In the case where C is positive semidefinite, C_m is the matrix of covariances between the first and the mth row in $\tilde{Y}_{\text{ext}}$ (when $\tilde{Y}_{\text{ext}}$ is arranged as an $M_{\text{ext}} \times N_{\text{ext}}$ matrix), $m = 0, \ldots, M_{\text{ext}} - 1$. For $L \in \mathbb{N}$, let

$$F_L = (\exp(-\hat{\imath} 2\pi lk/L)/\sqrt{L})_{l,k=0,\ldots,L-1}$$

denote the $L \times L$ discrete normalised Fourier transform matrix where $\hat{i}$ is the complex imaginary unit. For $m = 0, \ldots, M_{\text{ext}} - 1$, let c_m be the first row in C_m, and let

$$\Lambda_m = \text{diag}(\sqrt{N_{\text{ext}}} c_m \bar{F}_{N_{\text{ext}}})$$

be the diagonal matrix with diagonal entries given by $\sqrt{N_{\text{ext}}} c_m \bar{F}_{N_{\text{ext}}}$, where $\bar{}$ denotes complex conjugate. Further, let

$$\Omega = \text{diag}(1, w, \ldots, w^{M_{\text{ext}}-1})$$

where $w = \exp(\hat{i} 2\pi / M_{\text{ext}})$. Then by Theorem 5.8.1 in Davis (1979),

$$C = \bar{F}_\otimes \Lambda F_\otimes$$

where Λ is the diagonal matrix of eigenvalues λ_{mn} for C given by

$$\Lambda = \text{diag}(\lambda_{mn}; m = 0, \ldots, M_{\text{ext}} - 1, n = 0, \ldots, N_{\text{ext}} - 1) = \sum_{l=0}^{M_{\text{ext}}-1} \Omega^l \otimes \Lambda_l$$

and

$$F_\otimes = F_{M_{\text{ext}}} \otimes F_{N_{\text{ext}}}, \quad \bar{F}_\otimes = \bar{F}_{M_{\text{ext}}} \otimes \bar{F}_{N_{\text{ext}}},$$

where $\otimes$ is the Kronecker product. It can be shown that the eigenvalues λ_{mn} are real, so positive semidefiniteness of C is equivalent to that the eigenvalues are nonnegative. Combining the expressions for Λ and Λ_l yields that the $M_{\text{ext}} \times N_{\text{ext}}$ matrix $\tilde{\Lambda} = (\lambda_{mn})_{n,m=0}^{M_{\text{ext}}-1, N_{\text{ext}}-1}$ of the eigenvalues for C is given by

$$\tilde{\Lambda} = \sqrt{M_{\text{ext}} N_{\text{ext}}} \bar{F}_{M_{\text{ext}}} \tilde{C} \bar{F}_{N_{\text{ext}}}$$

where $\tilde{C}$ is the $M_{\text{ext}} \times N_{\text{ext}}$ matrix with rows $c_0, \ldots, c_{M_{\text{ext}}-1}$.

Suppose again that C is positive semidefinite. A simulation of $\tilde{Y}_{\text{ext}}$ regarded as a row vector is thus given by

$$U F_\otimes \Lambda^{1/2} \bar{F}_\otimes$$

where U is a standard normal vector of dimension $M_{\text{ext}} N_{\text{ext}}$. If we arrange $\tilde{Y}_{\text{ext}}$ and U as $M_{\text{ext}} \times N_{\text{ext}}$ matrices, this corresponds to calculating firstly,

$$\tilde{Z} = F_{M_{\text{ext}}} U F_{N_{\text{ext}}},$$

secondly,

$$Z_{mn} = \sqrt{\lambda_{mn}} \tilde{Z}_{mn}, \quad m = 0, \ldots, M_{\text{ext}} - 1, n = 0, \ldots, N_{\text{ext}} - 1,$$

and finally the simulation of $\tilde{Y}_{\text{ext}}$ is given by

$$\bar{F}_{M_{\text{ext}}} Z \bar{F}_{N_{\text{ext}}}.$$

If M_{ext} and N_{ext} are products of small primes, then the matrix $\tilde{\Lambda}$

SIMULATION OF GAUSSIAN FIELDS

and the simulation of $\tilde{Y}_{\text{ext}}$ can be computed efficiently using the two-dimensional *fast Fourier transform* (Press, Teukolsky, Vetterling & Flannery 1992), whereby a fast simulation algorithm is obtained. It is of course not guaranteed that C becomes positive semidefinite for given M_{ext} and N_{ext}. If negative eigenvalues are encountered, then increasing the values of M_{ext} and N_{ext} may help.

APPENDIX F

Nearest-neighbour Markov point processes

This appendix concerns nearest-neighbour Markov point processes (Baddeley & Møller 1989), a class of models that generalises Markov point processes. We restrict attention to the finite case where the processes are defined on $S \subset \mathbb{R}^d$ with $|S| < \infty$, cf. Section 6.1. Infinite Gibbs type nearest-neighbour Markov point processes are studied in Bertin, Billiot & Drouilhet (1999a) and Bertin, Billiot & Drouilhet (1999b).

F.1 Definition and characterisation

For a Markov point process X, two points $\xi, \eta \in X$ are neighbours or not regardless of the realisations of X. In a nearest-neighbour Markov point process X as defined below we change this so that the neighbourhood relation depends on the actual realisations $X = x$. We denote this relation by $\sim_x$.

Specifically, let $H \subset N_{\mathrm{f}}$ be a hereditary set, i.e. if $x \in H$ then $y \in H$ for $y \subset x$. Suppose that for any $x \in H$, $\sim_x$ is a symmetric and reflexive relation defined on x. The reason H is used instead of the entire state space N_{f} is that it may be impossible to define or to guarantee certain properties of $\sim_x$ on the whole of N_{f}, cf. Baddeley & Møller (1989) and Example F.2 below. For $x \in H$ and $\xi \in S \setminus x$ with $x \cup \xi \in H$, let $N_\xi(x) = \{\eta \in x : \eta \sim_{x \cup \xi} \xi\}$ denote the $x \cup \xi$-neighbourhood of ξ.

DEFINITION F.1 Let $h : N_{\mathrm{f}} \to [0, \infty)$ denote any hereditary function, and define λ^* in terms of h as in Section 6.3.1. Suppose that for all $x \in H$ with $h(x) > 0$ and all $\xi \in S \setminus x$ with $x \cup \xi \in H$, $\lambda^*(x, \xi)$ depends only on ξ, $N_\xi(x)$, and the restrictions of $\sim_x$ and $\sim_{x \cup \xi}$ to $N_\xi(x)$. Then h is said to be a *nearest-neighbour Markov function* (with respect to $\sim_x$). Particularly, if X is a point process with (unnormalised) density h with respect to Poisson$(S, 1)$ and $P(X \in H) = 1$, then X is called a *nearest-neighbour Markov point process (NNMPP)*.

Clearly, this definition covers the Ripley-Kelly case in Section 6.3.1 where $H = N_{\mathrm{f}}$ and $\sim_x = \sim$ does not depend on x. Other more interesting examples of nearest-neighbour relations and NNMPPs are given later in this appendix.

The class of nearest-neighbour Markov functions is characterised by the following Hammersley-Clifford type theorem. We need first to introduce some definitions. Consider any $y, x \in H$ with $y \subseteq x$. Let $\chi(y|x) = 1$ if $\xi \sim_x \eta$ for all $\xi, \eta \in y$, and $\chi(y|x) = 0$ otherwise. A function of the form $\Phi(y|x) = \varphi(y)^{\chi(y|x)}$ (taking $0^0 = 0$) is called an *interaction function* (with respect to $\sim_x$) if $\varphi : H \to [0, \infty)$ is hereditary and the following technical condition is satisfied:

$$\varphi(x) > 0, \ \varphi(\xi \cup N_\xi(x)) > 0 \ \Rightarrow \ \varphi(x \cup \xi) > 0 \quad \text{(F.1)}$$

for all $x \in H$ and $\xi \in S \setminus x$ with $x \cup \xi \in H$. Finally, for $y \subseteq z$ and $\xi, \eta \in S \setminus z$, we need two consistency conditions given by

$$\chi(y|z) \neq \chi(y|z \cup \xi) \ \Rightarrow \ y \subseteq N_\xi(z) \quad \text{(F.2)}$$

whenever $z \cup \xi \in H$, and

$$\xi \not\sim_x \eta \ \Rightarrow \ \chi(y|z \cup \xi) + \chi(y|z \cup \eta) = \chi(y|z) + \chi(y|x) \quad \text{(F.3)}$$

whenever $x = z \cup \{\xi, \eta\} \in H$. Both conditions are trivially satisfied in the Ripley-Kelly case when $H = N_f$, and in the case where $n(y) \leq 1$ (as $n(y) \leq 1$ implies $\chi(y|\cdot) = 1$). For a discussion of (F.1)-(F.3), see Baddeley & Møller (1989) and Section F.2.

THEOREM F.1 If the consistency condition (F.2) is satisfied and

$$h(x) = \prod_{y \subseteq x} \Phi(y|x), \quad x \in H, \quad \text{(F.4)}$$

where Φ is an interaction function, then h is a nearest-neighbour Markov function. If both consistency conditions (F.2)-(F.3) are satisfied and h is a nearest-neighbour Markov function, then there exists an interaction function $\Phi(\cdot|\cdot)$ such that h is given by (F.4).

Proof. Note that (F.2) and (F.4) imply

$$\lambda^*(x, \xi) = \prod_{y \subseteq N_\xi(x)} \varphi(y)^{\chi(y|x \cup \xi) - \chi(y|x)} \varphi(y \cup \xi)^{\chi(y|x \cup \xi)} \quad \text{(F.5)}$$

when $h(x) > 0$, $x \cup \xi \in H$, $\xi \in S \setminus x$. This verifies the first part in Theorem F.1. The other part is verified in Baddeley & Møller (1989). □

REMARK F.1 Baddeley & Møller (1989) consider a general state space S and extends Theorem F.1 to this case. Theorem F.1 reduces to Theorem 6.1 in the Ripley-Kelly case when $H = N_f$: if ϕ is an interaction function in the Ripley-Kelly sense, set $\Phi(y|x) = \varphi(y)^{\chi(y|x)}$ where $\varphi(y) = \phi(y)$ if $\phi(z) > 0$ for all $z \subseteq y$ and $\varphi(y) = 0$ otherwise.

EXAMPLES

If h is an unnormalised density of a point process, then (F.4) implies a spatial Markov property, see Kendall (1990) and Møller (1999a).

F.2 Examples

EXAMPLE F.1 (*Sequential neighbours on the real line*) Suppose $S \subset \mathbb{R}$. Let $x = \{x_1, \ldots, x_n\} \subset S$ where $x_1 < \cdots < x_n$, let $H = N_f$, and define the *sequential neighbour relation* by $x_i \sim_x x_j$ if and only if $|i - j| \leq 1$.

Since $\chi(y|\cdot) = 0$ when $n(y) \geq 3$, we need only to verify the consistency conditions (F.2)–(F.3) in the case where $y = \{y_1, y_2\}$ with $y_1 < y_2$. Further, as $\chi(y|z \cup \xi) = 1$ implies $\chi(y|z) = 1$, we need only to consider the case $\chi(y|z \cup \xi) = 0$ and $\chi(y|z) = 1$ in (F.2): then $(y_1, y_2) \cap z = \emptyset$ and $y_1 < \xi < y_2$, i.e. $y \subseteq N_\xi(z)$. Similarly, we need only to consider the case $\chi(y|x) = 0$ and $\chi(y|z) = 1$ in (F.3): then $(y_1, y_2) \cap z = \emptyset$ and $(y_1, y_2) \cap \{\xi, \eta\} \neq \emptyset$, and since $\xi \not\sim_x \eta$, some point in z lies between ξ and η but outside the interval $[y_1, y_2]$, so either $y_1 < \xi < y_2$ or $y_1 < \eta < y_2$ but not both. Thereby (F.2)–(F.3) are verified.

Consider

$$h(x) = \varphi(\emptyset) \prod_{i=1}^{n} \varphi(x_i) \prod_{i=1}^{n-1} \varphi(\{x_i, x_{i+1}\}) \tag{F.6}$$

where $\varphi > 0$. By Theorem F.1, h is a nearest-neighbour Markov function with respect to $\sim_x$. However, in general it is not a Markov function in the Ripley-Kelly sense, since for $\xi \in S \cap (x_i, x_{i+1})$,

$$\lambda^*(x, \xi) = \varphi(\xi) \varphi(\{\xi, x_i\}) \varphi(\{\xi, x_{i+1}\}) / \varphi(\{x_i, x_{i+1}\})$$

depends on the knowledge that ξ is falling between the ith and $(i+1)$th points in x.

By Theorem F.1, not all nearest-neighbour Markov functions (with respect to the sequential neighbour relation) need to be of the form (F.6). However, any positive nearest-neighbour Markov function $h > 0$ is of the form (F.6) where $\varphi > 0$. For example, the area-interaction process with density (6.27) and $S \subset \mathbb{R}$ is an NNMPP with respect to $\sim_x$, since we can rewrite the density as

$$f(x) = (\gamma^{-2R}/c) \beta^n \prod_{i=1}^{n-1} \gamma^{-\min\{x_{i+1} - x_i, 2R\}}$$

which is of the form (F.6) (here c denotes the normalising constant in (6.27)). As noticed in Example 6.5 (Section 6.3.2), this process is also Markov (in the Ripley-Kelly sense); but if we replace the interaction term in the product above by another nonconstant function $\varphi(\{x_i, x_{i+1}\})$, it is in general not Markov. Examples of this include renewal processes as discussed in Baddeley & Møller (1989).

EXAMPLE F.2 (*Dirichlet neighbours*) The sequential neighbour relation in Example F.1 can naturally be extended to the Dirichlet neighbour relation in any dimension d if $S \subset \mathbb{R}^d$, cf. Baddeley & Møller (1989). Here we just consider the planar case $d = 2$.

We say that $x \in N_f$ is in *quadratic general position* if no three points in x lie on the same line in $\mathbb{R}^2$, and no four points in x lie on the same circle in $\mathbb{R}^2$. Let H denote the set of finite point configurations in quadratic general position. Clearly H is hereditary and has probability one under the standard Poisson process. For $x \in H$, define the *Dirichlet* or *Voronoi cell* with "nucleus" $\xi \in x$ by

$$C(\xi|x) = \{\eta \in \mathbb{R}^2 : \|\xi - \eta\| \leq \|\kappa - \eta\| \text{ for all } \kappa \in x\},$$

i.e. the set of points in S that are not closer to another "nucleus" $\kappa \in x \setminus \xi$. These cells are convex polygons and constitute a subdivision of the plane called the *Dirichlet* or *Voronoi tessellation*, see Møller (1994b), Stoyan et al. (1995), and Okabe, Boots, Sugihara & Chiu (2000). Since x is in quadratic general position, a nonvoid intersection between two Dirichlet cells is a line segment called an edge, and these edges cannot be parallel; and a nonvoid intersection between three Dirichlet cells is a point called a vertex; while the intersection between four or more cells is always empty. Figure F.1 shows a Dirichlet tessellation generated by 100 independent and uniformly distributed points in the unit square.

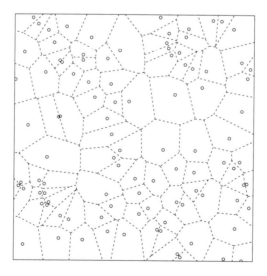

Figure F.1 *Dirichlet tessellation generated by 100 points uniformly dispersed in the unit square.*

Now, for $x \in H$, define $\xi, \eta \in x$ to be *Dirichlet neighbours* if $C(\xi|x) \cap$

CONNECTED COMPONENT MARKOV POINT PROCESSES   265

$C(\eta|x) \neq \emptyset$, i.e. if ξ and η share an edge in the Dirichlet tessellation generated by x. It is shown in Baddeley & Møller (1989) that the consistency conditions (F.2)–(F.3) are satisfied, and hence Theorem F.1 applies. Note that $\chi(y|x) = 1$ implies $n(y) \leq 3$. Examples of NNMPPs with respect to Dirichlet neighbours are given in Baddeley & Møller (1989); see also Bertin et al. (1999b).

EXAMPLE F.3 (*Ord's process*) Let H and $\sim_x$ be as in the previous example. Define $h : H \to [0, \infty)$ by

$$h(x) = \beta^{n(x)} \prod_{\xi \in x} k(|C(\xi|x)|)$$

where $\beta > 0$ and $k(\cdot) \geq 0$. Ord proposed (in the discussion of Ripley 1977) point process models with an unnormalised density of this form, where $k(a) = \mathbf{1}[a \geq a_0]$ and $a_0 > 0$ represents the minimum viable area of a cell. Note that h is not a nearest-neighbour Markov function with respect to the Dirichlet relation $\sim_x$, unless k is constant: $\lambda^*(x, \xi)$ depends on neighbours of neighbours of ξ.

However, under weak conditions on k, it is a nearest-neighbour Markov function with respect to another relation. Define the *iterated Dirichlet relation* $\sim_x^2$ by $\xi \sim_x^2 \eta$ if and only if there exists $\kappa \in x$ so that $\kappa \sim_x \xi$ and $\kappa \sim_x \eta$. This relation satisfies the consistency condition (F.2) but not (F.3). If k is positive or nondecreasing, then h is a nearest-neighbour Markov function with respect to $\sim_x^2$. For details, see Baddeley & Møller (1989).

F.3 Connected component Markov point processes

Let $\sim$ be a symmetric and reflexive relation on $S \subset \mathbb{R}^d$ and let $H = N_\mathrm{f}$. Define the *connected component relation* by

$$\xi \sim_x \eta \Leftrightarrow \exists \{x_1, \ldots, x_n\} \subseteq x : \xi = x_1 \sim x_2 \sim \cdots \sim x_n = \eta. \quad (\text{F.7})$$

For $y \subseteq x \in N_\mathrm{f}$, we call y a *connected component* in x if $\chi(y|x) = 1$ and $\chi(y \cup \xi|x) = 0$ for all $\xi \in x \setminus y$. For example, if $\sim$ is the R-close neighbourhood relation, $\xi \sim_x \eta$ if and only if ξ and η belong to the same connected component of the set given by the union of balls with centres in x and diameter R.

Note that $\chi(y|z) = 1$ implies $\chi(y|z \cup \xi) = 1$. So to verify (F.2) we need only to consider the case where $\chi(y|z) = 0$ and $\chi(y|z \cup \xi) = 1$. Obviously y is then connected to ξ, i.e. $y \subseteq N_\xi(z)$. To verify (F.3) we need only to consider the case where $\chi(y|z) = 0$ and $\chi(y|x) = 1$. Then y is connected to either ξ or η but not both, since $\xi \not\sim_x \eta$. So $\chi(y|z \cup \xi) + \chi(y|z \cup \eta) = 1$. Hence the two consistency conditions are verified.

For $x \in N_f$, let $C(x)$ denote the set of connected components in x. Theorem F.1 specialises to the following result from Baddeley, van Lieshout & Møller (1996).

COROLLARY F.1 h is a nearest-neighbour Markov function with respect to the connected component relation if and only if there exist a constant $\alpha \geq 0$ and a function $\psi : N_f \to [0, \infty)$ such that

$$h(x) = \alpha \prod_{y \in C(x)} \psi(y), \quad x \in N_f, \tag{F.8}$$

where $\psi(\cdot) \geq 0$ defined on the set of connected components $\{v \subset N_f : \chi(v|v) = 1\}$ is such that $\psi(y) > 0$ implies $\psi(z) > 0$ for all $z \subset y$.

Proof. Suppose that (F.8) holds. As

$$C(x \cup \xi) = \{y \in C(x) : y \not\subseteq N_\xi(x)\} \cup \{\xi \cup N_\xi(x)\}$$

we obtain

$$h(x) = \alpha \left[\prod_{y \in C(x):\, y \not\subseteq N_\xi(x)} \psi(y) \right] \left[\prod_{y \in C(x):\, y \subseteq N_\xi(x)} \psi(y) \right]$$

and

$$h(x \cup \xi) = \alpha \left[\prod_{y \in C(x):\, y \not\subseteq N_\xi(x)} \psi(y) \right] \psi(\xi \cup N_\xi(x)).$$

Hence h is easily seen to satisfy the conditions in Definition F.1.

Conversely, suppose that h is of the form (F.4) in Theorem F.1. Below we prove that this reduces to

$$h(x) = \alpha \prod_{y \in C(x)} \prod_{z \subseteq y:\, z \neq \emptyset} \varphi(z), \quad x \in N_f, \tag{F.9}$$

where $\alpha = \varphi(\emptyset)$ and $\varphi(\cdot) \geq 0$ is such that $\varphi(y) > 0$ implies $\varphi(z) > 0$ for $z \subset y$. From this we immediately obtain (F.8) with

$$\psi(y) = \prod_{z \subseteq y:\, z \neq \emptyset} \varphi(z).$$

Let $h(x)$ be given by (F.4). Note that $\chi(y|x) = 1$ if and only if $y \subseteq v$ for some $v \in C(x)$. Therefore (F.9) follows immediately if $\varphi(y) > 0$ for all $y \subseteq x$ or if $\varphi(y) = 0$ for some $y \subseteq x$ with $\chi(y|x) = 1$. In the remaining case, $\varphi(z) > 0$ for all $z \subseteq x$ with $\chi(z|x) = 1$ and $\varphi(y) = 0$ for some $y \subseteq x$ with $\chi(y|x) = 0$, and so the right hand sides in (F.4) and (F.9) do not agree. We show now that this case cannot occur, that is, $\varphi(z) > 0$ for all $z \subseteq x$ with $\chi(z|x) = 1$ implies $\varphi(y) > 0$ for all $y \subseteq x$ with $\chi(y|x) = 0$. To show this suppose that $v, w \in C(x)$ are disjoint. If $\xi \in v$ then $N_\xi(w) = \emptyset$, and by assumption $\varphi(\xi) > 0$ and $\varphi(w) > 0$, so

CONNECTED COMPONENT MARKOV POINT PROCESSES   267

(F.1) gives $\varphi(w \cup \xi) > 0$. If $\{\xi, \eta\} \subseteq v$ with $\xi \sim \eta$, then $N_\eta(w \cup \xi) = \xi$, and by assumption $\varphi(\{\xi, \eta\}) > 0$, so (F.1) gives $\varphi(w \cup \{\xi, \eta\}) > 0$ since $\varphi(w \cup \xi) > 0$. Continuing in this way we obtain that $\varphi(y) > 0$ whenever $\chi(y|x) = 0$. Finally, as φ is hereditary, the condition on φ stated below (F.9) is satisfied. □

REMARK F.2

- Like Theorem F.1, Corollary F.1 holds for an arbitrary space S.
- Because of the product form (F.8) we call a point process with a density of the form (F.8) for a *Markov connected component process*. An analog class of discrete random field models called Markov connected component fields are studied in Møller & Waagepetersen (1998). Extensions to point processes with interaction between the connected components are studied in Chin & Baddeley (2000). One such example is obtained by multiplying the right side in (F.8) by an indicator function so that different components are always separated by a given hard core parameter.
- One example is the *continuum random cluster model* given by

$$f(x) \propto \beta^{n(x)} \gamma^{-c(x)} \tag{F.10}$$

where $\beta, \gamma > 0$ are parameters and $c(x)$ denotes the number of connected components in x (Klein 1982, Møller 1994a, Chayes et al. 1995, Møller 1999a, Häggström et al. 1999). This is clearly not Markov in the Ripley-Kelly sense. The process is repulsive for $0 < \gamma \leq 1$ and attractive for $\gamma \geq 1$. See Häggström et al. (1999) and Møller (1999a) for examples of simulated point patterns and a further discussion of this model.
- In the symmetric case $\beta_1 = \beta_2 = \beta$ of the Widom-Rowlinson penetrable spheres mixture model (Example 6.8, Section 6.6.3), the marginal density for the superposition $Z = X_1 \cup X_2$ is a repulsive continuum random cluster model when $\sim$ in the definition (F.7) of the connected component relation is the R-close-neighbourhood relation: As

$P(Z \in F) \propto$

$$\sum_{m=0}^{\infty} \sum_{n=0}^{\infty} \frac{\beta^{m+n}}{m!n!} \int_T \cdots \int_T \int_T \cdots \int_T \mathbf{1}[d(\{\xi_1, \ldots, \xi_m\}, \{\eta_1, \ldots, \eta_n\}) > R,$$

$$\{\xi_1, \ldots, \xi_m, \eta_1, \ldots, \eta_n\} \in F] d\xi_1 \cdots d\xi_m d\eta_1 \cdots d\eta_n$$

$$= \sum_{l=0}^{\infty} \beta^l / l! \int_T \cdots \int_T 2^{c(\{\kappa_1, \ldots, \kappa_l\})} \mathbf{1}[\{\kappa_1, \ldots, \kappa_l\} \in F] d\kappa_1 \cdots \kappa_l,$$

the density of Z is of the form (F.10) with $\gamma = 1/2$.

- Though the area-interaction process (Examples 6.5, Section 6.3.2) is Markov, it seems more appealing to view it as a Markov connected component process with $\psi(y) = \beta^{n(y)} \gamma^{-|U_{y,R}|}$ in (F.8), cf. Remark 6.5 in Section 6.3.2.
- Baddeley et al. (1996) and Møller (2003b) show that shot noise Cox processes (and other related types of cluster processes where the mother process is Markov) are in general not Markov point processes but Markov connected component processes.

APPENDIX G

Ergodicity properties and other results for spatial birth-death processes

In this appendix we review those results in Preston (1977) for spatial birth-death processes which are relevant in this book, and establish in addition various useful results in the locally stable and constant death rate case. We use the same notation as in Section 11.1.

G.1 Jump processes

Spatial birth-death processes are special cases of jump processes, which in turn are special cases of continuous time Markov chains with piecewise constant right-continuous sample paths. Jump processes have been studied in e.g. Feller (1971) and Preston (1977). Before we study spatial birth-death processes in detail, it is useful to consider what is meant by a jump process defined on a general state space Ω.

Consider a stochastic process Y_t, $t \in [0, \infty)$, with state space Ω. Let $\alpha : \Omega \to [0, \infty)$ be a function and $P(x, F)$ a probability kernel defined for $x \in \Omega$ and $F \subseteq \Omega$, i.e. for each $x \in \Omega$, $P(x, \cdot)$ is a probability measure on Ω. Assume that conditional on $Y_t = x$ and the past history Y_s, $s < t$,

- the waiting time τ until the process jumps away from x is exponentially distributed with mean $1/\alpha(x)$, i.e.

$$t + \tau = \inf\{s > t : Y_s \neq x\}, \quad \tau \sim \mathrm{Exp}(\alpha(x))$$

(setting $\tau = \infty$ if $\alpha(x) = 0$);

- if we further condition on τ, then

$$Y_{t+\tau} \sim P(x, \cdot) \text{ if } \alpha(x) > 0, \quad Y_s = Y_t \text{ for all } s \geq t \text{ if } \alpha(x) = 0.$$

In the case $\alpha(x) > 0$, we assume that $P(x, \{x\}) = 0$. Then Y_t is called a *jump process* with *intensity* α and *transition kernel* $P(\cdot, \cdot)$. The process is uniquely determined if and only if it does not *explode*, i.e. if an infinite number of jumps in a finite time interval can never occur with a positive probability (Preston 1977, Section 3).

G.2 Coupling constructions

From now on we let Y_t, $t \geq 0$, be a spatial birth-death process with birth rate b and death rate $d(\cdot, \cdot)$. This is a jump process with state space $\Omega = N_f$, intensity $\alpha(x)$ given by (11.1), and transition kernel

$$P(x, F) = \left\{ \int_B \mathbf{1}[x \cup \xi \in F] b(x, \xi) d\xi + \sum_{\xi \in x} \mathbf{1}[x \setminus \xi \in F] d(x \setminus \xi, \xi) \right\} \bigg/ \alpha(x) \tag{G.1}$$

for $\alpha(x) > 0$, taking the sum equal to 0 if $x = \emptyset$. To check that this process does not explode and to study the ergodicity properties of the process, different coupling constructions turn out to be useful.

We consider first the case where b and d are given by (11.7) for a ϕ^*-locally stable unnormalised density h.

PROPOSITION G.1 The coupling constructions (Y_t, D_t), $t \geq 0$, considered in Algorithm 11.3 and Remark 11.1 (Section 11.1.4) are stochastic equivalent; Y_t is a spatial birth-death process with birth rate $b(y, \xi) = \lambda^*(y, \xi)$ and death rate $d(\cdot, \cdot) = 1$; and D_t is a spatial birth-death process with birth rate $b_D(x, \xi) = \phi^*(\xi)$ and death rate $d_D(\cdot, \cdot) = 1$.

Proof. We use the notation in Algorithm 11.3 and Remark 11.1.

Consider first the coupling construction in Algorithm 11.3. It follows immediately from Algorithm 11.3 that D_t is a spatial birth-death process with birth rate $b_D(x, \xi) = \phi^*(\xi)$ and death rate $d_D(\cdot, \cdot) = 1$. Further, (Y_t, D_t), $t \geq 0$, is seen to be a jump process with state space $\Omega = \{(y, x) \in E \times N_f : y \subseteq x\}$, intensity

$$\alpha(y, x) = c^* + n(x), \tag{G.2}$$

and transition kernel specified as follows, where $A \subseteq B$ and $\beta(y) = \int_B \lambda^*(y, \xi) d\xi$:

$$P((y, x), \{y\} \times \{x \cup \xi : \xi \in A\}) = \frac{1}{c^* + n(x)} \int_A (\phi^*(\xi) - \lambda^*(y, \xi)) d\xi,$$

$$P((y, x), \{(y \cup \xi, x \cup \xi) : \xi \in A\}) = \frac{1}{c^* + n(x)} \int_A \lambda^*(y, \xi) d\xi,$$

$$P((y, x), \{(y \setminus \eta, x \setminus \eta)\}) = \mathbf{1}[\eta \in x]/(c^* + n(x)). \tag{G.3}$$

Consider next the case where (Y_t, D_t) is obtained using the construction in Remark 11.1, and let T denote an arbitrary jump time of the process D_t. Suppose we condition on (Y_t, D_t), $t \leq T$, with $Y_T = y \in E$, $D_T = x = \{x_1, \ldots, x_n\} \in N_f$, $n \geq 0$, and $y \subseteq x$. Set $T_0 = \inf\{\tilde{T}_m - T : \tilde{T}_m > T\}$ and $T_i = \tau(x_i) - T$, where $\tau(x_i)$ is the life time associated to x_i in the marked Poisson process in Remark 11.1. Since the exponential distribution has no memory (Remark 3.3, Section 3.1.3), it follows from

COUPLING CONSTRUCTIONS 271

(11.15) that $T_0 \sim \text{Exp}(c^*)$ and $T_i \sim \text{Exp}(1)$, $i = 1,\ldots,n$, are mutually independent. The time for the first transition in D_t, $t > T$, is given by $T + \tilde{T}$ where $\tilde{T} = \min\{T_0,\ldots,T_n\}$. By Lemma 11.2 (Section 11.1.2), $\tilde{T} \sim \text{Exp}(c^* + n)$, $\tilde{T} = T_0$ with probability $c^*/(c^* + n)$, and $\tilde{T} = T_i$ with probability $1/(c^* + n)$ for $i = 1,\ldots,n$. If we further condition on that $\tilde{T} = T_0$, then the point $\xi \sim \bar{b}'$ added to D_t at time $T + \tilde{T}$ is also added to Y_t with probability $\lambda^*(y,\xi)/\phi^*(\xi)$. Hence, (Y_t, D_t), $t \geq 0$, is seen to be a jump process with the same intensity and transition kernel as given above for Algorithm 11.3.

Finally, along similar lines as in the proof of Proposition 11.1 and using the coupling construction in Remark 11.1, we obtain that Y_t is a spatial birth-death process with birth rate $b = \lambda^*$ and death rate $d(\cdot,\cdot) = 1$. □

COROLLARY G.1 In the locally stable and constant death rate case, the processes (Y_t, D_t), Y_t, and D_t do not explode.

Proof. Since a transition in Y_t implies a transition in D_t, it suffices to show that D_t does not explode. This follows from Remark 11.1, since

$$\#\{\text{transitions in } [0,t]\} \leq n(D_0) + 2\#\{\text{births in } [0,t]\}$$

which is finite with probability one, since the birth times follows a homogeneous Poisson process. □

Preston (1977) considers a more general setting, using a *coupling* of Y_t with a birth-death process defined on $\mathbb{N}_0$. For $n \in \mathbb{N}_0$, define

$$\beta_n = \sup\{\beta(x) : n(x) = n\}, \quad \delta_n = \inf\{\delta(x) : n(x) = n\},$$

and assume that all β_n are finite. Let N_t be a birth-death process with state space $\mathbb{N}_0$, intensity $\alpha_n = \beta_n + \delta_n$, and transition kernel p given by $p(n, n+1) = \beta_n/\alpha_n$ ($n \in \mathbb{N}_0$) and $p(n, n-1) = \delta_n/\alpha_n$ ($n \in \mathbb{N}$) for $\alpha_n > 0$. We can define (Y_t, N_t) as a jump process with state space $N_f \times \mathbb{N}_0$: The intensity is

$$\tilde{\alpha}(x,n) = \begin{cases} \alpha(x) + \alpha_n & \text{if } n(x) \neq n, \\ \delta(x) + \beta_n & \text{if } n(x) = n. \end{cases}$$

For $\tilde{\alpha}(x,n) > 0$, if $n(x) \neq n$, the transition kernel is given by

$$\tilde{P}((x,n), F \times \{n\}) = \alpha(x)P(x,F)/\tilde{\alpha}(x,n),$$
$$\tilde{P}((x,n), \{x\} \times \{n+1\}) = \beta_n/\tilde{\alpha}(x,n),$$
$$\tilde{P}((x,n), \{x\} \times \{n-1\}) = \delta_n/\tilde{\alpha}(x,n),$$

with $P(\cdot,\cdot)$ given by (G.1), while if $n(x) = n$, $F_{n-1} \subseteq \{y \in N_f : n(y) = $

$n-1\}$, and $F_{n+1} \subseteq \{y \in N_f : n(y) = n+1\}$,

$$\tilde{P}((x,n), F_{n+1} \times \{n+1\}) = \beta(x)P(x, F_{n+1})/\tilde{\alpha}(x,n),$$
$$\tilde{P}((x,n), \{x\} \times \{n+1\}) = (\beta_n - \beta(x))/\tilde{\alpha}(x,n),$$
$$\tilde{P}((x,n), F_{n-1} \times \{n-1\}) = \delta_n P(x, F_{n-1})/\tilde{\alpha}(x,n),$$
$$\tilde{P}((x,n), F_{n-1} \times \{n\}) = (\delta(x) - \delta_n)P(x, F_{n-1})/\tilde{\alpha}(x,n).$$

Note that as long as $n(Y_t) \neq N_t$, the two processes evolve independently of each other. In particular, the simple birth-death process *dominates* the spatial birth-death in the sense that

$$n(Y_s) \leq N_s \Rightarrow n(Y_t) \leq N_t, \quad t \geq s.$$

Using Lemmas 11.1–11.2 (Section 11.1.2), it is straightforwardly verified that Y_t and N_t are individual birth-death processes with the correct birth and death rates.

Preston (1977, Proposition 5.1) verifies that the process Y_t does not explode if the process N_t does not explode, and he gives sufficient conditions for nonexplosiveness of N_t. For our purpose, the following condition is usually sufficient.

PROPOSITION G.2 Suppose that either

$$\exists n_0 \in \mathbb{N}_0 : \beta_n = 0 \text{ for all } n \geq n_0, \tag{G.4}$$

or $\beta_n > 0$ for all $n \in \mathbb{N}_0$ and

$$\sum_{n=0}^{\infty} \left[\frac{1}{\beta_n} + \frac{\delta_n}{\beta_{n-1}\beta_n} + \ldots + \frac{\delta_1 \cdots \delta_n}{\beta_0 \cdots \beta_n} + \frac{\delta_0 \cdots \delta_n}{\beta_0 \cdots \beta_n} \right] = \infty. \tag{G.5}$$

Then the processes (Y_t, N_t), N_t, and Y_t do not explode.

The condition (G.4) is typically satisfied for simulation of point processes X with $P(n(X) \leq n_0) = 1$ for some $n_0 \geq 0$, e.g. X could be a hard core process. In the locally stable and constant death rate case, $\beta_{n_0} = 0$ is easily seen to imply (G.4), while if $\beta_n > 0$ for all $n \in \mathbb{N}_0$, then $\sum_{n=0}^{\infty} 1/\beta_n \geq \sum_{n=0}^{\infty} 1/c^* = \infty$, and so (G.5) is satisfied.

G.3 Detailed balance

The following proposition establishes that the detailed balance condition (11.5) implies reversibility.

PROPOSITION G.3 If (11.5) holds and $\mathbb{E}\beta(X) < \infty$ with $X \sim h$, then Y_t, $t \geq 0$, is reversible with respect to h, i.e. if $Y_0 \sim h$ then for any $t > 0$, the processes Y_s, $0 \leq s \leq t$, and Y_{t-s}, $0 \leq s \leq t$, are identically distributed.

DETAILED BALANCE

Proof. The processes Y_s, $0 \leq s \leq t$, and Y_{t-s}, $0 \leq s \leq t$, are identically distributed if for any $n \in \mathbb{N}$ and $0 \leq t_1 < \ldots < t_n \leq t$, $(Y_{t_1}, \ldots, Y_{t_n})$ and $(Y_{t_n}, \ldots, Y_{t_1})$ are identically distributed. Since Y_s, $s \geq 0$, is a time-homogeneous Markov process, we need only to verify that (Y_0, Y_t) and (Y_t, Y_0) are identically distributed if $Y_0 \sim h$. This is verified by a heuristic argument below, considering only what happens in an infinitesimally small time interval $[0, \mathrm{d}t]$; a formal proof is given in Preston (1977, Section 8).

For $x \in N_f$,

$$P(Y_s, \ 0 \leq s \leq \mathrm{d}t, \text{ has more than one transition} \mid Y_0 = x) = o(\mathrm{d}t) \tag{G.6}$$

where $o(\mathrm{d}t)$ denotes a term for which $o(\mathrm{d}t)/\mathrm{d}t \to 0$ when $\mathrm{d}t \to 0$. For $A \subseteq B$,

$$P(Y_{\mathrm{d}t} = x \cup \xi \text{ for some } \xi \in A \mid Y_0 = x) = \int_A b(x, \xi)\mathrm{d}\xi \times \mathrm{d}t + o(\mathrm{d}t), \tag{G.7}$$

and for $\eta \in x$,

$$P(Y_{\mathrm{d}t} = x \setminus \eta \mid Y_0 = x) = d(x \setminus \eta, \eta) \times \mathrm{d}t + o(\mathrm{d}t). \tag{G.8}$$

Suppose that $Y_0 \sim f$ where $f \propto h$ is the normalised density. Combining (G.6)–(G.8) with (6.1) and (11.5), we obtain that for $A \subseteq B$ and $F \subseteq N_f$,

$$P(Y_0 \in F, Y_{\mathrm{d}t} = Y_0 \cup \xi \text{ for some } \xi \in A)$$

$$= \sum_{n=0}^{\infty} \frac{\exp(-|B|)}{n!} \int_B \cdots \int_B \int_B \mathbf{1}[\{x_1, \ldots, x_n\} \in F, \xi \in A]$$

$$\qquad f(\{x_1, \ldots, x_n\}) b(\{x_1, \ldots, x_n\}, \xi) \mathrm{d}x_1 \cdots \mathrm{d}x_n \mathrm{d}\xi \times \mathrm{d}t + o(\mathrm{d}t)$$

$$= \sum_{n=0}^{\infty} \frac{\exp(-|B|)}{n!} \int_B \cdots \int_B \int_B \mathbf{1}[\{x_1, \ldots, x_n\} \in F, \xi \in A]$$

$$\qquad f(\{x_1, \ldots, x_n, \xi\}) d(\{x_1, \ldots, x_n\}, \xi) \mathrm{d}x_1 \cdots \mathrm{d}x_n \mathrm{d}\xi \times \mathrm{d}t + o(\mathrm{d}t)$$

$$= \sum_{n=1}^{\infty} \frac{\exp(-|B|)}{n!} \int_B \cdots \int_B \sum_{i=1}^{n} \mathbf{1}[\{x_1, \ldots, x_n\} \setminus x_i \in F, x_i \in A]$$

$$\qquad f(\{x_1, \ldots, x_n\}) d(\{x_1, \ldots, x_n\} \setminus x_i, x_i) \mathrm{d}x_1 \cdots \mathrm{d}x_n \times \mathrm{d}t + o(\mathrm{d}t)$$

$$= P(Y_{\mathrm{d}t} = Y_0 \setminus \xi \text{ for some } \xi \in Y_0 \cap A, Y_{\mathrm{d}t} \in F)$$

$$= P(Y_0 = Y_{\mathrm{d}t} \cup \xi \text{ for some } \xi \in A, Y_{\mathrm{d}t} \in F)$$

where the condition $\mathbb{E}\beta(X) < \infty$ ensures that the integrals above are finite. By an analogous argument,

$$P(Y_0 \in F, Y_{\mathrm{d}t} = Y_0 \setminus \xi \text{ for some } \xi \in Y_0)$$
$$= P(Y_0 = Y_{\mathrm{d}t} \setminus \xi \text{ for some } \xi \in Y_{\mathrm{d}t}, Y_{\mathrm{d}t} \in F).$$

Hence, for any $F, G \subseteq N_f$,

$$P(Y_0 \in F, Y_{dt} \in G, Y_0 \neq Y_{dt})$$
$$= P(Y_0 \in F, Y_{dt} \in G, n(Y_{dt}) = n(Y_0) + 1)$$
$$+ P(Y_0 \in F, Y_{dt} \in G, n(Y_{dt}) = n(Y_0) - 1) + o(dt)$$
$$= P(Y_0 \in G, Y_{dt} \in F, n(Y_0) = n(Y_{dt}) + 1)$$
$$+ P(Y_0 \in G, Y_{dt} \in F, n(Y_0) = n(Y_{dt}) - 1) + o(dt)$$
$$= P(Y_0 \in G, Y_{dt} \in F, Y_0 \neq Y_{dt})$$

and so
$$P(Y_0 \in F, Y_{dt} \in G) = P(Y_0 \in G, Y_{dt} \in F).$$
Thus (Y_0, Y_{dt}) and (Y_{dt}, Y_0) are identically distributed. □

G.4 Ergodicity properties

We now establish various limit results for the algorithms in Sections 11.1.2 –11.1.4. The first main result is Proposition G.4 which is concerned with the existence and properties of the limiting distribution $\lim_{t \to \infty} P(Y_t \in F | Y_0 = x)$ for $x \in N_f$ and $F \subseteq N_f$. The second main result is Proposition G.7 where V-uniform ergodicity and a central limit theorem is established in the locally stable and constant death rate case.

In order to establish the existence of $\lim_{t \to \infty} P(Y_t \in F | Y_0 = x)$ the notion of an *ergodic atom* becomes useful. We say that $\emptyset$ is an ergodic atom for Y_t if the mean return time to $\emptyset$ is finite:

$$\mathbb{E}[\inf\{t > 0 : Y_{t-} \neq \emptyset, \ Y_t = \emptyset\} | Y_0 = \emptyset] < \infty.$$

Similarly, 0 is an ergodic atom for N_t if the mean return time to 0 is finite, and $(\emptyset, 0)$ is an ergodic atom for (Y_t, N_t) if the mean return time to $(\emptyset, 0)$ is finite. Since N_t dominates Y_t, $\emptyset$ and $(\emptyset, 0)$ are ergodic atoms if 0 is an ergodic atom. If 0 is an ergodic atom, since the process (Y_t, N_t) regenerates each time N_t visits 0, we obtain by the renewal theorem (see e.g. Asmussen 1987) that $\lim_{t \to \infty} P((Y_t, N_t) \in F | (Y_0, N_0) = (\emptyset, 0))$ exists for all events $F \subseteq N_f \times \mathbb{N}_0$. For similar reasons, if $\emptyset$ is an ergodic atom and if the expected transition time for N_t from $n \in \mathbb{N}_0$ to 0 is finite, then $\lim_{t \to \infty} P(Y_t \in F | Y_0 = x)$ exists for $n(x) = n$ and $F \subseteq N_f$.

In fact 0 is an ergodic atom if and only if one of the following conditions holds:

(a) $\beta_0 = 0$;
(b) $\beta_0 > 0$, $\beta_m = 0$ for some $m \in \mathbb{N}$, and $\delta_n > 0$ for all $n \leq m$;
(c) $\beta_n > 0$ for all $n \in \mathbb{N}_0$, $\delta_n > 0$ for all $n \in \mathbb{N}$, and

$$\sum_{n=1}^{\infty} \frac{\beta_0 \cdots \beta_{n-1}}{\delta_1 \cdots \delta_n} < \infty.$$

ERGODICITY PROPERTIES

Case (a) is not interesting for the purpose of simulation, since it implies that $\emptyset$ is an absorbing state for Y_t. Below in Proposition G.4 we assume either (c) or a condition which implies (a) or (b). That the sum in (c) is finite simply means that the invariant density π_n for N_t given by the detailed balance condition

$$\pi_n \beta_n = \pi_{n+1} \delta_{n+1}, \quad n \in \mathbb{N}_0,$$

is well defined, since we obtain

$$\pi_n = \pi_0 \frac{\beta_0 \cdots \beta_{n-1}}{\delta_1 \cdots \delta_n}, \quad n \in \mathbb{N}.$$

The considerations above provide a sketch proof of Proposition G.4 below; a formal proof is given in Preston (1977, Theorem 7.1).

PROPOSITION G.4 Suppose that the process N_t does not explode, and that either (c) holds or that $\delta_n > 0$ for all $n \in \mathbb{N}$ and $\beta_m = 0$ for some $m \geq 0$. Then Y_t does not explode, and

(i) for $x \in N_f$ and $F \subseteq N_f$,

$$\kappa(F) = \lim_{t \to \infty} P(Y_t \in F | Y_0 = x)$$

exists and is independent of x;

(ii) κ is a probability measure on N_f, and κ is the unique *invariant distribution* for Y_t such that $Y_0 \sim \kappa$ implies $Y_t \sim \kappa$ for all $t \geq 0$;

REMARK G.1 If the conditions in Propositions G.3–G.4 are satisfied, then κ has invariant unnormalised density h (because reversibility implies invariance).

PROPOSITION G.5 The conditions in Propositions G.3–G.4 are satisfied in the locally stable and constant death rate case.

Proof. Clearly, (11.5) is satisfied, and since $\beta(\cdot) \leq c^*$ implies $\mathbb{E}\beta(X) < \infty$, the conditions in Proposition G.3 are satisfied. We have already noticed at the end of Section G.2 that N_t does not explode. Moreover, $\delta_n = n > 0$ for all $n \in \mathbb{N}$, and

$$\sum_{n=1}^{\infty} \frac{\beta_0 \cdots \beta_{n-1}}{\delta_1 \cdots \delta_n} \leq \sum_{n=1}^{\infty} c^{*n}/n! < \infty.$$

So the conditions in Proposition G.4 are satisfied. $\square$

Consider next the jump chain J_m, $m = 0, 1, 2, \ldots$, of Y_t, $t \geq 0$, assuming that the process Y_t does not explode. Its transition kernel is given by (G.1).

PROPOSITION G.6 Suppose that (11.5)–(11.6) are satisfied and that $\tilde{h}(x) = \alpha(x)h(x)$ is integrable with respect to Poisson$(B, 1)$. Then the jump chain is reversible with respect to $\tilde{h}$.

Proof. The conditions (11.5) and (11.6) imply that if $x \cup \xi \in E$ and $\xi \notin x$, then $\alpha(x) > 0$, $\alpha(x \cup \xi) > 0$, and

$$\tilde{h}(x)b(x,\xi)/\alpha(x) = \tilde{h}(x \cup \xi)d(x,\xi)/\alpha(x \cup \xi).$$

Combining this with (G.1) we easily obtain reversibility along similar lines as in the proof of Proposition 7.12 (Section 7.3.2). □

REMARK G.2 From Proposition G.6 we obtain

$$\mathbb{E}g(X) = \mathbb{E}(g(\tilde{X})/\alpha(\tilde{X}))/\mathbb{E}(1/\alpha(\tilde{X})) \tag{G.9}$$

whenever $X \sim h$ and $\tilde{X} \sim \tilde{h}$. This is in fact a general result for jump processes: if a jump process has invariant distribution Π and intensity α, if its jump chain has invariant distribution $\tilde{\Pi}$, and if $X \sim \Pi$ and $\tilde{X} \sim \tilde{\Pi}$, then (G.9) holds provided the expectations exist (see e.g. Norris 1997, Theorem 3.5).

Finally, we turn again to the locally stable and constant death rate case and consider the jump chain $(\tilde{J}_m, J'_m)$, $m = 0, 1, \ldots$, for the coupled process (Y_t, D_t), see Algorithm 11.3 (Section 11.1.4) and the discussion in Section 11.1.5. Recall that the state space is $\Gamma = \{(y,x) \in E \times N_\mathrm{f} : y \subseteq x\}$, and recall the definition of $G'_n(k)$ given by (11.21).

PROPOSITION G.7 For any constant $\beta > 1$, let $V(y,x) = \beta^{n(x)}$ for $(y,x) \in \Gamma$. For the $(\tilde{J}_m, J'_m)$-chain we have the following.

(i) Every set $C = \{(y,x) \in \Gamma : n(x) \leq \kappa\}$ is small for $\kappa > 0$.

(ii) The chain has a unique invariant distribution and is V-uniformly ergodic.

(iii) Let $X \sim h$ and let k be a real function defined on E so that either $k^2(y) \leq V(y,x)$ or $k \in \mathcal{L}^{2+\epsilon}(h)$ for some $\epsilon > 0$. Then regardless of the initial distribution, $\sqrt{n}(G'_n(k) - \mathbb{E}k(X))$ converges in distribution towards $N(0, (2c^*)^2\bar{\sigma}^2)$ as $n \to \infty$, where $\bar{\sigma}^2$ is the asymptotic variance of $(1/\sqrt{n})\sum_{m=0}^{n-1} k(\tilde{J}_m)/(c^* + n(J'_m))$.

Proof. It will become clear that (i) and (ii) follow effectively by domination and the ergodicity properties of the chain $N_m = n(J'_m)$, $m = 0, 1, \ldots$. Note that for $(y,x) \in \Gamma$ with $x \neq \emptyset$,

$$P^{n(x)}((y,x), \{(\emptyset, \emptyset)\}) = \prod_{m=1}^{n(x)} \frac{m}{m+c^*}, \tag{G.10}$$

ERGODICITY PROPERTIES

cf. (G.3), and
$$P^2((\emptyset,\emptyset),\{(\emptyset,\emptyset)\}) = \frac{1}{1+c^*}, \qquad (G.11)$$
since the chain always moves up from $(\emptyset,\emptyset)$ to some state $(y',x') \in \Gamma$ with $n(x') = 1$. Thus the chain is Ψ-irreducible, where $\Psi(F) = \mathbf{1}[(\emptyset,\emptyset) \in F]$, $F \subseteq \Gamma$.

(i) Let $F \subseteq \Gamma$, let $(y,x) \in \Gamma$ with $n(x) \leq \kappa$, and let $m \geq \kappa$ be an even integer. We claim that
$$P^m((y,x),F) \geq (1/(1+c^*))^m \Psi(F),$$
i.e. C is small. By (G.11), if $x = \emptyset$,
$$P^m((y,x),F) \geq P^2((\emptyset,\emptyset),\{(\emptyset,\emptyset)\})^{m/2} \mathbf{1}[(\emptyset,\emptyset) \in F] \geq (1/(1+c^*))^m \Psi(F)$$
(since $(1/(1+c^*))^m \leq (1/(1+c^*))^{m/2}$). By (G.10) and (G.11), if $m-n(x)$ is even and $x \neq \emptyset$,
$$P^m((y,x),F)$$
$$\geq P^{n(x)}((y,x),\{(\emptyset,\emptyset)\}) P^2((\emptyset,\emptyset),\{(\emptyset,\emptyset)\})^{(m-n(x))/2} \mathbf{1}[(\emptyset,\emptyset) \in F]$$
$$= \left(\prod_{m=1}^{n(x)} \frac{m}{m+c^*}\right) \left(\frac{1}{1+c^*}\right)^{(m-n(x))/2} \Psi(F) \geq (1/(1+c^*))^m \Psi(F).$$
Hence, if $m - n(x)$ is odd and $x \neq \emptyset$,
$$P^m((y,x),F)$$
$$\geq (n(x)/(n(x)+c^*))(1/(1+c^*))^{m-1} \Psi(F) \geq (1/(1+c^*))^m \Psi(F).$$
Thus C is small.

(ii) We verify that the conditions in Proposition 7.9 (Section 7.2.3) are satisfied. We have already established Ψ-irreducibility.

To establish existence of the invariant distribution, note first that the chain N_m is irreducible on $\mathbb{N}_0$ and has invariant density π' given by the detailed balance equation
$$\pi'_i \frac{c^*}{i+c^*} = \pi'_{i+1} \frac{i+1}{i+1+c^*}, \qquad i = 0,1,\ldots,$$
i.e.
$$\pi'_i = \frac{(c^*+i)(c^*)^i}{i! 2 c^*} \exp(-c^*).$$
Thus 0 is an ergodic atom for the N_m-chain, since the mean return time to 0 is finite (it is given by $1/\pi'_0$, see e.g. Theorem 1.7.7 in Norris 1997). Hence by domination, $(\emptyset,\emptyset)$ is an ergodic atom for the $(\tilde{J}_m, J'_m)$-chain. Since the $(\tilde{J}_m, J'_m)$-chain regenerates at $(\emptyset,\emptyset)$, it follows now by the renewal theorem (see e.g. Asmussen 1987) that the $(\tilde{J}_m, J'_m)$-chain has a limit distribution which is the unique invariant distribution.

To establish aperiodicity, we use Proposition 7.6 (Section 7.2.2) with $C = \{(y,x) \in \Gamma : n(x) \leq 1\}$: for $(y,x) \in C$ and $m \in \mathbb{N}$, considering each of the cases

$$P^m((y,x),C) = P^m((y,x),\{(y',x') : n(x') = n(x) \leq 1\})$$

if m is even, and

$$P^m((y,x),C) = P^m((y,x),\{(y',x') : n(x') \neq n(x), n(x') \leq 1\})$$

if m is odd, we easily obtain that $P^m((y,x),C) > 0$.

Finally, to establish the geometric drift condition, choose $\kappa > 0$ so large that

$$a \equiv \frac{\beta c^* + \kappa/\beta}{c^* + \kappa} < 1$$

(this is possible as $\beta > 1$), and set $C = \{(y,x) \in \Gamma : n(x) \leq \kappa\}$ and $b = \beta^{\kappa+1}$. Let $(y,x) \in \Gamma$ and $n = n(x)$. If $n > \kappa$ then by (G.3),

$$\mathbb{E}[V(\tilde{J}_1, J'_1)|(\tilde{J}_0, J'_0) = (y,x)] = \beta^{n+1} \frac{c^*}{c^* + n} + \beta^{n-1} \frac{n}{c^* + n}$$
$$\leq a\beta^n = aV(y,x),$$

while if $n \leq \kappa$ then

$$\mathbb{E}[V(\tilde{J}_1, J'_1)|(\tilde{J}_0, J'_0) = (y,x)] \leq b.$$

Thus (7.21) is verified.

(iii) This follows by a similar reasoning as in Remark 8.4 in Section 8.2.3, using that if J'_m follows the invariant distribution of the dominating jump chain,

$$\mathbb{E}1/(c^* + n(J'_m)) = 1/\mathbb{E}(c^* + n(D_0)) = 1/(2c^*),$$

cf. (G.9). $\square$

REMARK G.3 We can estimate $\bar{\sigma}^2$ by the method of batch means (Section 8.1.3). Since $(\tilde{J}_m, J'_m)$, $m = 1, 2, \ldots$, is in general a nonreversible Markov chain (Remark 11.2, Section 11.1.4), Geyer's initial series estimate does not apply.

REMARK G.4 Propositions G.4–G.7 extend to a more general setting with N_f replaced by a state space $\Omega = \sum_{n=0}^{\infty} \Omega_n$ where the Ω_n are disjoint sets, Ω_0 has only one element, and a jump from a state in Ω_n is to a state in either Ω_{n-1} or Ω_{n+1}; see Preston (1977).

We often have geometrically fast convergence towards κ, cf. Møller (1989). For instance, this is the case in the locally stable and constant death rate case.

The $(\tilde{J}_m, J'_m)$-chain is in general not uniformly ergodic. This follows by a similar reasoning as in Proposition 7.14 and Remark 7.9 (Section 7.3.2).

References

Adler, R. (1981). *The Geometry of Random Fields*, Wiley, New York.

Allen, M. P. & Tildesley, D. J. (1987). *Computer Simulation of Liquids*, Oxford University Press, Oxford.

Asmussen, S. (1987). *Applied Probability and Queues*, Wiley, Chichester.

Baddeley, A. J. & Gill, R. D. (1997). Kaplan-Meier estimators of distance distributions for spatial point processes, *Annals of Statistics* **25**: 263–292.

Baddeley, A. J., Møller, J. & Waagepetersen, R. (2000). Non- and semi-parametric estimation of interaction in inhomogeneous point patterns, *Statistica Neerlandica* **54**: 329–350.

Baddeley, A. J., Moyeed, R. A., Howard, C. V. & Boyde, A. (1993). Analysis of a three-dimensional point pattern with applications, *Applied Statistics* **42**: 641–668.

Baddeley, A. J. & van Lieshout, M. N. M. (1995). Area-interaction point processes, *Annals of the Institute of Statistical Mathematics* **46**: 601–619.

Baddeley, A. J., van Lieshout, M. N. M. & Møller, J. (1996). Markov properties of cluster processes, *Advances in Applied Probability (SGSA)* **28**: 346–355.

Baddeley, A. & Møller, J. (1989). Nearest-neighbour Markov point processes and random sets, *International Statistical Review* **2**: 89–121.

Baddeley, A. & Silverman, B. W. (1984). A cautionary example for the use of second-order methods for analysing point patterns, *Biometrics* **40**: 1089–1094.

Baddeley, A. & Turner, R. (2000). Practical maximum pseudolikelihood for spatial point patterns, *Australian and New Zealand Journal of Statistics* **42**: 283–322.

Barndorff-Nielsen, O. E. (1978). *Information and Exponential Families in Statistical Theory*, Wiley, Chichester.

Bartlett, M. S. (1963). The spectral analysis of point processes, *Journal of the Royal Statistical Society Series B* **29**: 264–296.

Bartlett, M. S. (1964). The spectral analysis of two-dimensional point processes, *Biometrika* **51**: 299–311.

Bartlett, M. S. (1975). *The Statistical Analysis of Spatial Pattern*, Chapman & Hall, London.

Bedford, T. & van den Berg, J. (1997). A remark on van Lieshout and Baddeley's J-function for point processes, *Advances in Applied Probability* **29**: 19–25.

Benes, V., Bodlak, K., Møller, J. & Waagepetersen, R. P. (2002). Bayesian analysis of log Gaussian Cox process models for disease mapping, *Technical Report R-02-2001*, Department of Mathematical Sciences, Aalborg University. Submitted.

Berman, M. & Turner, R. (1992). Approximating point process likelihoods with GLIM, *Applied Statistics* **41**: 31–38.

Berthelsen, K. K. & Møller, J. (2002a). A primer on perfect simulation for spatial point processes, *Bulletin of the Brazilian Mathematical Society* **33**: 351–367.

Berthelsen, K. K. & Møller, J. (2002b). Spatial jump processes and perfect simulation, *in* K. Mecke & D. Stoyan (eds), *Morphology of Condensed Matter*, Lecture Notes in Physics, Vol. 600, Springer-Verlag, Heidelberg, pp. 391–417.

Berthelsen, K. K. & Møller, J. (2003). Perfect simulation and inference for spatial point processes, *Scandinavian Journal of Statistics* **30**. To appear.

Bertin, E., Billiot, J.-M. & Drouilhet, R. (1999a). Existence of "nearest-neighbour" type spatial Gibbs models, *Advances in Applied Probability* **31**: 895–909.

Bertin, E., Billiot, J.-M. & Drouilhet, R. (1999b). Spatial Delaunay Gibbs point processes, *Stochastic Models* **15**: 181–199.

Besag, J. (1977a). Some methods of statistical analysis for spatial data, *Bulletin of the International Statistical Institute* **47**: 77–92.

Besag, J. & Clifford, P. (1989). Generalized Monte Carlo significance tests, *Biometrika* **76**: 633–642.

Besag, J. E. (1974). Spatial interaction and the statistical analysis of lattice systems (with discussion), *Journal of the Royal Statistical Society Series B* **36**: 192–236.

Besag, J. E. (1975). Statistical analysis of non-lattice data, *The Statistician* **24**: 179–195.

Besag, J. E. (1977b). Discussion of the paper by Ripley (1977), *Journal of the Royal Statistical Society Series B* **39**: 193–195.

Besag, J. E. (1994). Discussion of the paper by Grenander and Miller, *Journal of the Royal Statistical Society Series B* **56**: 591–592.

Besag, J. E. & Green, P. J. (1993). Spatial statistics and Bayesian computation (with discussion), *Journal of the Royal Statistical Society Series B* **55**: 25–37.

Besag, J., Milne, R. K. & Zachary, S. (1982). Point process limits of lattice processes, *Journal of Applied Probability* **19**: 210–216.

Best, N. G., Ickstadt, K. & Wolpert, R. L. (2000). Spatial Poisson regression for health and exposure data measured at disparate resolutions, *Journal of the American Statistical Association* **95**: 1076–1088.

Billingsley, P. (1995). *Probability and Measure*, Wiley, New York.

Bormann, F. H. (1953). The statistical efficiency of sample plot size and shape in forest ecology, *Ecology* **34**: 474–487.

Breyer, L. A. & Roberts, G. O. (2000). From Metropolis to diffusions: Gibbs states and optimal scaling, *Stochastic Processes and Their Applications* **90**: 181–206.

Brix, A. (1999). Generalized gamma measures and shot-noise Cox processes, *Advances in Applied Probability* **31**: 929–953.

Brix, A. & Chadoeuf, J. (2000). Spatio-temporal modeling of weeds and shot-

noise G Cox processes. Submitted.
Brix, A. & Diggle, P. J. (2001). Spatio-temporal prediction for log-Gaussian Cox processes, *Journal of the Royal Statistical Society Series B* **63**: 823–841.
Brix, A. & Kendall, W. S. (2002). Simulation of cluster point processes without edge effects, *Advances in Applied Probability* **34**: 267–280.
Brix, A. & Møller, J. (2001). Space-time multitype log Gaussian Cox processes with a view to modelling weed data, *Scandinavian Journal of Statistics* **28**: 471–488.
Brooks, S. & Gelman, A. (1998). General methods for monitoring convergence of iterative simulations, *Journal of Computational and Graphical Statistics* **7**: 434–455.
Brooks, S. & Roberts, G. O. (1998). Diagnosing convergence of Markov chain Monte Carlo algorithms, *Statistics and Computing* **8**: 319–335.
Brown, T. C., Silverman, B. W. & Milne, R. K. (1981). A class of two-type point processes, *Zeitschrift für Wahrscheinlichkeitstheorie und verwandte Gebicte* **58**: 299–308.
Buffon, G. (1777). Essai d'arithmétique morale, *Supplemént à l'Historie naturelle* **4**.
Burden, R. L. & Faires, D. J. (2001). *Numerical Analysis*, 7 edn, Brooks/Cole Publishing, Pacific Grove, CA.
Carter, D. S. & Prenter, P. M. (1972). Exponential spaces and counting processes, *Zeitschrift für Wahrscheinlichkeitstheorie und verwandte Gebiete* **21**: 1–19.
Chan, K. S. & Geyer, C. J. (1994). Discussion of the paper 'Markov chains for exploring posterior distributions' by Luke Tierney, *Annals of Statistics* **22**: 1747–1747.
Chayes, J. T., Chayes, L. & Kotecky, R. (1995). The analysis of the Widom-Rowlinson model by stochastic geometric methods, *Communications in Mathematical Physics* **172**: 551–569.
Chayes, L. & Machta, J. (1998). Graphical representations and cluster algorithms Part II, *Physica A* **254**: 477–516.
Chen, M.-H., Shao, Q.-M. & Ibrahim, J. G. (2000). *Monte Carlo Methods in Bayesian Computation*, Springer, New York.
Chin, Y. C. & Baddeley, A. J. (2000). Markov interacting component processes, *Advances in Applied Probability* **32**: 597–619.
Christakos, G. (1984). On the problem of permissible covariance and variogram models, *Water Resources Research* **20**: 251–265.
Christensen, N. L. (1977). Changes in structure, pattern and diversity associated with climax forest maturation in Piedmont, North Carolina, *American Midland Naturalist* **97**: 176–188.
Christensen, O. F., Diggle, P. J. & Ribeiro, P. (2003). An introduction to model-based geostatistics, *in* J. Møller (ed.), *Spatial Statistics and Computational Methods*, Lecture Notes in Statistics 173, Springer-Verlag, New York.
Christensen, O. F., Møller, J. & Waagepetersen, R. P. (2001). Geometric ergodicity of Metropolis-Hastings algorithms for conditional simulation in generalized linear mixed models, *Methodology and Computing in Applied Probability* **3**: 309–327.

Christensen, O. F., Roberts, G. O. & Rosenthal, J. S. (2003). Scaling limits for the transient phase of local Metropolis-Hastings algorithms. Submitted.

Christensen, O. F. & Waagepetersen, R. (2002). Bayesian prediction of spatial count data using generalized linear mixed models, *Biometrics* **58**: 280–286.

Clifford, P. (1990). Markov random fields in statistics, *in* G. R. Grimmett & D. J. A. Welsh (eds), *Disorder in Physical Systems. A Volume in Honour of J.M. Hammersley*, Clarendon Press, Oxford.

Clifford, P. & Nicholls, G. (1994). Comparison of birth-and-death and Metropolis-Hastings Markov chain Monte Carlo for the Strauss process. Department of Statistics, Oxford University. Manuscript.

Coles, P. & Jones, B. (1991). A lognormal model for the cosmological mass distribution, *Monthly Notices of the Royal Astronomical Society* **248**: 1–13.

Cowles, M. K. & Carlin, B. P. (1996). Markov chain Monte Carlo convergence diagnostics: a comparative review, *Journal of the American Statistical Association* **91**: 883–904.

Cox, D. R. (1955). Some statistical models related with series of events, *Journal of the Royal Statistical Society Series B* **17**: 129–164.

Cox, D. R. & Isham, V. (1980). *Point Processes*, Chapman & Hall, London.

Cressie, N. A. C. (1993). *Statistics for Spatial Data*, second edn, Wiley, New York.

Creutz, M. (1979). Confinement and the critical dimensionality of space-time, *Physical Review Letters* **43**: 553–556.

Daley, D. J. & Vere-Jones, D. (1988). *An Introduction to the Theory of Point Processes*, Springer-Verlag, New York.

Daley, D. J. & Vere-Jones, D. (2003). *An Introduction to the Theory of Point Processes. Volume I: Elementary Theory and Methods*, second edn, Springer-Verlag, New York.

Davis, P. J. (1979). *Circulant Matrices*, Wiley, Chichester.

Dellaportas, P. & Roberts, G. O. (2003). An introduction to MCMC, *in* J. Møller (ed.), *Spatial Statistics and Computational Methods*, Lecture Notes in Statistics 173, Springer-Verlag, New York, pp. 1–41.

Dietrich, C. R. & Newsam, G. N. (1993). A fast and exact method for multidimensional Gaussian stochastic simulation, *Water Resources Research* **29**: 2861–2869.

Diggle, P. J. (1983). *Statistical Analysis of Spatial Point Patterns*, Academic Press, London.

Diggle, P. J. (1985). A kernel method for smoothing point process data, *Applied Statistics* **34**: 138–147.

Diggle, P. J. (1986). Displaced amacrine cells in the retina of a rabbit: analysis of a bivariate spatial point pattern, *Journal of Neuroscience Methods* **18**: 115–125.

Diggle, P. J., Fiksel, T., Grabarnik, P., Ogata, Y., Stoyan, D. & Tanemura, M. (1994). On parameter estimation for pairwise interaction point processes, *International Statistical Review* **62**: 99–117.

Diggle, P. J., Gates, D. J. & Stibbard, A. (1987). A nonparametric estimator for pairwise interaction point processes, *Biometrika* **74**: 763–770.

Diggle, P. J. & Gratton, R. J. (1984). Monte Carlo methods of inference for implicit statistical models (with discussion), *Journal of the Royal Statistical Society Series B* **46**: 193 227.

Diggle, P. J., Lange, N. & Beneš, F. (1991). Analysis of variance for replicated spatial point patterns in clinical neuroanatomy, *Journal of the American Statistical Association* **86**: 618 625.

Diggle, P. J., Mateu, L. & Clough, H. E. (2000). A comparison between parametric and non-parametric approaches to the analysis of replicated spatial point patterns, *Advances of Applied Probability* **32**: 331 343.

Efron, B. & Tibshirani, R. J. (1993). *An Introduction to the Bootstrap*, Chapman & Hall, New York.

Feller, W. (1971). *An Introduction to Probability Theory and Its Aplications*, Vol. 2, Wiley, New York.

Fernández, R., Ferrari, P. A. & Garcia, N. L. (2002). Perfect simulation for interacting point processes, loss networks and Ising models, *Stochastic Processes and Their Applications* **102**: 63 88.

Fiksel, T. (1984a). Estimation of parameterized pair potentials of marked and nonmarked Gibbsian point processes, *Elektronische Informationsverarbeitung und Kypernetik* **20**: 270 278.

Fiksel, T. (1984b). Simple spatial-temporal models for sequences of geological events, *Elektronische Informationsverarbeitung und Kypernetik* **20**: 480–487.

Fiksel, T. (1988). Edge-corrected density estimators for point processes, *Statistics* **19**: 67 75.

Fill, J. A. (1998). An interruptible algorithm for perfect sampling via Markov chains, *Annals of Applied Probability* **8**: 131 162.

Fill, J. A., Machida, M., Murdoch, D. J. & Rosenthal, J. S. (2000). Extensions of Fill's perfect rejection sampling algorithm to general chains, *Random Structures and Algorithms* **17**: 290 316.

Foss, S. G. & Tweedie, R. L. (1998). Perfect simulation and backward coupling, *Stochastic Models* **14**: 187 203.

Gamerman, D. (1997). *Markov Chain Monte Carlo*, Chapman & Hall, London.

Gates, D. J. & Westcott, M. (1986). Clustering estimates for spatial point distributions with unstable potentials, *Annals of the Institute of Statistical Mathematics A* **38**: 123 135.

Gelfand, A. E. (1996). Model determination using sampling-based methods, *in* W. R. Gilks, S. Richardson & D. J. Spiegelhalter (eds), *Markov Chain Monte Carlo in Practice*, Chapman & Hall, London, pp. 145 161.

Gelfand, A. E. & Carlin, B. P. (1993). Maximum likelihood estimation for constrained or missing data models, *Canadian Journal of Statistics* **21**: 303–311.

Gelman, A. & Meng, X.-L. (1998). Simulating normalizing constants: from importance sampling to bridge sampling to path sampling, *Statistical Science* **13**: 163 185.

Geman, S. & Geman, D. (1984). Stochastic relaxation, Gibbs distributions and the Bayesian restoration of images, *IEEE Transactions on Pattern Analysis and Machine Intelligence* **6**: 721 741.

Georgii, H.-O. (1976). Canonical and grand canonical Gibbs states for continuum systems, *Communications of Mathematical Physics* **48**: 31 51.

Georgii, H.-O. (1988). *Gibbs Measures and Phase Transition*, Walter de Gruyter, Berlin.

Georgii, H.-O. (1994). Large deviations and the equivalence of ensembles for Gibbsian particle systems with superstable interaction, *Probability Theory and Related Fields* **99**: 171–195.

Georgii, H.-O. (1995). The equivalence of ensembles for classical systems of particles, *Journal of Statistical Physics* **80**: 1341–1378.

Georgii, H.-O. (2000). Phase transition and percolation in Gibbsian particle models, *in* K. Mecke & D. Stoyan (eds), *Statistical Physics and Spatial Statistics*, Lecture Notes in Physics, Springer-Verlag, Heidelberg, pp. 267–294.

Georgii, H.-O. & Haggström, O. (1996). Phase transition in continuum Potts models, *Communications in Mathematical Physics* **181**: 507–528.

Geyer, C. J. (1991). Estimating normalizing constants and reweighting mixtures in Markov chain Monte Carlo mixtures, *Technical Report 568r*, School of Statistics, University of Minnesota.

Geyer, C. J. (1992). Practical Markov chain Monte Carlo (with discussion), *Statistical Science* **8**: 473–483.

Geyer, C. J. (1994). On the convergence of Monte Carlo maximum likelihood calculations, *Journal of the Royal Society of Statistics Series B* **56**: 261–274.

Geyer, C. J. (1999). Likelihood inference for spatial point processes, *in* O. E. Barndorff-Nielsen, W. S. Kendall & M. N. M. van Lieshout (eds), *Stochastic Geometry: Likelihood and Computation*, Chapman & Hall/CRC, Boca Raton, Florida, pp. 79–140.

Geyer, C. J. & Møller, J. (1994). Simulation procedures and likelihood inference for spatial point processes, *Scandinavian Journal of Statistics* **21**: 359–373.

Geyer, C. J. & Thompson, E. A. (1992). Constrained Monte Carlo maximum likelihood for dependent data, *Journal of the Royal Society of Statistics Series B* **54**: 657–699.

Geyer, C. J. & Thompson, E. A. (1995). Annealing Markov chain Monte Carlo with applications to pedigree analysis, *Journal of the American Statistical Association* **90**: 909–920.

Gilks, W. R., Richardson, S. & Spiegelhalter, D. J. (1996). *Markov Chain Monte Carlo in Practice*, Chapman & Hall, London.

Goulard, M., Särkkä, A. & Grabarnik, P. (1996). Parameter estimation for marked Gibbs point processes through the maximum pseudo-likelihood method, *Scandinavian Journal of Statistics* **23**: 365–379.

Grandell, J. (1976). *Doubly Stochastic Poisson Processes*, Springer Lecture Notes in Mathematics 529. Springer-Verlag, Berlin.

Green, P. J. (1995). Reversible jump MCMC computation and Bayesian model determination, *Biometrika* **82**: 711–732.

Häggström, O. & Nelander, K. (1998). Exact sampling from anti-monotone systems, *Statistica Neerlandia* **52**: 360–380.

Häggström, O., van Lieshout, M. N. M. & Møller, J. (1999). Characterization results and Markov chain Monte Carlo algorithms including exact simulation for some spatial point processes, *Bernoulli* **5**: 641–659.

Heikkinen, J. & Arjas, E. (1998). Non-parametric Bayesian estimation of a

REFERENCES

spatial Poisson intensity, *Scandinavian Journal of Statistics* **25**: 435–450.

Heikkinen, J. & Penttinen, A. (1999). Bayesian smoothing in the estimation of the pair potential function of Gibbs point processes, *Bernoulli* **5**: 1119–1136.

Jensen, E. B. V. (1998). *Local Stereology*, World Scientific, Singapore.

Jensen, E. B. V. & Nielsen, L. S. (2000). Inhomogeneous Markov point processes by transformation, *Bernoulli* **6**: 761–782.

Jensen, E. B. V. & Nielsen, L. S. (2001). A review on inhomogeneous spatial point processes, *in* I. V. Basawa, C. C. Heyde & R. L. Taylor (eds), *Selected Proceedings of the Symposium on Inference for Stochastic Processes*, Vol. 37, IMS Lecture Notes & Monographs Series, Beachwood, Ohio, pp. 297–318.

Jensen, J. L. (1993). Asymptotic normality of estimates in spatial point processes, *Scandinavian Journal of Statistics* **20**: 97–109.

Jensen, J. L. & Künsch, H. R. (1994). On asymptotic normality of pseudo likelihood estimates for pairwise interaction processes, *Annals of the Institute of Statistical Mathematics* **46**: 475–486.

Jensen, J. L. & Møller, J. (1991). Pseudolikelihood for exponential family models of spatial point processes, *Annals of Applied Probability* **3**: 445–461.

Kallenberg, O. (1975). *Random Measures*, Akademie-Verlag, Berlin.

Kallenberg, O. (1984). An informal guide to the theory of conditioning in point processes, *International Statistical Review* **52**: 151–164.

Karr, A. F. (1991). *Point Processes and Their Statistical Inference*, Marcel Dekker, New York.

Kelly, F. P. & Ripley, B. D. (1976). A note on Strauss' model for clustering, *Biometrika* **63**: 357–360.

Kelsall, J. E. & Wakefield, J. C. (2002). Modeling spatial variation in disease risk: a geostatistical approach, *Journal of the American Statistical Association* **97**: 692–701.

Kendall, W. S. (1990). A spatial Markov property for nearest-neighbour Markov point processes, *Journal of Applied Probability* **28**: 767–778.

Kendall, W. S. (1997). Perfect simulation for spatial point processes, *Proceedings of ISI 51st session*, Istanbul, pp. 163–166.

Kendall, W. S. (1998). Perfect simulation for the area-interaction point process, *in* L. Accardi & C. Heyde (eds), *Probability Towards 2000*, Springer Lecture Notes in Statistics 128, Springer Verlag, New York, pp. 218–234.

Kendall, W. S. & Møller, J. (2000). Perfect simulation using dominating processes on ordered spaces, with application to locally stable point processes, *Advances in Applied Probability* **32**: 844–865.

Kerscher, M. (2000). Statistical analysis of large-scale structure in the Universe, *in* K. R. Mecke & D. Stoyan (eds), *Statistical Physics and Spatial Statistics*, Lecture Notes in Physics, Springer, Berlin, pp. 36–71.

Kerstan, J., Matthes, K. & Mecke, J. (1974). *Unbegrenzt teilbare Punktprozesse*, Akademie-Verlag, Berlin.

Kingman, J. F. C. (1993). *Poisson Processes*, Clarendon Press, Oxford.

Klein, W. (1982). Potts-model formulation of continuum percolation, *Physical Review B* **26**: 2677–2678.

Kong, A., McCullagh, P., Nicolae, D., Tan, Z. & Meng, X.-L. (2003). A theory of statistical models for Monte Carlo integration (with discussion), *Journal of the Royal Statistical Society Series B* **65**. To appear.

Kutoyants, Y. A. (1984). *Parameter estimation for stochastic processes*, Heldermann Verlag, Berlin.

Lieshout, M. N. M. van (2000). *Markov Point Processes and Their Applications*, Imperial College Press, London.

Lieshout, M. N. M. van & Baddeley, A. J. (1996). A nonparametric measure of spatial interaction in point patterns, *Statistica Neerlandica* **50**: 344–361.

Lieshout, M. N. M. van & Baddeley, A. J. (1999). Indices of dependence between types in multivariate point patterns, *Scandinavian Journal of Statistics* **26**: 511–532.

Lieshout, M. N. M. van & Baddeley, A. J. (2002). Extrapolating and interpolating spatial patterns, *in* A. B. Lawson & D. Denison (eds), *Spatial Cluster Modelling*, Chapman & Hall/CRC, Boca Raton, Florida, pp. 61–86.

Liu, J. S. (2001). *Monte Carlo Strategies in Scientific Computing*, Springer, New York.

Lotwick, H. W. (1982). Simulation of some spatial hard core models, and the complete packing problem, *Journal of Statistical Computation and Simulation* **15**: 295–314.

Lotwick, H. W. & Silverman, B. W. (1982). Methods for analysing spatial point processes of several types of points, *Journal of the Royal Statistical Society Series B* **44**: 406–413.

Lund, J., Penttinen, A. & Rudemo, M. (1999). Bayesian analysis of spatial point patterns from noisy observations, *Preprint 1999:57*, Department of Mathematical Statistics, Chalmers University of Technology.

Lund, J. & Rudemo, M. (2000). Models for point processes observed with noise, *Biometrika* **87**: 235–249.

Lund, J. & Thönnes, E. (2004). Perfect simulation for point processes given noisy observations, *Computational Statistics*. To appear.

Marinari, E. & Parisi, G. (1992). Simulated tempering: a new Monte Carlo scheme, *Europhysics Letters* **19**: 451–458.

Mase, S. (1991). Asymptotic equivalence of grand canonical MLE and canonical MLE of pair potential functions of Gibbsian point process models, *Technical Report 292*, Statistical Research Group, Hiroshima University.

Mase, S. (1992). Uniform LAN condition of planar Gibbsian point processes and optimality of maximum likelihood estimators of soft-core potential functions, *Probability Theory and Related Fields* **92**: 51–67.

Mase, S. (1995). Consistency of the maximum pseudo-likelihood estimator of continuous state space Gibbs processes, *Annals of Applied Probability* **5**: 603–612.

Mase, S. (1999). Marked Gibbs processes and asymptotic normality of maximum pseudo-likelihood estimators, *Mathematische Nachrichten* **209**: 151–169.

Mase, S. (2002). Asymptotic properties of MLE's of Gibbs models on $\mathbb{R}^d$. Manuscript.

Mase, S., Møller, J., Stoyan, D., Waagepetersen, R. P. & Döge, G. (2001). Packing densities and simulated tempering for hard core Gibbs point processes,

REFERENCES

Annals of the Institute of Statistical Mathematics **53**: 661–680.

Matérn, B. (1960). *Spatial Variation*, Meddelanden från Statens Skogforskningsinstitut, **49** (5).

Matérn, B. (1986). *Spatial Variation*, Lecture Notes in Statistics 36, Springer-Verlag, Berlin.

Matheron, G. (1975). *Random Sets and Integral Geometry*, Wiley, New York.

Matsumoto, M. & Nishimura, T. (1998). Mersenne Twister: a 623-dimensionally equidistributed uniform pseudorandom number generator, *ACM Transactions on Modeling and Computer Simulation* **8**: 3–30.

McKeague, I. W. & Loizeaux, M. (2002). Perfect sampling for point process cluster modelling, *in* A. B. Lawson & D. Denison (eds), *Spatial Cluster Modelling*, Chapman & Hall/CRC, Boca Raton, Florida, pp. 62–107.

Mecke, J. (1967). Stationäre zufällige Maße auf lokalkompakten Abelschen Gruppen, *Zeitschrift für Wahrscheinlichkeitstheorie und verwandte Gebiete* **9**: 36–58.

Mecke, J. (1999). On the relationship between the 0-cell and the typical cell of a stationary random tessellation, *Pattern Recognition* **232**: 1645–1648.

Mecke, J., Schneider, R. G., Stoyan, D. & Weil, W. R. (1990). *Stochastische Geometrie*, Birkhäuser Verlag, Basel.

Meester, R. & Roy, R. (1996). *Continuum Percolation*, Cambridge University Press, New York.

Meng, X.-L. & Wong, W. H. (1996). Simulating ratios of normalizing constants via a simple identity: a theoretical exploration, *Statistica Sinica* **6**: 831–860.

Mengersen, K. L., Robert, C. P. & Guihenneuc-Jouyaux, C. (1999). MCMC convergence diagnostics: a review, *in* J. O. Berger, J. M. Bernado, A. P. Dawid & A. F. M. Smith (eds), *Bayesian Statistics 6*, Oxford University Press, Oxford, pp. 415–440.

Metropolis, N., Rosenbluth, A. W., Rosenbluth, M. N., Teller, A. H. & Teller, E. (1953). Equations of state calculations by fast computing machines, *Journal of Chemical Physics* **21**: 1087–1092.

Meyn, S. P. & Tweedie, R. L. (1993). *Markov Chains and Stochastic Stability*, Springer-Verlag, London.

Molchanov, I. (1997). *Statistics of the Boolean Model for Practitioners and Mathematicians*, Wiley, Chichester.

Møller, J. (1989). On the rate of convergence of spatial birth-and-death processes, *Annals of the Institute of Statistical Mathematics* **3**: 565–581.

Møller, J. (1994a). Contribution to the discussion of N.L. Hjort and H. Omre (1994): Topics in spatial statistics, *Scandinavian Journal of Statistics* **21**: 346–349.

Møller, J. (1994b). *Lectures on Random Voronoi Tessellations*, Lecture Notes in Statistics 87, Springer-Verlag, New York.

Møller, J. (1999a). Markov chain Monte Carlo and spatial point processes, *in* O. E. Barndorff-Nielsen, W. S. Kendall & M. N. M. van Lieshout (eds), *Stochastic Geometry: Likelihood and Computation*, Monographs on Statistics and Applied Probability 80, Chapman & Hall/CRC, Boca Raton, Florida, pp. 141–172.

Møller, J. (1999b). Perfect simulation of conditionally specified models, *Journal* of the *Royal Statistical Society Series B* **61**: 251–264.

Møller, J. (2001). A review of perfect simulation in stochastic geometry, *in* I. V. Basawa, C. C. Heyde & R. L. Taylor (eds), *Selected Proceedings of the Symposium on Inference for Stochastic Processes*, Vol. 37, IMS Lecture Notes & Monographs Series, Beachwood, Ohio, pp. 333–355.

Møller, J. (2003a). A comparison of spatial point process models in epidemiological applications, *in* P. J. Green, N. L. Hjort & S. Richardson (eds), *Highly Structured Stochastic Systems*, Oxford University Press, Oxford, pp. 264–268.

Møller, J. (2003b). Shot noise Cox processes, *Advances in Applied Probability* **35**. To appear.

Møller, J. (ed.) (2003c). *Spatial Statistics and Computational Methods*, Springer Lecture Notes in Statistics 173, Springer-Verlag, New York.

Møller, J. & Nicholls, G. (1999). Perfect simulation for sample-based inference, *Technical Report R-99-2011*, Department of Mathematical Sciences, Aalborg University. *Statistics and Computing*. Conditionally accepted.

Møller, J. & Schladitz, K. (1999). Extensions of Fill's algorithm for perfect simulation, *Journal of the Royal Statistical Society Series B* **61**: 955–969.

Møller, J. & Sørensen, M. (1994). Parametric models of spatial birth-and-death processes with a view to modelling linear dune fields, *Scandinavian Journal of Statistics* **21**: 1–19.

Møller, J., Syversveen, A. R. & Waagepetersen, R. P. (1998). Log Gaussian Cox processes, *Scandinavian Journal of Statistics* **25**: 451–482.

Møller, J. & Waagepetersen, R. P. (1998). Markov connected component fields, *Advances in Applied Probability* **30**: 1–35.

Møller, J. & Waagepetersen, R. P. (2002). Statistical inference for Cox processes, *in* A. B. Lawson & D. Denison (eds), *Spatial Cluster Modelling*, Chapman & Hall/CRC, Boca Raton, Florida, pp. 37–60.

Møller, J. & Waagepetersen, R. P. (2003). An introduction to simulation-based inference for spatial point processes, *in* J. Møller (ed.), *Spatial Statistics and Computational Methods*, Lecture Notes in Statistics 173, Springer-Verlag, New York, pp. 143–198.

Moyeed, R. A. & Baddeley, A. J. (1991). Stochastic approximation of the MLE for a spatial point pattern, *Scandinavian Journal of Statistics* **18**: 39–50.

Murdoch, D. J. & Green, P. J. (1998). Exact sampling from a continuous state space, *Scandinavian Journal of Statistics* **25**: 483–502.

Newman, M. E. J. & Barkema, G. T. (2000). *Monte Carlo Methods in Statistical Physics*, Oxford University Press, Oxford.

Neyman, J. & Scott, E. L. (1958). Statistical approach to problems of cosmology, *Journal of the Royal Statistical Society Series B* **20**: 1–43.

Nguyen, X. X. & Zessin, H. (1979). Integral and differential characterizations of Gibbs processes, *Mathematische Nachrichten* **88**: 105–115.

Nielsen, L. S. (2000). Modelling the position of cell profiles allowing for both inhomogeneity and interaction, *Image Analysis and Stereology* **19**: 183–187.

Nielsen, L. S. & Jensen, E. B. V. (2003). Statistical inference for transformation inhomogeneous Markov point processes. Manuscript submitted for publication.

REFERENCES

Norman, G. E. & Filinov, V. S. (1969). Investigations of phase transition by a Monte-Carlo method, *High Temperature* **7**: 216–222.

Norris, J. R. (1997). *Markov Chains*, Cambridge University Press, Cambridge.

Nummelin, E. (1984). *General Irreducible Markov Chains and Non-negative Operators*, Cambridge University Press, Cambridge.

Ogata, Y. (1998). Space-time point-process models for earthquake occurences, *Annals of the Institute of Statistical Mathematics* **50**: 379–402.

Ogata, Y. & Tanemura, M. (1981). Estimation of interaction potentials of spatial point patterns through the maximum likelihood procedure, *Annals of the Institute of Statistical Mathematics B* **33**: 315–338.

Ogata, Y. & Tanemura, M. (1984). Likelihood analysis of spatial point patterns, *Journal of the Royal Statistical Society Series B* **46**: 496–518.

Ogata, Y. & Tanemura, M. (1985). Estimation of interaction potentials of marked spatial point patterns through the maximum likelihood method, *Biometrics* **41**: 421–433.

Ogata, Y. & Tanemura, M. (1986). Likelihood estimation of interaction potentials and external fields of inhomogeneous spatial point patterns, *in* I. S. Francis, B. F. J. Manly & F. C. Lam (eds), *Proceedings of the Pacific Statistical Congress*, Elsevier, Amsterdam, pp. 150–154.

Ohser, J. & Mücklich, F. (2000). *Statistical Analysis of Microstructures in Materials Science*, Wiley, New York.

Okabe, A., Boots, B., Sugihara, K. & Chiu, S. N. (2000). *Spatial Tessellations. Concepts and Applications of Voronoi Diagrams*, second edn, Wiley, Chichester.

Peebles, P. J. E. (1974). The nature of the distribution of galaxies, *Astronomy and Astrophysics* **32**: 197–202.

Peebles, P. J. E. & Groth, E. J. (1975). Statistical analysis of extragalactic objects. V: three-point correlation function for the galaxy distribution in the Zwicky catalog, *Astrophysical Journal* **196**: 1–11.

Penttinen, A. (1984). *Modelling Interaction in Spatial Point Patterns: Parameter Estimation by the Maximum Likelihood Method*, Number 7 in Jyväskylä Studies in Computer Science, Economics, and Statistics, Univeristy of Jyväskylä.

Penttinen, A., Stoyan, D. & Henttonen, H. M. (1992). Marked point processes in forest statistics, *Forest Science* **38**: 806–824.

Peskun, P. H. (1973). Optimum Monte Carlo sampling using Markov chains, *Biometrika* **60**: 607–612.

Press, W. H., Teukolsky, S. A., Vetterling, W. T. & Flannery, B. P. (1992). *Numerical Recipes in C: The Art of Scientific Computing*, second edn, Cambridge University Press, Cambridge.

Preston, C. (1976). *Random Fields*, Lecture Notes in Mathematics 534. Springer-Verlag, Berlin.

Preston, C. J. (1977). Spatial birth-and-death processes, *Bulletin of the International Statistical Institute* **46**: 371–391.

Priestley, M. B. (1981). *Spectral Analysis and Time Series*, Academic Press, London.

Propp, J. G. & Wilson, D. B. (1996). Exact sampling with coupled Markov chains and applications to statistical mechanics, *Random Structures and Algorithms* **9**: 223–252.

Quine, M. P. & Watson, D. F. (1984). Radial simulation of n-dimensional Poisson processes, *Journal of Applied Probability* **21**: 548–557.

Rathbun, S. L. & Cressie, N. (1994a). Asymptotic properties of estimators for the parameters of spatial inhomogeneous Poisson processes, *Advances in Applied Probability* **26**: 122–154.

Rathbun, S. L. & Cressie, N. (1994b). Space-time survival point processes: longleaf pines in southern Georgia, *Journal of the American Statistical Association* **89**: 1164–1174.

Reiss, R.-D. (1993). *A Course on Point Processes*, Springer-Verlag, New York.

Richardson, S. (2003). Spatial models in epidemiological applications, *in* P. J. Green, N. L. Hjort & S. Richardson (eds), *Highly Structured Stochastic Systems*, Oxford University Press, Oxford, pp. 237–259.

Ripley, B. D. (1976). The second-order analysis of stationary point processes, *Journal of Applied Probability* **13**: 255–266.

Ripley, B. D. (1977). Modelling spatial patterns (with discussion), *Journal of the Royal Statistical Society Series B* **39**: 172–212.

Ripley, B. D. (1979). Simulating spatial patterns: dependent samples from a multivariate density. Algorithm AS 137, *Applied Statistics* **28**: 109–112.

Ripley, B. D. (1981). *Spatial Statistics*, Wiley, New York.

Ripley, B. D. (1987). *Stochastic Simulation*, Wiley, New York.

Ripley, B. D. (1988). *Statistical Inference for Spatial Processes*, Cambridge University Press, Cambridge.

Ripley, B. D. (1989). Gibbsian interaction models, *in* D. A. Griffiths (ed.), *Spatial Statistics: Past, Present and Future*, Image, New York, pp. 1–19.

Ripley, B. D. & Kelly, F. P. (1977). Markov point processes, *Journal of the London Mathematical Society* **15**: 188–192.

Robert, C. P. & Casella, G. (1999). *Monte Carlo Statistical Methods*, Springer-Verlag, New York.

Roberts, G. O. & Rosenthal, J. S. (1997). Geometric ergodicity and hybrid Markov chains, *Electronic Communications in Probability* **2**: 13–25.

Roberts, G. O. & Rosenthal, J. S. (1998). Optimal scaling of discrete approximations to Langevin diffusions, *Journal of the Royal Statistical Society Series B* **60**: 255–268.

Roberts, G. O. & Tweedie, R. L. (1996). Exponential convergence of Langevin diffusions and their discrete approximations, *Bernoulli* **2**: 341–363.

Roberts, G. O. & Tweedie, R. L. (2003). *Understanding MCMC*. In preparation.

Rosenthal, J. S. (1995). Minorization conditions and convergence rates for Markov chain Monte Carlo, *Journal of the American Statistical Association* **90**: 558–566.

Rossky, P. J., Doll, J. D. & Friedman, H. L. (1978). Brownian dynamics as smart Monte Carlo simulation, *Journal of Chemical Physics* **69**: 4628–4633.

Ruelle, D. (1969). *Statistical Mechanics: Rigorous Results*, W.A. Benjamin, Reading, Massachusetts.

Ruelle, D. (1971). Existence of a phase transition in a continuous classsical system, *Physical Review Letters* **27**: 1040–1041.

REFERENCES

Santaló, L. (1976). *Integral Geometry and Geometric Probability*, Addison-Wesley, Reading, MA.

Särkkä, A. (1993). *Pseudo-likelihood Approach for Pair Potential Estimation of Gibbs Processes*, Ph.D. thesis, Department of Statistics, University of Jyväskylä.

Schladitz, K. & Baddeley, A. J. (2000). A third-order point process characteristic, *Scandinavian Journal of Statistics* **27**: 657–671.

Schlater, M. (1999). Introduction to positive definite functions and unconditional simulation of random fields, *Technical Report ST 99-10*, Lancaster University.

Schlather, M. (2001). On the second-order characteristics of marked point processes, *Bernoulli* **7**: 99–117.

Schoenberg, F. P., Brillinger, D. R. & Guttorp, P. M. (2002). Point processes, spatial-temporal, *in* A. El-Shaarawi & W. Piegorsch (eds), *Encyclopedia of Environmetrics*, Vol. 3, Wiley, New York, pp. 1573–1577.

Snethlage, M. (2000). *Über die statistische Analyse von Clusterpunktprozessen durch die Paarkorrelationsfunktion*, Ph.D. thesis, Technische Universität Freiberg.

Stoyan, D. & Grabarnik, P. (1991). Second-order characteristics for stochastic structures connected with Gibbs point processes, *Mathematische Nachrichten* **151**: 95–100.

Stoyan, D., Kendall, W. S. & Mecke, J. (1995). *Stochastic Geometry and Its Applications*, second edn, Wiley, Chichester.

Stoyan, D. & Stoyan, H. (1994). *Fractals, Random Shapes and Point Fields*, Wiley, Chichester.

Stoyan, D. & Stoyan, H. (1998). Non-homogeneous Gibbs process models for forestry—a case study, *Biometrical Journal* **40**: 521–531.

Stoyan, D. & Stoyan, H. (2000). Improving ratio estimators of second order point process characteristics, *Scandinavian Journal of Statistics* **27**: 641–656.

Strauss, D. J. (1975). A model for clustering, *Biometrika* **63**: 467–475.

Student (1907). On the error of counting with a haemacytometer, *Biometrika* **10**: 179–180.

Suomela, P. (1976). *Construction of Nearest Neighbour Systems*, Dissertation 10, Department of Mathematics, University of Helsinki.

Swendsen, R. H. & Wang, J.-S. (1987). Nonuniversal critical dynamics in Monte Carlo simulations, *Physical Review Letters* **58**: 86–88.

Takacs, R. (1986). Estimator for the pair-potential of a Gibbsian point process, *Statistics* **17**: 429–433.

Thomas, M. (1949). A generalization of Poisson's binomial limit for use in ecology, *Biometrika* **36**: 18–25.

Thönnes, E. (1999). Perfect simulation of some point processes for the impatient user, *Advances in Applied Probability* **31**: 69–87.

Tierney, L. (1994). Markov chains for exploring posterior distributions, *Annals of Statistics* **22**: 1701–1728.

Tierney, L. (1998). A note on Metropolis-Hastings kernels for general state spaces, *Annals of Applied Probability* **8**: 1–9.

Torrie, G. M. & Valleau, J. P. (1977). Nonphysical sampling distributions in Monte Carlo free-energy estimation: umbrella sampling, *Journal of Chemical Physics* **23**: 187–199.

Waagepetersen, R. (2002). Convergence of posteriors for discretized log Gaussian Cox processes, *Research Report R-02-2024*, Department of Mathematics, Aalborg University. Submitted.

Waagepetersen, R. & Sorensen, S. (2001). A tutorial on reversible jump MCMC with a view toward applications in QTL-mapping, *International Statistical Review* **69**: 49–61.

Wackernagel, H. (1995). *Multivariate Geostatistics*, Springer-Verlag, Berlin.

Wässle, H., Boycott, B. B. & Illing, R.-B. (1981). Morphology and mosaic of on- and off-beta cells in the cat retina and some functional considerations, *Proceedings of the Royal Society London Series B* **212**: 177–195.

Widom, B. & Rowlinson, J. S. (1970). A new model for the study of liquid-vapor phase transitions, *Journal of Chemical Physics* **52**: 1670–1684.

Wilson, D. B. (2000a). How to couple from the past using a read-once source of randomness, *Random Structures and Algorithms* **16**: 85–113.

Wilson, D. B. (2000b). Layered multishift coupling for use in perfect sampling algorithms (with a primer to CFTP), *in* N. Madras (ed.), *Monte Carlo Methods*, Vol. 26, Fields Institute Communications Series, American Mathematical Society, Providence, pp. 141–176.

Wolpert, R. L. & Ickstadt, K. (1998). Poisson/gamma random field models for spatial statistics, *Biometrika* **85**: 251–267.

Wood, A. T. A. & Chan, G. (1994). Simulation of stationary Gaussian processes in $[0,1]^d$, *Journal of Computational and Graphical Statistics* **3**: 409–432.

Zhung, J., Ogata, Y. & Vere-Jones, D. (2002). Stochastic declustering of space-time earthquake occurences, *Journal of the American Statistical Association* **97**: 369–380.

Subject index

absolutely continuous, 24
absorption, 207
acceptance probability, 109, 113
aggregation, 34, 36
anisotropy, 34
anti-monotonicity, 222
aperiodicity, 121
area-interaction process, 102, 263, 268
asymptotic variance, 138
attraction, 34
attractive point process, 83
auto-correlation, 137
auto-covariance, 138

balanced Cox process, 78
batch means, 138
binomial point process, 14
birth process, 20
birth rate, 206
birth-death Metropolis-Hastings algorithm, 113
birth-death-move Metropolis-Hastings algorithm, 115
Boolean model, 26
bootstrap, 40
boundary, 93
bridge sampling formula, 144
burn-in, 136

Campbell measure, 248
Campbell-Mecke theorem, 249
canonical ensemble, 91
canonical sufficient statistic, 141
central limit theorem, 124
centre process, 63
CFTP, 216

CFTP algorithm, 219
Choleski decomposition, 77
circulant embedding, 77, 257
clan of ancestors, 232
cluster, 63
cluster process, 63
clustering, 34, 36
coalescence, 222
complete spatial randomness, 17
conditional approach, 94, 160
conditional intensity, 178
conditional log likelihood, 160
conditional pairwise interaction point process, 85
conditional simulation for an LGCP, 190
conditional simulation for an SNCP, 186
conditional simulation of a Neyman-Scott process, 185
conditional simulation of a Poisson-gamma process, 188
conditional Strauss process, 86
confidence interval, 40
connected component, 265
connected component relation, 265
constant death rate case, 210
continuous time Markov process, 207
continuum Ising model, 104, 116, 223
continuum random cluster model, 267
convergence diagnostics, 136
correlation function, 73
coupling construction, 211, 217, 218, 271
coupling from the past, 216
Cox process, 58

Cox process driven by a χ^2 field, 73, 76
critical value, 96
cross moment measure, 48
cross pair correlation function, 48
cross second order intensity reweighted stationary, 49
cross summary statistics, 48
cross-correlations, 137
cross-over trick, 222

daughter process, 63
death rate, 206
density for a finite marked point process, 100
density for a multivariate point process, 100
density for a Poisson process, 25
dependent thinning, 213
detailed balance condition, 209
differential characterisation of Gibbs point processes, 95, 250
diffuse measure, 13
directional cross K function, 49
directional K-function, 34
Dirichlet cell, 264
Dirichlet neighbours, 264
Dirichlet tessellation, 264
distribution at a typical point, 252
distribution of a point process, 9, 242
DLR-equations, 95
domination, 212, 272
doubling scheme, 221, 231
doubly stochastic Poisson process, 57
drift criterion for recurrence, 121

edge correction factor, 37, 39
edge effects, 94, 159
effective sample size, 138
efficiency of Monte Carlo estimates, 138
EM algorithm, 154
empty space function, 35
energy, 91
Epanečnikov kernel, 37
ergodic atom, 274
ergodic average, 119

ergodicity, 122
exact Monte Carlo p-value, 151
exact simulation, 216
existence of Gibbs distributions, 95
explosion, 207, 269
exponential correlation function, 74
exponential family, 141
extended Slivnyak-Mecke theorem, 21

F-function, 35
fast Fourier transform, 77, 259
Fill's algorithm, 236
finite Gibbs point process, 91
finite Gibbs state, 91
finite point configurations, 82
finite range of interaction, 91, 94
first order inhomogeneity, 98
first order stationarity, 30
Fisher information, 146
fixed number of points Metropolis-Hastings algorithm, 109
forward coupling time, 220
free boundary condition, 159
free boundary likelihood, 159
free boundary pseudo likelihood, 172
funnelling, 222

G-function, 35
Gaussian correlation function, 74
generating functional for a point process, 9
generating functional for a Poisson process, 17
geometric drift condition, 123
geometric ergodicity, 122
germ-grain model, 8
Gibbs point process, 95
Gibbs sampler, 111, 116
grand canonical ensemble, 91

Hammersley-Clifford-Ripley-Kelly theorem, 89
hard core, 86, 96
hard core Gibbs process, 96
hard core process, 86

SUBJECT INDEX

Harris recurrence, 119
Hastings ratio, 109, 113
Hausdorff metric, 244
hereditary, 83
higher order summary statistics, 36
homogeneous pairwise interaction point process, 84
homogeneous point process, 30
homogeneous Poisson point process, 14
horizontal backward coupling time, 219
horizontal CFTP, 219, 227

identifiability, 141
importance sampling approximation of log likelihood, 147
importance sampling distribution, 143
importance sampling estimator, 143
importance sampling formula, 142
importance sampling parameter, 143
independence of components, 50
independent scattering, 16
independent thinning, 22, 32, 33, 49
infinite order of interaction, 91
infinite range of interaction, 88
inhomogeneous Markov point processes, 98
inhomogeneous point process, 30
inhomogeneous Poisson point process, 14
initial distribution, 118
initial sequence estimate, 139
integral characterisation of Gibbs point processes, 95
integrated auto-correlation time, 138
intensity, 14, 30
intensity function, 13, 30
intensity measure, 13, 30, 247
intensity of a jump process, 269
interaction, 17
interaction function, 84, 89, 94, 262
invariant distribution, 118
invariant distribution for a spatial birth-death process, 275
irreducibility, 108, 119

isotropy, 15

J-function, 35
jump chain, 207
jump process, 269
jump times, 206

K-function, 33
Kaplan-Meier estimators, 46
kernel, 36, 61, 62

L-function, 33
Langevin-Hastings algorithm, 191
Lennard-Jones process, 88
LGCP, 72
likelihood equation, 146
linked Cox process, 78
local dominated integrability, 141
local finiteness, 7
local integrability, 13
local Markov property, 89
local specification, 95
local stability, 83
locally finite measure, 13
locally stable and constant death rate case, 210
locally stable version of the birth-death Metropolis-Hastings algorithm, 114
log Gaussian Cox process, 72, 199
log likelihood function, 146
log likelihood ratio, 147, 150
logistic process, 58
lower chain, 222
lower envelope, 41
lower process, 222

mark distribution, 26, 54
mark space, 8
marked Markov point process, 99
marked pairwise interaction point process, 101
marked point process, 8
marked Poisson process, 26
Markov chain, 107
Markov chain Monte Carlo, 107

SUBJECT INDEX

Markov connected component process, 267
Markov density function, 89
Markov function, 89
Markov point process, 89
Markov random field, 101
Matérn cluster process, 62
maximum likelihood estimate, 146
maximum profile likelihood estimate, 163
maximum pseudo likelihood estimate, 172
MCMC, 107
measure, 9
Metropolis algorithm, 110
Metropolis-adjusted Langevin algorithm, 191
minimal canonical sufficient statistic, 141
minus sampling, 39, 46
missing data, 151
missing data approach, 159
missing data likelihood, 159
missing data toroidal log likelihood, 160
mixed Poisson process, 58
mixing behaviour, 108, 137
MLE, 146
moment measure, 247
monotonicity, 221
Monte Carlo EM algorithm, 154
Monte Carlo error, 138
Monte Carlo methods, 135
Monte Carlo MLE, 149
Monte Carlo variance, 138
mother process, 63
MPLE, 172
multiscale process, 86
multitype point process, 8
multitype Poisson process, 26
multivariate Cox process, 78
multivariate LGCP, 79
multivariate log Gaussian Cox process, 79
multivariate Markov point process, 99
multivariate point process, 8

multivariate Poisson process, 27
multivariate shot noise Cox process, 80
multivariate SNCP, 80

nearest-neighbour function, 35, 50
nearest-neighbour Markov function, 261
nearest-neighbour Markov point process, 261
neighbourhood, 89, 93, 261
neighbours, 89
Newton-Raphson, 147
Neyman-Scott process, 61
NNMPP, 261
no memory, 19
normalising constant, 82

observed Fisher information, 146
offspring, 63
Ord's process, 265
overlap area process, 87
overlapping disc relation, 101

pair correlation function, 31
pairwise interaction point process, 84
Papangelou conditional intensity, 95
Papengelou conditional intensity for a marked point process, 100
parametric bootstrap, 151
parent process, 63
partition function, 82, 91
path sampling identity, 145
paving, 241
perfect simulation, 216
perfect simulation algorithm for SNCPs, 255
periodicity, 121
phase transition, 96
point configuration, 7
point process, 242
Poisson cluster process, 63
Poisson point process, 14
Poisson-gamma process, 65, 197
Polish space, 244
power exponential family, 74

SUBJECT INDEX

product space, 241
proposal density, 109
Propp-Wilson's CFTP algorithm, 221
Propp-Wilson's monotone CFTP algorithm, 222
pseudo likelihood equation, 173
pseudo likelihood for point processes, 171

quadratic general position, 264

R-close-neighbourhood relation, 89
R-close-neighbours, 88
radial simulation procedure, 24
random counting measure, 242
random independent displacements, 27
random independent thinning, 58
random labelling, 28, 50, 54
random updating scheme, 110
range of interaction, 84, 91
rate, 14
ratio-unbiasedness, 39
read-once algorithm, 227
reduced Campbell measure, 248
reduced moment measure, 247
reduced Palm distribution, 249
reduced-sample estimate, 40, 46
refined doubling scheme, 231
regularity, 34, 36
rejection sampling, 207, 234
repulsion, 34
repulsive point process, 83
retention probability, 22
reverse logistic regression, 144
reversibility, 108, 118
Ripley's K-function, 34
Ruelle stability, 84

sandwiching, 222
score function, 146
second order factorial moment measure, 30
second order intensity reweighted stationary, 32
second order product density, 30

second order reduced moment measure, 32
separable space, 243
sequential neighbours on the real line, 263
shot noise Cox process, 63
shot noise G Cox process, 65
simple border correction estimate, 40
simple point process, 242
simulated tempering, 112, 115, 128
Slivnyak-Mecke's theorem, 20
small set, 120
SNCP, 63
spacing, 139
spatial birth-death process, 206
spatial Markov property, 93
spatial point process, 7
spatstat, 174
spherical contact distribution, 35
SRS, 217
stable correlation function, 74
standard Poisson point process, 14
standard proof, 10
stationary case of an SNCP, 64
stationary multivariate point process, 49
stationary point process, 14
stochastic approximation, 148
stochastic recursive sequence, 217
stopping time, 219
Strauss process, 85, 141
Strauss type disc process, 101
strong laws of large numbers, 119
subsampling, 139
superposition, 22
Swendsen-Wang algorithm, 117
systematic updating scheme, 110

Takacs-Fiksel estimation, 177
thinning of a homogeneous Markov point process, 98
Thomas process, 62, 192
time-homogeneous Markov chain, 118
toroidal approach, 160
toroidal approximation, 159
toroidal log likelihood, 160

toroidal pseudo likelihood, 172
total variation norm, 121
trace plots, 136
transformation of a homogeneous
        Markov point process, 98
transformation of standard Poisson
        process, 20
transition kernel, 118
transition kernel of a jump process,
        269
translation invariant pairwise
        interaction functions, 95
triplet process, 92
truncated Langevin-Hastings
        algorithm, 192
trust factor, 148
trust region, 148
typical point, 54

umbrella sampling, 144
uniform ergodicity, 123
unit rate Poisson point process, 14
upper chain, 222
upper envelope, 41
upper process, 222

vertical backward coupling time, 219
vertical CFTP, 219
very-soft-core process, 88
void event, 9, 243
void probability, 9
Voronoi cell, 264
Voronoi tessellation, 264

Wald test, 150
Widom-Rowlinson penetrable
        spheres mixture model,
        101, 224
Widom-Rowlinson penetrable
        spheres model, 91, 96

Notation index

$C^!$, 248
E, 114
E_n, 110
$\text{Exp}(\cdot)$, 18
F, 35
F_2, 258
G, 35
G_X, 10
$\text{Geo}(\cdot)$, 111
H, 261
J, 35
K, 33
L, 33
M, 8
$N(B)$, 9, 242
$N_d(0, \omega^2 I_d)$, 62
N_f, 82
N_{lf}, 7, 242
$N_\varepsilon(x)$, 261
$P(x, F)$, 118
$P'''(x, F)$, 118
$P_0^!$, 251
$P_\xi^!$, 249, 251
$\text{Poisson}(S, \rho)$, 14
S, 7
T, 8
$\text{Uniform}(\cdot)$, 14
$V_{\theta, x}(y)$, 153
$V_\theta(x)$, 141
W, 7
$W_{\ominus r}$, 40
X, 7, 242
X_B, 7
$Y_m^n(x)$, 217
$b(\cdot, \cdot)$, 206
$b(\xi, r)$, 8
$\text{binomial}(\cdot, \cdot, \cdot)$, 14
c, 82

$c(\xi, \eta)$, 73
c^*, 84
c_θ, 140
$d(\cdot, \cdot)$, 206
$d(\xi, B)$, 46
$d(\xi, \eta)$, 8
$d(x_1, x_2)$, 101, 267
f, 81
f_n, 82
$g(\xi)$, 31
$g(\xi, \eta)$, 31
$g(r)$, 36
h, 82
h_n, 82
$j(\theta)$, 146
$k(\cdot)$, 61
$k(c, \cdot)$, 62
$\bar{k}_n$, 119
$l(\theta)$, 146
$m(\xi)$, 73
$n(x)$, 7
$p(x|\rho)$, 185
$\text{po}(\cdot)$, 14
$t-$, 211
$u(\theta)$, 146
$x \cup \{\xi\}$, 7
$x \setminus \eta$, 7
x_B, 7
$\mathbb{E}_0^!$, 252
$\mathbb{E}_\theta$, 141
$\mathbb{N}$, 14
$\mathbb{N}_0$, 118
$\mathbb{R}^d$, 7
$\mathbb{T}$, 218
$\mathbb{Z}$, 218
$\mathcal{A}$, 241
$\mathcal{B}$, 8
$\mathcal{B}_0$, 9

$\mathcal{K}$, 32, 252
$\mathcal{L}^a(\Pi)$, 119
$\mathcal{L}^a(h)$, 119
$\mathcal{N}_{\mathrm{lf}}$, 8, 242
$\Gamma(\cdot,\cdot)$, 208
Φ, 63
$\Phi(\cdot|\cdot)$, 262
Π, 118
$\Pi(k)$, 119
$\alpha^{(2)}$, 30
$\alpha^{(n)}$, 247
$\beta(\cdot)$, 206
$\delta(\cdot)$, 206
λ^*, 83, 100
μ, 13, 30, 247
$\mu(A)$, 241
$\mu^{(n)}$, 247
ω_d, 17
∂_B, 93
$\phi(\cdot)$, 84, 89
ϕ^*, 84
$\varphi(\cdot)$, 262
$\varphi(\cdot,\cdot)$, 217
ρ, 13, 30
$\sigma(\mathcal{A})$, 241
σ_d, 17
$\mathbf{1}[\cdot]$, 9
$\|\cdot\|$, 8, 244
$\|\cdot\|_{\mathrm{TV}}$, 121
$\|\cdot\|_V$, 122
$\otimes$, 241
$\sim$, 14, 89, 108
$\sim_x$, 261
$\sum^{\neq}$, 22
$\emptyset$, 7